Dietary Phytochemicals

Chukwuebuka Egbuna • Sadia Hassan
Editors

Dietary Phytochemicals

A Source of Novel Bioactive Compounds for the Treatment of Obesity, Cancer and Diabetes

Springer

Editors
Chukwuebuka Egbuna
Faculty Natural Science
Department of Biochemistry
Chukwuemeka Odumegwu
Ojukwu University
Uli, Nigeria

Sadia Hassan
Department of Nutritional Sciences
Faculty of Science and Technology
Government College Women University
Faisalabad, Pakistan

ISBN 978-3-030-72998-1 ISBN 978-3-030-72999-8 (eBook)
https://doi.org/10.1007/978-3-030-72999-8

This Springer imprint is published by the registered company Springer Nature Switzerland AG
The registered company address is: Gewerbestrasse 11, 6330 Cham, Switzerland

Preface

The proper understanding of the interplay between obesity, cancer, and diabetes and how to manage them is the main focus of this book. This book entitled *Dietary Phytochemicals – A Source of Novel Bioactive Compounds for the Treatment of Obesity, Cancer and Diabetes* is a collection of chapters that were collaboratively written by authors from key institutions across the globe. An effort was made to highlight the link between obesity, diabetes, and cancer and the potentials of functional phytochemicals for their prevention and treatment. In Part I, the etiology and pathophysiology of obesity, diabetes, and cancer were fully discussed. Also, the role of peptides and the link between insulin, obesity, and cancer were presented. In Part II, the functions and the sources of important dietary phytochemicals were presented. Special chapters were dedicated to MicroRNAs as targets of dietary phytochemicals and the potentials of phytochemicals for breast cancer treatment.

Uli, Nigeria — Chukwuebuka Egbuna
Faisalabad, Pakistan — Sadia Hassan

Contents

Etiology of Obesity, Cancer, and Diabetes

Iqra Yasmin, Wahab Ali Khan, Saima Naz, Muhammad Waheed Iqbal, Chinaza G. Awuchi, Chukwuebuka Egbuna, Sadia Hassan, Kingsley C. Patrick-Iwuanyanwu, and Chukwuemelie Zedech Uche

1 Introduction

Obesity is of public health concern globally. The prevalence of obesity is on a constant increase. The effects of this non-communicable disease on the overall health of affected individual is devastating. Obesity worsens the symptoms of several diseases such as osteoarthritis, cancer, type 2 diabetes mellitus (T2DM), cardiovascular disease, and psychological disturbances (Dixon, 2010), and, as a result, significantly contribute to economic burden worldwide (Anyanwu et al., 2020; Wang et al. 2011; Withrow and Alter 2011). Factors responsible for obesity include biological, behavioral, and environmental factors. Obesity has been associated with cancer and diabetes. Cancer can be broadly classified into three categories, namely, carcinomas, lymphomas, and sarcomas. Approximately 1 in 3 cancer deaths are due

I. Yasmin (✉)
Center of Excellence for Olive Research and Training, Barani Agricultural Research Institute, Chakwal, Pakistan

W. A. Khan
District Food Laboratory, Technical Wing, Punjab Food Authority, Lahore, Pakistan

S. Naz
Department of Clinical Nutrition, Nur International University, Lahore, Pakistan

M. W. Iqbal
School of Food and Biological Engineering, Jiangsu University, Zhenjiang, China

Riphah College of Rehabilitation and Allied Health Sciences, Riphah International University, Faisalabad, Pakistan

C. G. Awuchi
School of Natural and Applied Sciences, Kampala International University, Kampala, Uganda
e-mail: awuchi.chinaza@kiu.ac.ug

C. Egbuna
Africa Centre of Excellence in Public Health and Toxicological Research (ACE-PUTOR), University of Port Harcourt, Port Harcourt, Choba, Nigeria

Department of Biochemistry, Faculty of Science, University of Port Harcourt, Port Harcourt, Rivers State, Nigeria

C. Egbuna, S. Hassan (eds.), *Dietary Phytochemicals*,
https://doi.org/10.1007/978-3-030-72999-8_1

to high BMI, low intake of fruits and vegetables, coupled with sedentary lifestyle or lack of physical activities, tobacco addiction, and alcoholism. Diabetes mellitus is a pancreas disorder which results in high glucose levels in the blood. There are different biological, environmental, and behavioral factors which metabolically interact to potentiate these diseases in healthy individuals. Mitigating obesity, cancer, and diabetes, which are often caused by several factors, require comprehensive approach. This chapter presents the etiologies and risk factors of obesity, cancer, and diabetes.

2 Obesity

Obesity is considered a public health issue which require immediate control measures. To better understand the etiology of obesity, it is pertinent to understand its underlying causes, which include excessive calorie intake, insufficient physical activity or sedentary lifestyle, poor food choices, genetic factors, and so on. Proper understanding of the role of genetics in obesity is also very important (Pereira-Lancha et al., 2012). The prevalence of obesity is influenced by various socio-economic, environmental, and behavioral factors which result in the accumulation of fats in the adipose tissue. Obesity is prevalent in both developing and developed countries with more than 500 million people affected globally. Primarily, obesity occurs when calorie intake is more than energy expenditure with resultant weight gain. Obesity can also develop from the interaction of genes with other biological and environmental factors, such as diet, and lifestyle.

The treatment of obesity is not as simple as the affected individuals may desire. Reduced metabolic rate and gradual increase in appetite due to increased activities of appetite-stimulating hormones may undermine the effort of conventional methods used in treating obesity (Hemmingsson 2014; Johansson et al. 2014).

However, prevention is an important and crucial strategy in the management of obesity. Identifying different risk factors of obesity and how these factors are linked up with each other is vital. Children between 1–5 years can also be overweight or

Department of Biochemistry, Faculty of Natural Sciences, Chukwuemeka Odumegwu Ojukwu University, Uli Campus, Ihiala, Anambra State, Nigeria

S. Hassan
Department of Nutritional Sciences, Faculty of Science and Technology, Government College Women University, Faisalabad, Pakistan

K. C. Patrick-Iwuanyanwu
Africa Centre of Excellence in Public Health and Toxicological Research (ACE-PUTOR), University of Port Harcourt, Port Harcourt, Choba, Nigeria

Department of Biochemistry, Faculty of Science, University of Port Harcourt, Port Harcourt, Rivers State, Nigeria
e-mail: kc.patrick-iwuanyanwu@uniport.edu.ng

C. Z. Uche
Department of Medical Biochemistry and Molecular Biology, Faculty of Basic Medical Sciences, University of Nigeria, Enugu Campus, Nsukka, Nigeria
e-mail: chukwuemelie.uche@unn.edu.ng

obese (Cunningham et al., 2017; Cunningham et al., 2014; Matthews et al., 2017). If a person is overweight in childhood, obesity is highly likely in adolescence (Cunningham et al. 2017; Ward et al. 2017). Increasing awareness on the risk factors associated with infant overweight is critical to preventing obesity in adolescence (Baidal and Taveras 2012; Danese and Tan 2014).

The drastic increase in obesity in adolescents is mostly due to behavioral changes directly caused by overeating and consistent weight gain. Clinical studies have shown that depression in adulthood can lead to obesity in adults (Luppino et al. 2010; Wurtman and Wurtman 2015). Additionally, mood swings are also associated with adult obesity, including chronic anxiety, etc. (Dallman et al. 2005), severe premenstrual syndrome (late luteal phase dysphoric disorder; LLPDD), and post-traumatic stress syndromes (PTSD) (Kubzansky et al. 2014). The tendency of obesity linked with mood disorders may occur in patients receiving behavioral medicines and antidepressant (Arjona et al. 2004; Fava et al. 2000; Wurtman and Wurtman 2015). In this condition, drug gradually affects the patient's eating behavior in two ways; either (a) reduced satiety, leading to the urge to eat more, or (b) increased appetite for calorie-rich foods (Wurtman and Wurtman 2015). Both conditions can be drastic and ultimately lead to weight gain (Fava et al. 2000). Similarly, in the case of some psychotropic drugs, the physical activity of the patient is limited, leading to decreased burning of calories; and, is one of the main causes of weight gain in such patients (Paluska and Schwenk 2000). Other environmental and psychological factors that contribute to weight gain in depressed individuals, include unemployment, low income, low self-esteem, disturbance in relation, family issues, educational challenges, isolation, etc., all of which can lead to abnormal food consumption and/or intake of high-calorie foods (Wurtman and Wurtman 2015). Weight gain is not only observed in individuals treated with psychotropic drugs, it is also observed in the population with hormonal disturbance, excessive intake of supplementation, anxiety, medical condition, etc. (Gariepy et al. 2010).

2.1 *Etiology of Obesity*

Ingesting calories in excess more than the body needs results in the storage of extra calories in adipose tissue (O'Rahilly 2009). When energy intake is more than energy expenditure, the result is weight gain, which when consistent can lead to overweight (BMI within 25–30), followed by obesity (BMI of 30 and above). Multiple factors responsible for obesity, include biological, behavioral, and environmental factors, among others. These factors are responsible for energy storage and weight gain (Aldhoon-Hainerová et al. 2014; Apalasamy and Mohamed 2015; Rao et al. 2014). Energy is mainly stored due to calorie intake in the form of macronutrients, such as carbohydrate, protein, fat, and alcohol. Calories are utilized via three metabolic processes; (1) thermic effect of food (TEF) (2) physical activity in which calories are burned (3) resting (basal) metabolic rate. Approximately 8–10% energy is utilized in TEF (energy required to digest and metabolized food). Weight maintenance

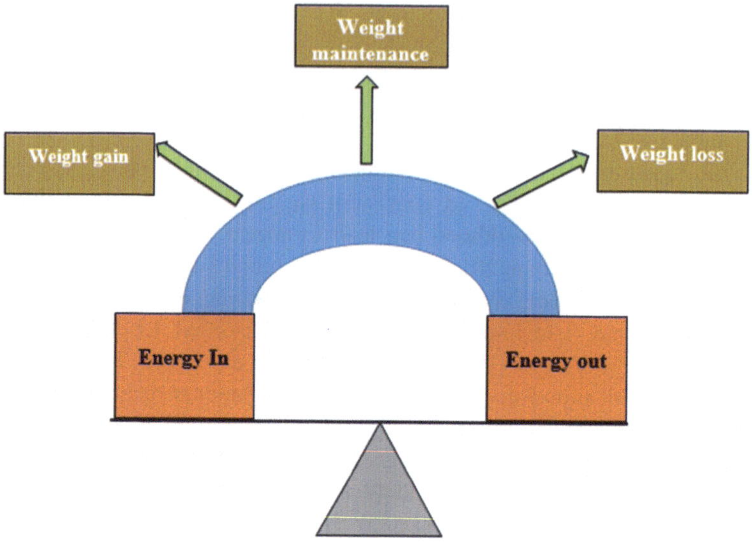

Fig. 1 Energy balance. Depicting an association between intake of food (Energy In) and the energy expenditure through the body during physical activity and metabolism (Energy Out)

is achieved when energy intake and energy expenditure are equal (Fig. 1). Changes in weight occur when energy intake is more than energy out or energy expenditure is more than energy intake. Both conditions lead to an imbalance in body weight. A healthy person become obese when energy intake is consistently more than energy expenditure for a prolonged period of time, usually months or years. Different factors such as genetic, environmental, and behavioral factors which directly contribute to high energy intake may result in weight gain (Fig. 2) (Hill et al. 2012).

2.2 *Biological Factors*

Biological factors associated with obesity include brain-gut axis, genetics, pregnancy, prenatal determinants, neuroendocrine conditions, menopause, medications, gut microbiome, physical disability, and viruses. If any of these factors is present and interacts with other factors (such as environmental and behavioral factors), it can lead to obesity.

2.2.1 Genetics

Most people with obesity have multiple genes which predisposes them to excessive weight gain. There is strong evidence that genetic makeup can contribute to the development of obesity. Studies on clones and adoptees showed that >70% of the

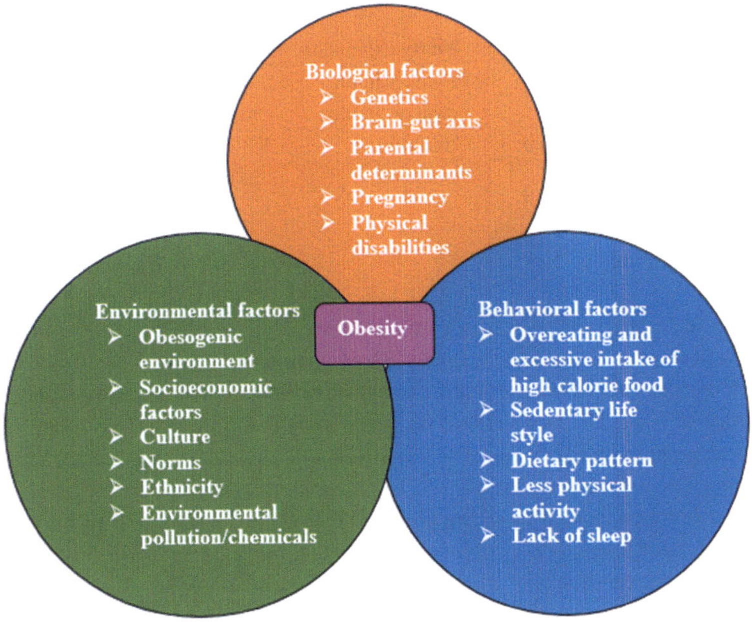

Fig. 2 Etiology of obesity

individual fat difference are due to genetic variations (Farooqi and O'rahilly 2008). Genetic obesity is associated with all aspects of energy homeostasis, including energy intake, physical activity, TEF, basal metabolic rate, food intake, etc. (Pereira-Lancha et al. 2012). The ob (Lep) gene that encodes the peptide leptin, and its alteration, causes obesity, and has been the main focus of researchers working on heritability of obesity (Zhang et al. 1994). Genetic studies of obesity showed that there are more than 300 genes and gene markers which are associated with obesity, and their interactions with environmental factors (Chagnon et al. 2003; Friedman 2003). There are 3 subgroups of genetic-induced obesity, including;

1. Monogenic obesity (due to single gene defect, e.g., leptin)
2. Syndromic obesity (chromosomal abnormalities, e.g., Prader-Willi syndrome)
3. Polygenic obesity (due to more than one gene variants) (Rao et al. 2014)

Monogenic Obesity

Monogenic obesity is caused by single gene mutation/variation. To date, several forms of early human obesity have been identified, which mostly occur due to gene mutation controlling the leptin-melanocortin path. This path regulates the intake of food. The genes responsible for this kind of obesity are listed in Table 1 (Farooqi and O'rahilly 2008; Rao et al. 2014). Continued efforts are made to identify and discover other potential candidate genes that are responsible for obesity in affected individuals.

Table 1 Genes associated with obesity

Gene	Encoded protein	Obesity expression	Cases identified
LEPR	Leptin receptor	Few days in life	2–3% of severe early-obesity
LEP	Leptin	Few days in life	<50 patient globally
MC4R	Melanocortin-4 receptor	Childhood	2–3% obesity in children and adults
SIM1	Single-minded 1	Childhood	<50 patient globally

Syndromic Obesity

Currently, at least 25 forms of syndromic obesity are known. Some of the most common syndromic obesity are; Bardet-Biedl syndrome, Prader-Willi syndrome, Alstrom syndrome, aniridia, genitourinary anomalies, Wilms tumor and mental retardation (Chung 2012). Prader-Willi is the most common syndromic obesity and a complex genetic disease caused by modifications in gene expression on the paternally inherited chromosome 15q11.2-q13 region. Under this syndrome, the affected individual is characterized by being obese, short stature, cognitive disability, excess food intake, and behavioral problems. Obesity is the main cause of morbidity and mortality among these sets of individuals (Angulo et al. 2015).

Polygenic Obesity

Genetic alteration with polygenic effects on energy intake is clinically significant, as this is the most vital human cases of obesity (Hinney et al. 2010). Some of the obesogenic traits arise partly due to the mutation in more than one genes. Polygenic variants are the alleles at different gene loci that collectively affect the heritage, quality, and quantity of the expression of phenotypes. The polygenic basis of obesity depends on individuals; every affected individual has a unique set of polygenic genes linked with the disease (Hinney et al. 2010). Genome-wide link and Metabochip studies showed single nucleotide polymorphisms with polygenic effects on body mass index. In meta-analysis of genome-wide link studies, as well as Metabochip studies, involving 339,224 individuals, 97 loci (new and previously identified) were found to be linked with BMI. The study demonstrated that these known loci are responsible for 2.7% of deviation in BMI (Robinson et al. 2015). Although, extensive research is still required to discover a wider range of susceptible loci that are present and associated with the environment in the expression of obesity (Rao et al. 2014).

2.2.2 Brain-gut Axis

The brain-gut axis (Fig. 3) regulates the utilization of food in response to homeostatic and non-homeostatic signals. In case of homeostatic (energy balance) system, the arcuate nucleus of hypothalamus integrates nutrient, neural, and hormonal

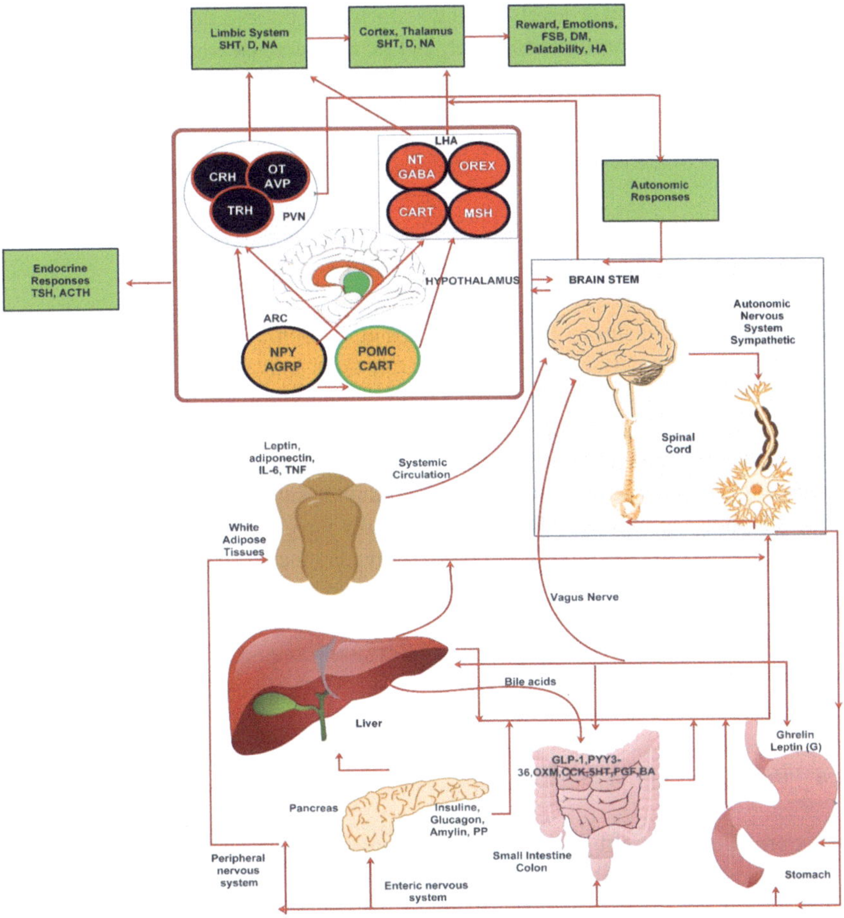

Fig. 3 Brain-gut axis and regulation of food intake

signals to regulate satiation, hunger, and satiety through higher cortical centers that control these signals to stimulate the sympathetic and parasympathetic nervous system, gastric function, and hormone secretion (Hussain and Bloom 2013). These signals are the means of communication between the brain, gastrointestinal system and adipose tissue. Individual characteristics associated with the brain, gastrointestinal system, and adipose tissue may cause variations in energy intake and storage among different individuals (French et al. 2012). The gastric function also affects hunger, satiation, and satiety. Similarly, the intake of food may also cause alterations in gastric function (Acosta and Camilleri 2014).

In case of obese individuals, there is difference in the brain-gut axis compared to healthy people. Obesity is directly related to higher fasting gastric volume, higher gastric emptying of solids and liquids, decreased satiation (a difference of 50 calories/5 kgm^{-2} of BMI), low level of satiation-linked hormone peptide YY (PYY)

(Acosta et al. 2015), lower levels of glucagon-like peptide-1 and cholecystokinin, higher levels of acyl-ghrelin (Sato et al. 2014), and increased levels of leptin with leptin resistance (receptor downregulation) (Myers et al. 2008). Furthermore, most of the biological etiologies of obesity disturb the brain-gut axis homeostasis. There is need to study whether change in brain gut-axis with intake of food is an effect or cause.

2.2.3 Prenatal Determinants

Several research have shown that nutritional and environmental factors during fetal and newborn life can permanently affect the functions and structure of many organs, which can lead to obesity. It has been noted that high pre-pregnancy and gestational weight can affect the body composition of newborn, which is the major cause of childhood obesity. Gestational diabetes is also associated with childhood overweight (Baptiste-Roberts et al. 2012). In addition, smoking during pregnancy appears to predispose the infant to excessive BMI and eventually weight gain. Interestingly, it has been observed that the method of childbirth may also increase the risk of obesity in offspring. Also, epidemiologic data suggest a direct relationship between newborn delivery and child adiposity. A systematic review and meta-analysis of 15 studies (approximately 163,796 participants) found a positive connection between childbirth and more BMI in adulthood (Darmasseelane et al. 2014).

2.2.4 Pregnancy

During pregnancy, weight gain is directly associated with fear of postpartum retention of excess weight. Different epidemiologic studies suggest that gestational weight gain is positively related to weight retention after pregnancy but not to a greater extent (Mannan et al. 2013). Studies have suggested that that the tendency of 2–3 kg weight gain and increase in waist-hip ratio after first pregnancy during 5 years is more as compared to non-pregnant women. After pregnancy, most of the women retained 4–5 kg, which often lead to obesity. A study conducted on 540 women, recorded their weight during pregnancy, after 6 months of pregnancy, and after10 years. The research showed that an average increase of 6.3 kg in weight from pre-pregnancy to follow-up (Rooney et al. 2005). Thus it is clear from the result that there is a positive correlation between pregnancy-induced weight gain and risk of obesity after pregnancy.

2.2.5 Menopause

Another important cause of obesity in women is menopause. According to a study conducted on American women, 48% of postmenopausal women (>50 years of age) are overweight and >25% are at high risk of obesity compared to premenopausal women (Lambrinoudaki et al. 2010). Menopause is associated with changes in body composition and structure, such as accumulation of more fat mass, increased BMI, increased body weight, less physical activity, slow basal metabolic rate, and increased risk of other health disorders. Most of these conditions usually lead to less energy expenditure and, sometimes. Perimenopausal hormonal changes are directly linked with drastic increase in abdominal adiposity and more hip to waist ratio. Estrogen therapy may prevent these body composition changes and their associated metabolic consequences (Davis et al. 2012).

2.2.6 Physical Disability

Epidemiological studies showed that individuals with disabilities are at high risk of obesity compared to normal individuals. Individuals with disabilities are often unable to do physical activity and their body composition changes due to less energy expenditure. A study was conducted on the change of body composition in patients with acute spinal cord injury. Results showed a significant decrease in bone-mineral density and lean body mass during the first year of injury, due to less energy expenditure and increase in adiposity (Singh et al. 2014). One of the main causes of obesity in children is physical disability. The study also showed that children and teenagers with spastic quadriplegic cerebral palsy had a significantly low total energy expenditure and non-basal energy expenditure as compared to their healthy counterparts (Singh et al. 2014). It is conclusive to state that individuals with disabilities, either physically or mentally, are at high risk of developing obesity and other disorders.

2.2.7 Gut Microbiome

Different animals and human studies reported that gut microflora affects body weight by influencing energy metabolism and inflammation (Murphy et al. 2013). The human gut consists of 10–100 trillion microorganisms, indicating a 10-fold more microbial load compared to human cells in the human body (Cani and Delzenne 2011). Gut microflora breaks down large indigestible molecules into small digestible molecules and plays an important role in energy metabolism in the human body (Flint et al. 2012). In fact, the metabolites produce in the gut after fermentation of carbohydrates into simple sugar is important in metabolism. A study reported that gut microbiota isolated from an obese rat can transfer the obesity phenotype when transplanted into a germ-free rat. One of the advanced findings has seen in the diet-induced obese rat as well as the genetic mouse models of obesity, ob/ob and toll-like receptor 5 (TLR5) knockout mice (Vijay-Kumar et al. 2010). After that tremendous discoveries, it opens a new horizon for human studies to better understand the link between the gut microbiome and obesity in microbial diversity among obese and healthy individuals. However, bacterial species or genus also directly associated with obesity. There is need to further explore the microbial diversity in relation to obesity and energy expenditure.

2.3 *Environmental Factors*

The environment is one of the critical factors in the etiology of obesity. In most cases, biological factors interact with environmental factors to express obesity. These dynamics are required to be treated at first by endocrinologists, policy makers, and health practitioners involved in the treatment of obesity. Some of the major environmental determinants are discussed below.

2.3.1 Obesogenic Environment

Nowadays, food has been abundant and readily available for consumption. Most people eat food for pleasure and entertainment rather than for survival. The food industries and food supply chain make food readily available by creating a favorable environment for fast food, ready to drink beverages, and the foods rich in calories are among the major causes of weight gain (Bray 2014). Consequently, in this way, environment has become one of the major contributing factors in the development of obesity. Another cause of weight gain is modernization and the trend of fast food among the population. The new generation consumes a lot of fast and fried food items and alcoholic drinks that are considered one of the pillars of the obesogenic environment. Nowadays, these food supply chains attract their customers by providing different offers (such as buy one get one free) and celebrating different days (such as Mothers' Day, Fathers' Day, Independent Day, etc.). In this fast developing life, many people and families spend less time in the kitchen.

2.3.2 Socio-Economic Factors

There is an inverse relationship between socio-economic status and obesity all around the globe. This link is assumed to arise from the effects of society, status, gender, ethnicity, culture, economics, education, norms, etc. (Dubowitz et al., 2013). Each family has different traditions, culture, and foods. Culture consists of different beliefs, characteristics, and behaviors of different age groups, gender, society, and ethnicities that are followed by different families. Consequently, due to cultural variations, we have different norms and values that directly influence our dietary habits, health, and body weight.

Society directly affects a person's mental and physical health. Generally, it has been observed that obese individuals are immensely affected by stigma, biases, body shaming, and discrimination compared to healthy persons. The bias they usually face in their daily routine comes not only from society but also from their own family. All the discriminations received from the society influence his/her emotional and psychological behavior, and consequently affect his/her physical well-being. In our society today, overweight youth who encounter body shaming, disrespect, and weight-based teasing are more prone to unhealthy eating, poor weight management, disturbed dietary lifestyle, binge eating, and minimum physical activity (Puhl and Brownell 2001).

2.3.3 Environmental Chemicals and Obesity

Nowadays, some of the alarming concerns are environmental pollution and the role of environmental chemicals in obesity. Different studies showed that certain endocrine-disrupting chemicals, or "obesogens," are potential risks to obesity. For example, phthalates are well-recognized obesogens, as they react with peroxisome proliferator-activated receptors (PPARs), which are responsible for the development of fat. There is need to investigate and identify more obesogens to prevent environmental health issues (Thayer et al. 2012).

2.4 Behavioral Factors

The behavioral factors are the monogenic factors or etiologies responsible for energy intake, less energy expenditure, overeating, etc. Individual choices and lifestyles essentially interplay between biological and environmental factors to express obesity. We are in an obesogenic food environment with foods in abundance, readily available, affordable, lots of variety, and convenience. In spite of all these conditions, the individual choice plays a major role in the consumption of food. Similarly, other behaviors, such as less physical activity, overeating, lifestyle, insufficient sleep, drinking habit, and smoking are self-inflicted behaviors that are not affected by the environment.

2.4.1 Increased Calorie Intake and Eating Habits

Currently, the trend of per capita consumption of refined cereal grains, sugar, and fats is dramatically increasing in form of processed, ready-to-eat fast foods, resulting in more calorie intake. Studies have demonstrated that diets high in carbohydrates and fat are linked with more energy intake. Mozaffarian et al. (2011) examined the relationship between lifestyle and change in weight in 3 groups of people (120,877 healthy women and men) over a period of 4 years. The results reported that weight gain was due to the consumption of potatoes, potatoes chips, sweetened beverages, unprocessed red meat, and processed foods, and less consumption of fruits, vegetables, whole grains, nuts, and yogurt (Mozaffarian et al. 2011). In other studies, it has been reported that excessive intake of liquid calories is a significant risk factor for the development of obesity. Studies showed that liquid beverages are rich sources of sugar, which is associated with high intake of calories and subsequent weight gain (Woodward-Lopez et al. 2011).

Another important issue is that weight gain is linked with eating patterns. Two eating disorders are linked with obesity: (1) night eating syndrome (NES) and (2) binge eating disorder (BED) (Beksinska et al. 2010). According to the Diagnostic and Statistical Manual of Mental Disorders (DSM-V), BED is defined as "frequent overeating in a short period of time marked by feelings of lack of control". On the other hand, NES is characterized by night eating, night snacking, eating after awakening from sleep, or more food consumption after having an evening meal. In studies exploring the relationship between NES and obesity, NES was reported as one of the major causes of obesity, such as increase in BMI due to more calorie intake at night with no physical activity (de Zwaan et al. 2014).

2.4.2 Sedentary Lifestyle and Less Physical Activity

Physical activity plays a vital role in maintaining personal fitness. It can constitute 20–30% of total energy expenditure per day. The energy expended during physical activity varies from person to person based on body shape, composition, structure, weight, lifestyle, dietary patterns, etc. Sedentary lifestyle such as sustained watching of TV, sitting at one place for a long time, using laptops, etc. are related to weight gain and obesity, despite exercise and diet (Hu 2003). Physiologically, a sedentary lifestyle constitutes less energy expenditure, thus promotes obesity due to reduced physical activity (Hamilton et al. 2007). There is an inverse relationship between physical activity and tendency to become obese. Many studies showed that physically active children are less likely to become overweight and obese in childhood and puberty, and have less chance to become obese in adulthood (Hills et al. 2011). Unfortunately, in this technological era, it is not rare for a person to spend all day in sitting position, with little or no physical activity (Hamilton et al. 2007). The situation is alarming because less physical activity leads to risks of several chronic disease, such as diabetes, cardiovascular diseases, and overall mortality. There is need to promote and spare at least 30 min per day for physical activity to control obesity and other health-related disorders (Fox and Hillsdon 2007).

2.4.3 Insufficient Sleep

In behavioral etiologies of obesity, another concerning factor related to modernization is insufficient sleep and exhaustion, usually due to night shift jobs, long working hours, extra work, entertainment, etc. Research findings strongly recommend sleeping at least 8 h per day to keep the mind and body healthy and active in both children and adults (Cappuccio et al. 2008). Research conducted on 12 healthy men observed the effect of sleep on their appetite and BMI. The subjects were kept 2 days without sleep and 2 days of sleep extension spaced 6 weeks apart with controlled physical activity and food intake. Results showed that sleep deprivation was linked with decrease in leptin and increase in ghrelin associated with hunger and appetite. Epidemiological studies suggest that more food consumption and adiposity may contribute to insufficient sleep. Several studies have attempted to explain the association between insufficient sleep and weight gain.

2.4.4 Quitting Smoking

There is a concept that smoking cause weight reduction. Epidemiologic studies support this statement, with a study stating that 4–5 kg weight reduction was observed in smokers compared to nonsmokers (Audrain-McGovern and Benowitz 2011). Smokers usually gain weight when they withdraw from smoking. This weight gain may be due to nicotine withdrawal, increased food intake, and less energy expenditure (Filozof et al. 2004). Proper diet and adequate physical activity are helpful strategies for smokers planning to quit smoking.

3 Cancer

Cancer is a disease characterized by abnormal cell division, whereby the cell divides in an uncontrolled manner which eventually ends up with formation of a tumor (Saravanan et al., 2020). Further uncontrolled division results in the spread of the tumor, a process termed as metastases. The local or metastatic cancer causes suffering to the affected individuals, and may even cause death. There are various forms of cancer. This uncontrolled cell division can arise from any tissue or organ of the body. According to recent reports of World Health Organization (WHO), an estimated 9.6 million deaths were due to cancer in 2018, which makes cancer the second leading cause of deaths worldwide. This turns out that 1 in every 6 deaths is due to cancer. It has also been reported by WHO (2018) that approximately one third of cancer patients are due to five most important dietary and behavioral risk factors; including high body mass index (BMI), low fruits and vegetables consumption, sedentary life style, use of alcohol, and tobacco usage (Table 2).

Table 2 Prevalence of different cancers

S/No.	Cancer type	Number of cases (Millions)	Number of deaths
1.	Lung cancer	2.09	1.76 million
2.	Breast cancer	2.09	627,000
3.	Colorectal cancer	1.80	862,000
4.	Prostate cancer	1.28	–
5.	Skin cancer (non-melanoma)	1.04	–
6.	Stomach cancer	1.03	783,000

Source: (WHO 2018)

3.1 *Major Types of Cancer*

Cancer can be broadly classified into three categories, namely, carcinomas, lymphomas, and sarcomas.

1. **Carcinomas:** Carcinomas accounts for about 85% of the total cases diagnosed worldwide. Carcinomas develop in the cells/tissues of epithelial layers. It has four subcategories; adenocarcinomas (e.g. Lung Cancer), Squamous cell carcinomas (e.g. Oral Cancer), Transitional cell carcinoma (e.g. Bladder Cancer), and Basal cell carcinomas (e.g. Skin Cancer) (Vanita et al. 2011).
2. **Sarcomas**: The cancer which originates in the mesodermal tissues falls under the category of sarcomas (Lau et al. 2011). This category includes the tumors of bone, muscle, fat tissues, and hematopoietic tumors.
3. **Lymphomas**: This type of cancer accounts for about 9% of the total cases diagnosed in the world (Bali et al. 2010). As the name implies, lymphomas occur in the lymph cell of the human immune system. Individuals suffering from cancer caused by lymphomas has enlarged lymph nodes and modification of lumps. Lymph nodes are present in the lymphatic system of human body along with blood vessels of circulatory system (Bali et al. 2010).

3.2 *Causes of Cancer*

Technically it is a difficult task to find out the basic cause of any type of cancer. The reason behind this difficulty is that cancerous cells are actually affected by several cultural and extracellular microenvironment condition (Nagy 2011). However there are certain factors which have strong evidence in increasing the risks of cancer, including poor dietary habits, tobacco use, alcohol consumption, obesity, UV light, sedentary life style, and hereditary factors (López-Lázaro 2016). There are certain genes which act as tumor suppressors (e.g. BRCA1 and BRCA2); any mutation in these genes will be responsible for causing hereditary cancers, such as ovarian cancer, breast cancer, etc. (Kurioka et al. 2011). There are numerous studies which showed that there are high levels of mRNA in the tumors of metastatic stage, which can be cured with chemotherapy (Liu, 2010).

3.2.1 Diet and Physical Activity

According to the WHO, one of the most important factor accountable for cancer is diet. High BMI, less consumption of fruits and vegetables (low-energy-dense foods), lack of physical activity, tobacco use, and alcohol use are prominent risk factors, as they account for about one third cause of cancer deaths. Although, it has been estimated that around 30% of cancers are due to dietary factors in industrialized regions, only a few definite relationships between specific nutrient-related factors and cancer have been established scientifically.

There is large research data which supports that overweight and obesity increase the risk of cancer, especially in esophagus, colorectal, breast in post-menopausal women, endometrium, and kidney, while high amount of alcohol consumption increases the risk of cancers such as esophagus, pharynx, oral cavity, larynx, breast, and liver cancers. Liver cancer has increased due to exposure to aflatoxins, while nasopharynx cancer is on the rise due to some types of fermented and salted fish. Risk of colon cancer can be reduced with physical activity. Similarly, there are other risk factors which probably cause different types of cancer. For example; colorectal cancer is associated with high intake of preserved meat; very hot drinks and spicy food leads to oral cavity, esophagus, and pharynx cancer; stomach cancer is probably due to salt and salt-preserved products.

Healthy eating patterns such as increased portions of fruit and vegetables in the diet are helpful against stomach, esophagus, colorectal, and oral cavity cancers. Healthy lifestyle that includes sufficient physical activity has been reported to be protective against breast cancer. The American Institute for Cancer Research (AICR) and World Cancer Research Fund (WCRF) state that high intake of fiber reduces the risk of colorectal cancer, while intake of red meat and processed meat increase the risk of colorectal cancer. The dietary recommendations to prevent cancers include high fiber foods and reduced intake of energy-dense foods, salted items and red meat (Fig. 4).

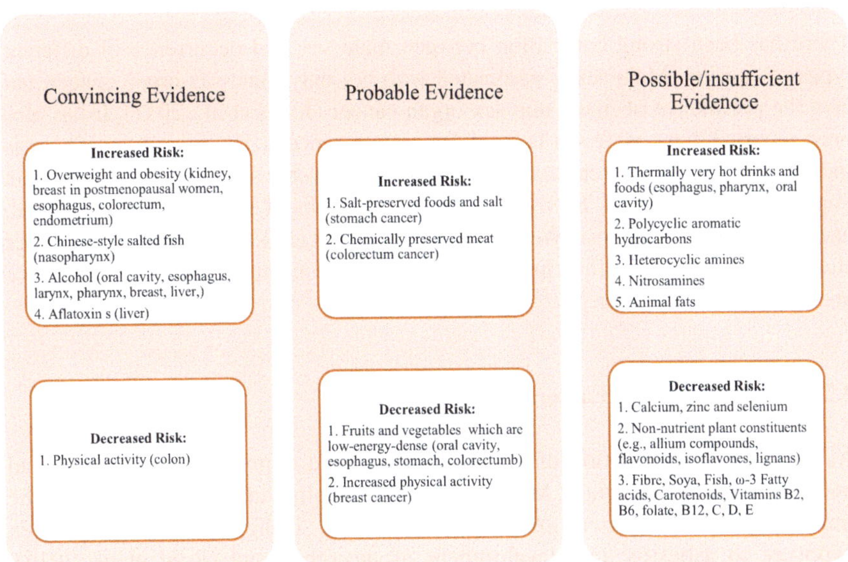

Fig. 4 Different risk factors for cancer

3.2.2 Use of Addictive Substances

The extensive use of tobacco is now recognized globally as a leading cause of different cancers and is currently estimated to account for 27% of cancer deaths (WHO 2018). Smoking tobacco increases the risk of developing many types of cancer such as cancers of the lung, larynx (voice box), mouth, throat, esophagus, kidney, stomach, pancreas, bladder, and cervix. The occurrence of lung cancer is about 20 times in chain smokers as compared to the nonsmokers (Furrukh 2013). However, nonsmoking individual who are exposed to secondhand smoke, termed as "Secondhand smokers", are also at risk of developing cancers due to the presence of tobacco smoke in the environment. However, many lung cancer cases have been reported in people who have never been smokers (Sun et al. 2007); many smokers never develop lung cancer.

Studies report that risk of developing prostate cancer is linked with the dose of alcohol being consumed by an individual (Fowke et al. 2015). All types of alcohol, including wine, beer, and spirits, increase cancer risk (Scheideler and Klein 2018). Alcohol increases the risk of seven types of cancers which includes highly prevalent cancers such as breast and bowel cancers (Corrao et al. 2004), along with being a major cause of liver cancer (Turati et al. 2014). Breast cancer risk is also relatively higher among those who consume relatively small amounts of alcohol (Seitz et al. 2012).

3.2.3 Sex and Reproductive Health

There has been strong correlation between male sex and occurrence of different types of cancers. Male sex is associated with not only male sex organ cancer, but also the prevalence of many non-sex organ cancer (Jones et al., 2011). It has also been reported that unsafe sex is a risk for certain cancers. The role of unsafe sex is indirect in causing different sexual organ cancers. Unsafe sex is a source of Human Papillioma Virus spread. This virus increases the risks of vulvar, vaginal, cervical, and anal cancers in its host (Weiderpass 2010; Vanita et al. 2011). However, further studies are needed to sufficiently establish the relationship between sex difference and occurrence of cancer.

3.2.4 Environmental Factors

Asbestos is a combination of certain minerals and is present in industrial and housing building materials. Mesothelioma is a malignant tumor formed on the lining of lungs, abdomen, and heart. There is a strong relationship between exposure to asbestos and development of mesothelioma (Reid et al. 2014). Benzene is a chemical found in gasoline, smoking, and pollution. According to American Cancer Society, people who are regularly exposed to high amount of benzene are at risk of developing cancers, such as acute lymphocytic leukemia

(ALL), acute myeloid leukemia (AML), chronic lymphocytic leukemia (CLL), non-Hodgkin lymphoma, and multiple myeloma. Skin cancer is often caused by exposure to the UV rays of the sun. Sunburn or a tan is actually the result of cell damage caused by the sun. Skin cancer can be prevented in most cases (Watson et al. 2016).

3.2.5 Genetics

Genetics is one of the key factors of cancer, and plays major role in cancer development. If there is any history of cancer in one's family, like breast carcinoma, taking extra precautions is significant. When cancer is genetic, a mutated gene can be passed down to the future descendants (Pomerantz and Freedman 2011). The metabolism of cancerous cells is controlled by several tumor suppressor genes and oncogenes. Any mutations in these genes can result in suppression or expression of metabolic enzymes which can develop cancer. Humans differ greatly with respect to their genetic makeup and similarly, as does the human populations of different regions; environmental factors are very much diverse around the globe. As a result, wide variation is seen in the prevalence of certain types of cancer in different regions. This can be illustrated by the risk of colon cancer and its associated mortality rate, which is different in various regions of Europe. Germany has reported the highest colon cancer cases and deaths, while, in comparison, Greece shows almost 50% less colon cancer cases and deaths (Carlberg et al. 2016). Of note is the epigenetic mechanisms of individuals which are more affected by the environmental factors, compared to genes. Several population based studies showed that migrant people are more prone to developing cancers which are highly prevalent in their host country (Carlberg et al. 2016).

4 Diabetes Mellitus

Diabetes mellitus is a pancreas disorder (an insulin producing gland) and results in high blood glucose level. The pancreas is located between the stomach and the spine, and plays a vital role in the process of digestion by releasing insulin into the blood (Diamond 2003; Wannamethee et al. 2001). Insulin is a hormone that regulates the blood glucose levels. Due to some physiological abnormalities, the release of this hormone may not function properly, leading to higher levels of glucose in the blood (Diamond 2003).

Diabetes mellitus is generally divided into four categories: type 1, type 2, gestational diabetes, and other categories of diabetes (Dungan et al. 2000). Type I diabetes may happen at any age. Type I diabetes happens because of autoimmune destruction of beta cells of pancreas. Many diabetes patients have markers of autoimmune destruction at the time of diagnose, comprising of antibodies of islet cells, tyrosine phosphatases IA-2, IA-2b, glutamic acid decarboxylase, and insulin. In younger individuals, there tends to be a rapid rate of beta-cell destruction and development of ketoacidosis. Older individuals tend to have a more indolent course and later development of ketoacidosis, leading to that them often being labeled as having latent autoimmune diabetes (Opara and Dagogo-Jack 2019). Type 2 diabetes manifests as insulin resistance. In majority of the patients, initial plasma insulin concentration tends to increase, although insufficient to obtain glucose homeostasis due to insulin receptor resistance (Maina 2018). Over time, due to progressive failure of the beta-cell, absolute insulin deficiency tends to develop. A small number of patients diagnosed with type 2 diabetes will have severe insulinopenia with normal or near-normal insulin sensitivity at the time of diagnosis (Dungan et al. 2000).

Gestational diabetes occurs when glucose intolerance is identified during pregnancy. Diabetes can also be caused by other factors, including genetic defects, diseases of the exocrine pancreas, endocrinopathies, infections, and drugs (Dungan et al. 2000). The genetic defects include mature-onset diabetes of the young (MODY), which presents as impaired insulin secretion with little or no insulin resistance (Opara and Dagogo-Jack 2019). It can occur within the first 6 months of life (neonatal) and can be transient or permanent (Dungan et al. 2000). It can also occur at later age, although many exhibit mild hyperglycemia at early age. It has autosomal-dominant inheritance, with the varying natural history depending on the underlying genetic defect (Hasler et al. 1995). There are various genetic disorders that can involve mutations of the insulin receptor, leading to insulin resistance (Dungan et al. 2000). Additionally, there are several genetic disorders that could lead to diabetes whose mechanisms are still controversial (Opara and Dagogo-Jack 2019). Fibrotic changes to the pancreatic parenchyma, due to diseases such as cystic fibrosis or chronic pancreatitis, can lead to diabetes. In addition, diseases such as pancreatic cancer or pancreatic trauma can also lead to diabetes. In the presence of pre-existing failure of beta-cell, other endocrinopathies, such as cortisol, growth hormone, glucagon, and epinephrine, can exacerbate insulin resistance. Infectious disease, such as congenital rubella, can predispose individual to development of diabetes. Other infectious agents are thought to predispose individuals to diabetes (Hasler et al. 1995). Various drugs can also induce insulin resistance and impair beta-cell function (Dungan et al. 2000).

With such a disease as diabetes, it is imperative that patient learns about this illness and how to manage it. Such an undertaking is significant in ensuring that, in circumstances where diabetic patients experience complications from this disease and are not close to a hospital or a medically trained professional, they can respond to such situations in an effective manner. Diamond (2003) argues that, in such a case, the knowledge the diabetic patients have on diabetes might be the only thing that will save their life. It is also vital to note that diabetes is lifestyle illness and the routine or day-to-day undertakings of an individual with diabetes is of great concern, especially in relation to having adequate knowledge about this disease (Maina, 2018).

4.1 *Causes of Type 1 Diabetes Mellitus*

To realize the fundamentals of type I diabetes mellitus (T1DM), one needs to know the activities that lead to the establishment of this disease (Fig. 5). Hitherto, the etiology of diabetes have been classified into 3 categories, which include genetic susceptibility, viral, and environmental factors.

4.1.1 Genetic Susceptibility Factors

Type 1 diabetes mellitus can be hereditary. Epidemiological studies have revealed an elevated prevalence (6% in siblings, compared to 0.4% in common population) of T1DM among the families of T1DM patients due to genetic susceptibility (Pociot and McDermott 2002). The immunological studies of T1DM was developed on the hypothesis that insulin is known as non-self-substance in people suffering from T1DM. Therefore, it is rational to accept that either an impairment in insulin structure or in recognition site can be responsible for T1DM (Acharjee et al. 2013).

In mammals, each nucleated cell has distinct marker molecule that is expressed at the surface of each cell, which helps in cell identification. Major histocompatibility complex (MHC) is a group of genes responsible for the synthesis of these marker molecules (Willey et al. 2009). Human leukocyte antigen complex is responsible for the same task in human beings. Two chromosomal sections in the human species genome have presented with significant and consistent evidence of a link with T1DM. These including HLA at the insulin gene region at chromosome 11 and at chromosome 6. In this context, HLA genes are categorized into 3 distinct classes (class I, II, and III) according to the type of molecules they produced (Pociot and McDermott 2002).

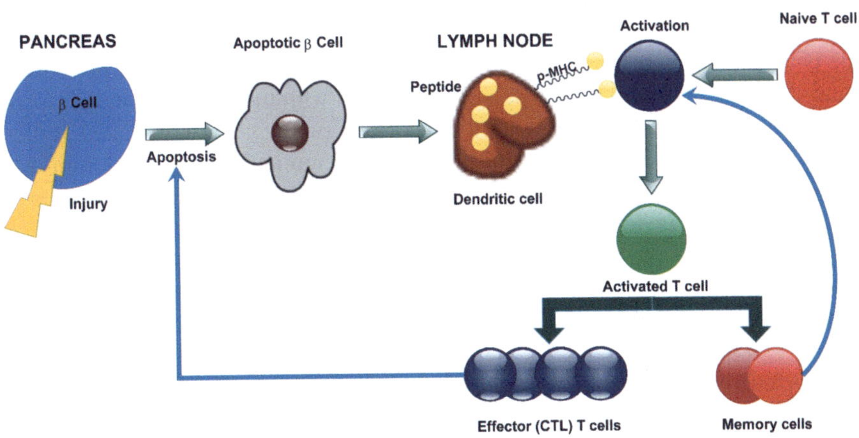

Fig. 5 Etiology of type I diabetes mellitus

Class I marker molecules are established at the surface of all nucleated cells and these cells are recognized as self. Class II molecules are found only on surface of particular cell that vigorously participate in body defense system including dendritic cells, epithelial cells, macrophages, and T cells. In T1DM, class II genes are of chief significance and can be further separated into 3 subclasses, which include HLA-DR, HLA-DP, and HLA-DQ. Several genetic studies have depicted that few variants of HLA-DR and DQ genes (HLA-DRB1, DQA1, and DQB1) take part in the genetic predisposition to T1DM. Amongst the HLA-DQ is a fierce vulnerability (Redondo and Concannon 2020). The insulin genetic region on chromosome 11 is the second significant genetic susceptibility for T1DM. Genetic studies on mice and humans have also shown the presence of insulin as autoantigen and antibody in the onset of T1DM (Pihoker et al. 2005). Some other chromosomal regions of human genome have also shown possible roles in onset of T1DM, such as interleukin-2-α- chain receptor (IL-2RA), protein tyrosine phosphatase non-receptor type 22 (PTPN-22), and cytotoxic T-lymphocyte antigen-4 (Espino-Paisán et al. 2009).

Recently, genome wide association studies have been applied to categorize genetic loci-associated T1DM. In addition to the traditional methods of evaluating chromosome, genome wide association studies analyze the whole genome for single nucleotide polymorphism (SNP). SNP frequently take place in T1DM patients, and usually linked to the disease. These SNPs are used to mark the vulnerability loci. By applying this technology, several susceptible loci were identified for T1DM, including BACH2, CD69, CD226, CCR5, CLEC16A, CI1QTNF6, CTSH, ERBB3, GLIS3, IFIH1, IL-2, IL-2RA, IL-7R, IL-10, IL-18RAP, IL-19, IL-20, IL-27, PRKCQ, and SH2B3 (Barrett et al. 2009; Cooper et al. 2012).

The occasional complex cascades that lead to T1DM have been streamlined in a model pattern developed by Mahaffy and Edelstein-Keshet (2007). The study pointed out that any impairment of insulin producing β-cells can initiate the onset and the triggering of T cells against self-antigens of the human body system. The impaired β-cells undergo apoptosis which make self-antigen with short peptides. In the pancreatic lymph nodes, these peptides are present on the antigen presenting dendritic cells. The native T cells, when in contact with these kind of antigens, are unable to recognize this as self-system protein and usually get discriminated, resulting in the identification of these substances as foreign antigens. A portion of discriminated T cells persist as memory cells, while other fractions actively take part in the cell apoptosis, leading to T1DM (Acharjee et al. 2013).

4.1.2 Virus-Related Contagions

Studies have shown the association of T1DM with different strains of enteroviruses. Coxsackie virus B-4 has a protein 2C identical to the glutamic-acid-decarboxylase enzyme, which is present on the islets of Langerhans. Due to molecular simulation, protein 2C erroneously assemble as a self-molecule, thus not confronted by the T lymphocytes. Generally, T cells attack the protein envelope, including VP1, VP2, and VP3 of Coxsackie virus; the T cell response decrease significantly in patients with T1DM, compared to normal persons, which ultimately results in destruction of β-cells (Varela-Calvino et al. 2000).

4.1.3 Role of Environment

Epidemiological studies have indicated the role of environment in the development of T1DM. Research conducted on people with same ethnic background but different geographical regions (Estonia vs. Finland), showed changes in incidence of diabetes mellitus, including T1DM. The probability of developing T1DM was 3 times in Finland compared to Estonia (Zimmet et al. 2001). Along with other environmental factors, contact to antigenic substances in early period of life is believed to play role in this development. Some constituents, like serum albumin, gluten, and β-casein are involved in the development of T1DM. Albumin and β-casein from cow milk appear to act via synthesis of T lymphocytes that explicitly attack the β-cells-specific glucose transporter molecule (GLUT-21) (Pozzilli 1998). Gluten sometimes initiated subclinical inflammation of the intestine, which increases the fraction of antagonistic T cells. The state of the β-cells plays significant role in the pathogenesis of T1DM. In addition, high intake of food with high glycemic index increases insulin demand, thus pushing the β-cells to synthesize more insulin that speed up the conversion and temporary storage of glucose. Consequently, the increased weight gain in people may increase the onset of T1DM (Buschard 2011). Strachan (1989) presented a hypothesis which contributes to the knowledge about diabetes mellitus.

4.2 Etiology of Type 2 Diabetes Mellitus

The evidence of the regularization of insulin secretion after surgery helps in the understanding that pancreas and liver behavior in the duration of hypocaloric dieting could lead to pathogeneses and etiology of type 2 diabetes mellitus (T2DM) (Camastra et al. 2007). The deposition of fat in liver and pancreas increase the risk of T2DM (Fig. 6). Fatty liver decreases fasting glucose metabolism and raises export of VLDL triacylglycerol, which enhances fat delivery to all tissues in the body, including the islets (Adiels et al. 2006). Observational studies related to reversal of T2DM showed that if the primary impact of high caloric balance is removed then the whole process is reversible (Lim et al. 2011). Prolonged control of blood glucose level in formerly diagnosed diabetic patients after bariatric surgery shows that diabetes may not occur up to ten years, except significant weight gain take place (Sjöström et al. 2004). For a patient with T2DM, reduction in pancreas and liver fat content will result in relief from fatty acid mediated dysfunctions. Individual tolerance to fat exposure differs. If an individual has T2DM, there could be abnormal fat in the pancreas and liver (Taylor 2013).

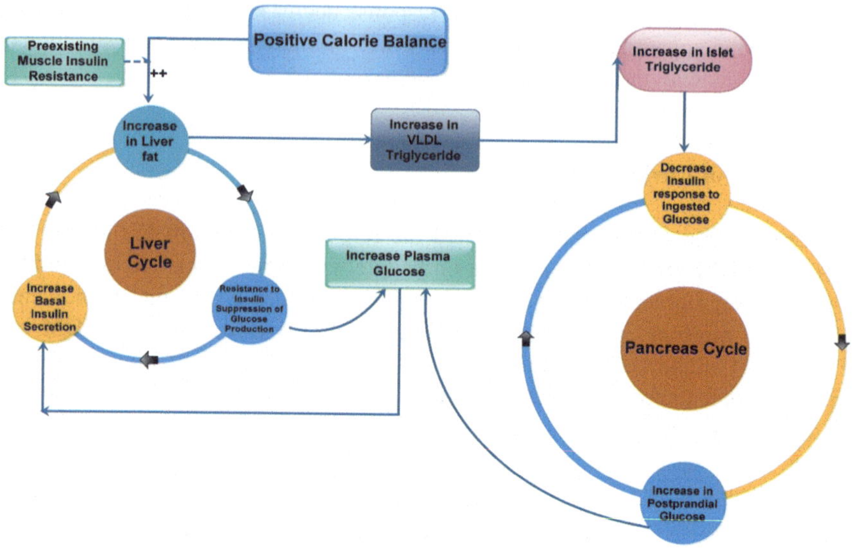

Fig. 6 Etiology of type 2 diabetes

5 Conclusion

Obesity is a heterogenic disease caused by several metabolic factors with the principal cause resulting from high-calorie foods with little or no physical activity. Factors responsible for obesity, include biological, behavioral, and environmental factors. Obesity has been associated with cancer and diabetes. There are different biological, environmental, and behavioral factors which metabolically interact to potentiate these diseases in healthy individuals. Mitigating the problem of obesity, cancer, and diabetes require comprehensive approach. Well-designed studies are required to explore other potential causes and treatments of these diseases.

References

Acharjee, S., Ghosh, B., Al-Dhubiab, B. E., & Nair, A. B. (2013). Understanding type 1 diabetes: Etiology and models. *Canadian Journal of Diabetes, 37*(4), 269–276.

Acosta, A., & Camilleri, M. (2014). Gastrointestinal morbidity in obesity. *Annals of the New York Academy of Sciences, 1311*, 42.

Acosta, A., Camilleri, M., Shin, A., Vazquez-Roque, M. I., Iturrino, J., Burton, D., & Zinsmeister, A. R. (2015). Quantitative gastrointestinal and psychological traits associated with obesity and response to weight-loss therapy. *Gastroenterology, 148*(3), 537–546.

Adiels, M., Taskinen, M. R., Packard, C., Caslake, M. J., Soro-Paavonen, A., & Westerbacka, J. (2006). Overproduction of large VLDL particles is driven by increased liver fat content in man. *Diabetologia, 49*(4), 755–765.

Aldhoon-Hainerová, I., Včelák, J., & Zamrazilová, H. (2014). Monogenic obesity-current status of molecular genetic research and clinical importance. *Casopis lekaru ceskych, 153*(4), 200–206.

Angulo, M., Butler, M., & Cataletto, M. (2015). Prader-Willi syndrome: A review of clinical, genetic, and endocrine findings. *Journal of Endocrinological Investigation, 38*(12), 1249–1263.

Anyanwu, G. O., Kolb, A. F., & Bermano, G. (2020). Antiobesity functional leads and targets for drug development. In Egbuna et al. (Eds.), *Phytochemicals as lead compounds for new drug discovery* (pp. 143–160). https://doi.org/10.1016/b978-0-12-817890-4.00009-3.

Apalasamy, Y. D., & Mohamed, Z. (2015). Obesity and genomics: Role of technology in unraveling the complex genetic architecture of obesity. *Human Genetics, 134*(4), 361–374.

Arjona, A. A., Zhang, S. X., Adamson, B., & Wurtman, R. J. (2004). An animal model of antipsychotic-induced weight gain. *Behavioural Brain Research, 152*(1), 121–127.

Audrain-McGovern, J., & Benowitz, N. (2011). Cigarette smoking, nicotine, and body weight. *Clinical Pharmacology & Therapeutics, 90*(1), 164–168.

Baidal, J. A. W., & Taveras, E. M. (2012). Childhood obesity: Shifting the focus to early prevention. *Archives of Pediatrics & Adolescent Medicine, 166*(12), 1179–1181.

Bali, A., Singh, M. P., Padmavathi, K. M., & Ahmed, J. (2010). Malignant Fibrous Histiocytoma - An Unusual Transformation from Benign to Malignant. *Journal of Cancer Science and Therapy, 2*, 053–057.

Baptiste-Roberts, K., Nicholson, W. K., Wang, N.-Y., & Brancati, F. L. (2012). Gestational diabetes and subsequent growth patterns of offspring: The National Collaborative Perinatal Project. *Maternal and Child Health Journal, 16*(1), 125–132.

Barrett, J. C., Clayton, D. G., Concannon, P., Akolkar, B., Cooper, J. D., & Erlich, H. A. (2009). Genome-wide association study and meta-analysis find that over 40 loci affect risk of type 1 diabetes. *Nature Genetics, 41*(6), 703–707.

Beksinska, M. E., Smit, J. A., Kleinschmidt, I., Milford, C., & Farley, T. M. (2010). Prospective study of weight change in new adolescent users of DMPA, NET-EN, COCs, nonusers and discontinuers of hormonal contraception. *Contraception, 81*(1), 30–34.

Bray, G. A. (2014). *Handbook of obesity—Volume 1: Epidemiology, etiology, and physiopathology* (Vol. 1). CRC Press.

Buschard, K. (2011). What causes type 1 diabetes? Lessons from animal models. *Apmis, 119*, 1–19.

Camastra, S., Manco, M., Mari, A., Greco, A. V., Frascerra, S., Mingrone, G., & Ferrannini, E. (2007). β-cell function in severely obese type 2 diabetic patients: Long-term effects of bariatric surgery. *Diabetes Care, 30*(4), 1002–1004.

Cani, P. D., & Delzenne, N. M. (2011). The gut microbiome as therapeutic target. *Pharmacology & Therapeutics, 130*(2), 202–212.

Cappuccio, F. P., Taggart, F. M., Kandala, N.-B., Currie, A., Peile, E., Stranges, S., & Miller, M. A. (2008). Meta-analysis of short sleep duration and obesity in children and adults. *Sleep, 31*(5), 619–626.

Carlberg, C., Ulven, S. M., & Molnár, F. (2016). *Nutrigenomics*. Springer.

Chagnon, Y. C., Rankinen, T., Snyder, E. E., Weisnagel, S. J., Pérusse, L., & Bouchard, C. (2003). The human obesity gene map: The 2002 update. *Obesity Research, 11*(3), 313–367.

Chung, W. K. (2012). An overview of mongenic and syndromic obesities in humans. *Pediatric blood & cancer, 58*(1), 122–128.

Cooper, J., Howson, J., Smyth, D., Walker, N., Stevens, H., Yang, J. H. M., She, J.-X., Eisenbarth, G. S., Rewers, M., Todd, J. A., Akolkar, B., Concannon, P., Erlich, H. A., Julier, C., Morahan, G., Nerup, J., Nierras, C., Pociot, F., & Rich, S. S. (2012). Confirmation of novel type 1 diabetes risk loci in families. *Diabetologia, 55*(4), 996–1000.

Corrao, G., Bagnardi, V., Zambon, A., & La Vecchia, C. (2004). A meta-analysis of alcohol consumption and the risk of 15 diseases. *Preventive Medicine, 38*(5), 613–619.

Cunningham, S. A., Datar, A., Narayan, K. V., & Kramer, M. R. (2017). Entrenched obesity in childhood: Findings from a national cohort study. *Annals of Epidemiology, 27*(7), 435–441.

Cunningham, S. A., Kramer, M. R., & Narayan, K. (2014). Incidence of childhood obesity in the United States. *The New England Journal of Medicine, 370*, 403–411.

Dallman, M. F., Pecoraro, N. C., & la Fleur, S. E. (2005). Chronic stress and comfort foods: Self-medication and abdominal obesity. *Brain, Behavior, and Immunity, 19*(4), 275–280.

Danese, A., & Tan, M. (2014). Childhood maltreatment and obesity: Systematic review and meta-analysis. *Molecular psychiatry, 19*(5), 544–554.

Darmasseelane, K., Hyde, M. J., Santhakumaran, S., Gale, C., & Modi, N. (2014). Mode of delivery and offspring body mass index, overweight and obesity in adult life: A systematic review and meta-analysis. *PLoS One, 9*(2), e87896.

Davis, S. R., Castelo-Branco, C., Chedraui, P., Lumsden, M. A., Nappi, R. E., Shah, D., & Writing Group of the International Menopause Society for World Menopause Day. (2012). Understanding weight gain at menopause. *Climacteric, 15*(5), 419–429.

de Zwaan, M., Müller, A., Allison, K. C., Brähler, E., & Hilbert, A. (2014). Prevalence and correlates of night eating in the German general population. *PLoS One, 9*(5), e97667.

Diamond, J. (2003). The double puzzle of diabetes. *Nature, 423*(6940), 599–602.

Dixon, J. B. (2010). The effect of obesity on health outcomes. *Molecular and Cellular Endocrinology, 316*(2), 104–108.

Dubowitz, T., Ghosh-Dastidar, M., Steiner, E., Escarce, J. J., & Collins, R. L. (2013). Are our actions aligned with our evidence? The skinny on changing the landscape of obesity. *Obesity, 21*(3), 419–420.

Dungan, K., Grossman, A., Hershman, J., Koch, C., McLachlan, R., & New, M. (2000). *SourceEndotext [Internet]*. South Dartmouth (MA): MDText. com. Inc..

Espino-Paisán, L., Urcelay, E., & Santiago, J. L. (2009). An insight into the genetics of type 1 diabetes. *Inmunología, 28*(4), 173–181.

Farooqi, I. S., & O'rahilly, S. (2008). Mutations in ligands and receptors of the leptin-melanocortin pathway that lead to obesity. *Nature Clinical Practice Endocrinology & Metabolism, 4*(10), 569–577.

Fava, M., Judge, R., Hoog, S. L., Nilsson, M. E., & Koke, S. C. (2000). Fluoxetine versus sertraline and paroxetine in major depressive disorder: Changes in weight with long-term treatment. *The Journal of Clinical Psychiatry, 61*(11), 863–867.

Filozof, C., Fernandez Pinilla, M., & Fernández-Cruz, A. (2004). Smoking cessation and weight gain. *Obesity Reviews, 5*(2), 95–103.

Flint, H. J., Scott, K. P., Duncan, S. H., Louis, P., & Forano, E. (2012). Microbial degradation of complex carbohydrates in the gut. *Gut Microbes, 3*(4), 289–306.

Fowke, J. H., McLerran, D. F., Gupta, P. C., He, J., Shu, X. O., Ramadas, K., et al. (2015). Associations of body mass index, smoking, and alcohol consumption with prostate cancer mortality in the Asia Cohort Consortium. *American Journal of Epidemiology, 182*(5), 381–389.

Fox, K., & Hillsdon, M. (2007). Physical activity and obesity. *Obesity Reviews, 8*(Suppl. 1), 115–121.

French, S. A., Epstein, L. H., Jeffery, R. W., Blundell, J. E., & Wardle, J. (2012). Eating behavior dimensions. Associations with energy intake and body weight. A review. *Appetite, 59*(2), 541–549.

Friedman, J. M. (2003). A war on obesity, not the obese. *Science, 299*, 856–858.

Furrukh, M. (2013). Tobacco smoking and lung cancer: Perception-changing facts. *Sultan Qaboos University Medical Journal, 13*(3), 345.

Gariepy, G., Nitka, D., & Schmitz, N. (2010). The association between obesity and anxiety disorders in the population: A systematic review and meta-analysis. *International Journal of Obesity, 34*(3), 407–419.

Hamilton, M. T., Hamilton, D. G., & Zderic, T. W. (2007). Role of low energy expenditure and sitting in obesity, metabolic syndrome, type 2 diabetes, and cardiovascular disease. *Diabetes, 56*(11), 2655–2667.

Hasler, W. L., Soudah, H. C., Dulai, G., & Owyang, C. (1995). Mediation of hyperglycemia-evoked gastric slow-wave dysrhythmias by endogenous prostaglandins. *Gastroenterology, 108*(3), 727–736.

Hemmingsson, E. (2014). A new model of the role of psychological and emotional distress in promoting obesity: Conceptual review with implications for treatment and prevention. *Obesity Reviews, 15*(9), 769–779.

Hill, J. O., Wyatt, H. R., & Peters, J. C. (2012). Energy balance and obesity. *Circulation, 126*(1), 126–132.

Hills, A. P., Andersen, L. B., & Byrne, N. M. (2011). Physical activity and obesity in children. *British Journal of Sports Medicine, 45*(11), 866–870.

Hinney, A., Vogel, C. I., & Hebebrand, J. (2010). From monogenic to polygenic obesity: Recent advances. *European Child & Adolescent Psychiatry, 19*(3), 297–310.

Hu, F. B. (2003). Sedentary lifestyle and risk of obesity and type 2 diabetes. *Lipids, 38*(2), 103–108.

Hussain, S., & Bloom, S. (2013). The regulation of food intake by the gut-brain axis: Implications for obesity. *International Journal of Obesity, 37*(5), 625–633.

Johansson, K., Neovius, M., & Hemmingsson, E. (2014). Effects of anti-obesity drugs, diet, and exercise on weight-loss maintenance after a very-low-calorie diet or low-calorie diet: A systematic review and meta-analysis of randomized controlled trials. *The American Journal of Clinical Nutrition, 99*(1), 14–23.

Jones, L. P., Buelto, D., Tago, E., & Owusu-Boaitey, K. E. (2011). Abnormal mammary adipose tissue environment of Brca1 mutant mice show a persistent deposition of highly vascularized multilocular adipocytes. *Journal of Cancer Science & Therapy, 1*(Suppl. 2), 4.

Kubzansky, L. D., Bordelois, P., Jun, H. J., Roberts, A. L., Cerda, M., Bluestone, N., & Koenen, K. C. (2014). The weight of traumatic stress: A prospective study of posttraumatic stress disorder symptoms and weight status in women. *JAMA Psychiatry, 71*(1), 44–51.

Kurioka, D., Takagi, A., Yoneda, M., Hirokawa, Y., Shiraishi, T., & Watanabe, M. (2011). Multicellular spheroid culture models: Applications in prostate cancer research and therapeutics. *Journal of Cancer Science and Therapy, 3*(3), 060–065.

Lambrinoudaki, I., Brincat, M., Erel, C. T., Gambacciani, M., Moen, M. H., Schenck-Gustafsson, K., et al. (2010). EMAS position statement: Managing obese postmenopausal women. *Maturitas, 66*(3), 323–326.

Lau, G. S. K., Chan, J. Y. W., & Wei, W. I. (2011). Role of surgery in the treatment of radiation-induced sarcomas of the head and neck. *Journal of Cell Science & Therapy, S2*, 002.

Lim, E. L., Hollingsworth, K., Aribisala, B. S., Chen, M., Mathers, J., & Taylor, R. (2011). Reversal of type 2 diabetes: Normalisation of beta cell function in association with decreased pancreas and liver triacylglycerol. *Diabetologia, 54*(10), 2506–2514.

Liu, Z. L. (2010). Unification of gene expression data applying mRNA quantification references for comparable analyses. *Journal of Microbial and Biochemical Technology, 2*, 124–126.

López-Lázaro, M. (2016). What is the main cause of cancer? *Cancer Studies and Terapeutics, 1*(1), 1–4.

Luppino, F. S., de Wit, L. M., Bouvy, P. F., Stijnen, T., Cuijpers, P., Penninx, B. W., & Zitman, F. G. (2010). Overweight, obesity, and depression: A systematic review and meta-analysis of longitudinal studies. *Archives of General Psychiatry, 67*(3), 220–229.

Mahaffy, J. M., & Edelstein-Keshet, L. (2007). Modeling cyclic waves of circulating T cells in autoimmune diabetes. *SIAM Journal on Applied Mathematics, 67*(4), 915–937.

Maina, J. W. (2018). Understanding the types and causes of diabetes mellitus. *International Journal of Biology, 3*(1), 202–207.

Mannan, M., Doi, S. A., & Mamun, A. A. (2013). Association between weight gain during pregnancy and postpartum weight retention and obesity: A bias-adjusted meta-analysis. *Nutrition Reviews, 71*(6), 343–352.

Matthews, E., Wei, J., & Cunningham, S. (2017). Relationship between prenatal growth, postnatal growth and childhood obesity: A review. *European Journal of Clinical Nutrition, 71*(8), 919–930.

Mozaffarian, D., Hao, T., Rimm, E. B., Willett, W. C., & Hu, F. B. (2011). Changes in diet and lifestyle and long-term weight gain in women and men. *New England Journal of Medicine, 364*(25), 2392–2404.

Murphy, E. F., Clarke, S. F., Marques, T. M., Hill, C., Stanton, C., Ross, R. P., & Cotter, P. D. (2013). Antimicrobials: Strategies for targeting obesity and metabolic health? *Gut Microbes, 4*(1), 48–53.

Myers, M. G., Cowley, M. A., & Münzberg, H. (2008). Mechanisms of leptin action and leptin resistance. *Annual Review of Physiology, 70*, 537–556.

Nagy, M. A. (2011). HIF-1 is the Commander of gateways to cancer. *Journal of Cancer Science and Clinical Therapies, 3*, 035–040.

O'Rahilly, S. (2009). Human genetics illuminates the paths to metabolic disease. *Nature, 462*(7271), 307–314.

Opara, E. C., & Dagogo-Jack, S. (2019). *Nutrition and diabetes: Pathophysiology and management*. CRC Press.

Paluska, S. A., & Schwenk, T. L. (2000). Physical activity and mental health. *Sports Medicine, 29*(3), 167–180.

Pereira-Lancha, L. O., Campos-Ferraz, P. L., & Lancha, A. H. (2012). Obesity: Considerations about etiology, metabolism, and the use of experimental models. *Diabetes, Metabolic Syndrome and Obesity: Targets and Therapy, 5*, 75.

Pihoker, C., Gilliam, L. K., Hampe, C. S., & Lernmark, Å. (2005). Autoantibodies in diabetes. *Diabetes, 54*(2), 52–61.

Pociot, F., & McDermott, M. (2002). Genetics of type 1 diabetes mellitus. *Genes & Immunity, 3*(5), 235–249.

Pomerantz, M. M., & Freedman, M. L. (2011). The genetics of cancer risk. *Cancer Journal (Sudbury, Mass.), 17*(6), 416.

Pozzilli, P. (1998). Prevention of insulin-dependent diabetes mellitus. *Diabetes/Metabolism Reviews, 14*(1), 69–84.

Puhl, R., & Brownell, K. D. (2001). Bias, discrimination, and obesity. *Obesity Research, 9*(12), 788–805.

Rao, K. R., Lal, N., & Giridharan, N. (2014). Genetic & epigenetic approach to human obesity. *The Indian Journal of Medical Research, 140*(5), 589.

Redondo, M. J., & Concannon, P. (2020). Genetics of type 1 diabetes comes of age. *Diabetes Care, 43*(1), 16–18.

Reid, A., De Klerk, N. H., Magnani, C., Ferrante, D., Berry, G., Musk, A. W., & Merler, E. (2014). Mesothelioma risk after 40 years since first exposure to asbestos: A pooled analysis. *Thorax, 69*(9), 843–850.

Robinson, M. R., Hemani, G., Medina-Gomez, C., Mezzavilla, M., Esko, T., Shakhbazov, K., & Gustafsson, S. (2015). Population genetic differentiation of height and body mass index across Europe. *Nature Genetics, 47*(11), 1357–1362.

Rooney, B. L., Schauberger, C. W., & Mathiason, M. A. (2005). Impact of perinatal weight change on long-term obesity and obesity-related illnesses. *Obstetrics & Gynecology, 106*(6), 1349–1356.

Saravanan, K., Egbuna, C., Averal, H. I., Kannan, S., Elavarasi, S., & Bahadur, B. (Eds.). (2020). *Drug development for cancer and diabetes: A path to 2030* (1st ed.). New York: Apple Academic Press. https://doi.org/10.1201/9780429330490.

Sato, T., Ida, T., Nakamura, Y., Shiimura, Y., Kangawa, K., & Kojima, M. (2014). Physiological roles of ghrelin on obesity. *Obesity Research & Clinical Practice, 8*(5), 405–413.

Scheideler, J. K., & Klein, W. M. (2018). Awareness of the link between alcohol consumption and cancer across the world: A review. *Cancer Epidemiology and Prevention Biomarkers, 27*(4), 429–437.

Seitz, H. K., Pelucchi, C., Bagnardi, V., & Vecchia, C. L. (2012). Epidemiology and pathophysiology of alcohol and breast cancer. *Alcohol and Alcoholism, 47*(3), 204–212.

Singh, R., Rohilla, R. K., Saini, G., & Kaur, K. (2014). Longitudinal study of body composition in spinal cord injury patients. *Indian Journal of Orthopaedics, 48*, 168–177.

Sjöström, L., Lindroos, A. K., Peltonen, M., Torgerson, J., Bouchard, C., & Carlsson, B. (2004). Lifestyle, diabetes, and cardiovascular risk factors 10 years after bariatric surgery. *New England Journal of Medicine, 351*(26), 2683–2693.
Strachan, D. P. (1989). Hay fever, hygiene, and household size. *British Medical Journal, 299*(6710), 1259.
Sun, S., Schiller, J. H., & Gazdar, A. F. (2007). Lung cancer in never smokers—A different disease. *Nature Review Cancer, 7*, 778–790.
Taylor, R. (2013). Type 2 diabetes: Etiology and reversibility. *Diabetes Care, 36*(4), 1047–1055.
Thayer, K. A., Heindel, J. J., Bucher, J. R., & Gallo, M. A. (2012). Role of environmental chemicals in diabetes and obesity: A National Toxicology Program workshop review. *Environmental Health Perspectives, 120*(6), 779–789.
Turati, F., Galeone, C., Rota, M., Pelucchi, C., Negri, E., Bagnardi, V., & La Vecchia, C. (2014). Alcohol and liver cancer: A systematic review and meta-analysis of prospective studies. *Annals of Oncology, 25*(8), 1526–1535.
Vanita, P., Subrahmanyam, V., & Jhansi, K. (2011). A short note on cancer. *Journal of Carcinogene Mutagene, 2*, 128.
Varela-Calvino, R., Sgarbi, G., Arif, S., & Peakman, M. (2000). T-Cell reactivity to the P2C non-structural protein of a diabetogenic strain of coxsackievirus B4. *Virology, 274*(1), 56–64.
Vijay-Kumar, M., Aitken, J. D., Carvalho, F. A., Cullender, T. C., Mwangi, S., Srinivasan, S., & Gewirtz, A. T. (2010). Metabolic syndrome and altered gut microbiota in mice lacking Toll-like receptor 5. *Science, 328*(5975), 228–231.
Wang, K., Li, W.-D., Zhang, C. K., Wang, Z., Glessner, J. T., Grant, S. F., & Price, R. A. (2011). A genome-wide association study on obesity and obesity-related traits. *PLoS One, 6*(4), e18939.
Wannamethee, S. G., Shaper, A. G., & Perry, I. J. (2001). Smoking as a modifiable risk factor for type 2 diabetes in middle-aged men. *Diabetes Care, 24*(9), 1590–1595.
Ward, Z. J., Long, M. W., Resch, S. C., Giles, C. M., Cradock, A. L., & Gortmaker, S. L. (2017). Simulation of growth trajectories of childhood obesity into adulthood. *New England Journal of Medicine, 377*, 2145–2153.
Watson, M., Holman, D. M., & Maguire-Eisen, M. (2016). Ultraviolet radiation exposure and its impact on skin cancer risk. *Seminars in Oncology Nursing, 32*(3), 241–254.
Weiderpass, E. (2010). Lifestyle and cancer risk. *Journal of Preventive Medicine and Public Health, 43*(6), 459–471.
Willey, J. M., Sherwood, L., & Woolverton, C. J. (2009). *Prescott's principles of microbiology* (pp. 126–152). New York: McGraw-Hill.
Withrow, D., & Alter, D. A. (2011). The economic burden of obesity worldwide: A systematic review of the direct costs of obesity. *Obesity Reviews, 12*(2), 131–141.
Woodward-Lopez, G., Kao, J., & Ritchie, L. (2011). To what extent have sweetened beverages contributed to the obesity epidemic? *Public Health Nutrition, 14*(3), 499–509.
World Health Organization. (2018). https://www.who.int/news-room/fact-sheets/detail/cancer
Wurtman, J. J., & Wurtman, R. J. (2015). Depression can beget obesity can beget depression. *Journal of Clinical Psychiatry, 76*(12), 1619–1621.
Zhang, W. M., Kuchar, S., & Mozes, S. (1994). Body fat and RNA content of the VMH cells in rats neonatally treated with monosodium glutamate. *Brain Research Bulletin, 35*(4), 383–385.
Zimmet, P., Alberti, K., & Shaw, J. (2001). Global and societal implications of the diabetes epidemic. *Nature, 414*(6865), 782–787.

Pathophysiology of Obesity and Diabetes

Tabussam Tufail, Aiman Ijaz, Sana Noreen, Muhammad Umair Arshad, Syed Amir Gilani, Shahid Bashir, Ahmad Din, Muhammad Zia Shahid, Ammar Ahmad Khan, Anees Ahmed Khalil, and Chinaza Godswill Awuchi

1 Introduction

Obesity or overweight is characterized by extra amount of adipose tissues to the degree that influences the physiological, psychosocial, physical health, and wellbeing of an individual. The prevalence of obesity is increasing dramatically in many countries around the world. The incidence of obesity can be reduced when economic cost, physical awareness, dietary patterns, socioeconomic hazards, morbidity, and disease mortality are taken into consideration (Boulton et al. 2005). The

T. Tufail (✉)
University Institute of Diet & Nutrition Sciences, Faculty of Allied Health Sciences, The University of Lahore, Lahore, Pakistan

Institute of Home & Food Sciences, Government College University Faisalabad, Faisalabad, Pakistan
e-mail: tabussam.tufail@dnsc.uol.edu.pk

A. Ijaz · S. Noreen · S. A. Gilani · S. Bashir · M. Z. Shahid · A. A. Khan · A. A. Khalil
University Institute of Diet & Nutrition Sciences, Faculty of Allied Health Sciences, The University of Lahore, Lahore, Pakistan
e-mail: aiman.ijaz@dnsc.uol.edu.pk; shahid.bashir@rsmi.uol.edu.pk; zia.shahid@dnsc.uol.edu.pk; ammar.ahmad@rsmi.uol.edu.pk; anees.khalil@dnsc.uol.edu.pk

M. U. Arshad
Institute of Home & Food Sciences, Government College University Faisalabad, Faisalabad, Pakistan

A. Din
National Institute of Food Science & Technology, University of Agriculture Faisalabad, Faisalabad, Pakistan

C. G. Awuchi
School of Natural and Applied Sciences, Kampala International University, Kampala, Uganda
e-mail: awuchi.chinaza@kiu.ac.ug

C. Egbuna, S. Hassan (eds.), *Dietary Phytochemicals*,
https://doi.org/10.1007/978-3-030-72999-8_2

high prevalence of Type 2 diabetes mellitus (T2DM) among adults and children can be correlated with the prevalence of obesity (Bhupathiraju and Hu 2016). Among the adult population in many developed nations, such as the United States, approximately 60–70% of the population falls under the category of overweight or obese. About 171 million people worldwide are suffering from T2DM, which is expected to increase to at least 360 million or more by 2030 (Katsarou et al. 2017; Egbuna et al. 2019; Saravanan et al. 2020; Tupas et al. 2020). Obesity and T2DM are linked to insulin tolerance.

β-cells of the islet of Langerhans of the pancreas produces insulin to control the reduction in levels of insulin, in order to maintain normal glucose tolerance. History of T2DM shows that dysfunction of endothelial cells occurs along with obesity or resistance of insulin in people suffering from DM. β-cells of the pancreas may not compensate adequately for a reduction in insulin sensitivity to initiate resistance of insulin and obesity, thus leading to T2DM. In obese people, "non-esterified fatty acids" (NEFAs) secreted from adipose tissue may enhance the resistance of insulin; and dysfunction of β-cell is mostly correlated (Care 2019). β-cells play vital role in ensuring that the levels of glucose in the blood lie within a normal range in healthy people, which may contribute to an increase in the availability of insulin to the β-cells when the fat tissues, muscles, liver, and kidneys need glucose. The alterations in insulin sensitivity must be matched by an equal change in circulating levels of insulin when the levels of glucose require stability. Failure in this procedure ultimately causes high glucose levels in the blood (Czech 2017).

There is an effective response to insulin resistance if a person has healthy β-cells. On the other hand, excessive glucose resistance or elevated fasting glucose may be present if pancreatic β-cells become impaired, and the development in of T2DM can ensue. A continuous decrease in β-cell activity is a major factor in the onset of T2DM. It has already been shown through various studies that dysfunction of β-cell causes high post-prandial blood glucose levels and inadequate insulin production. There would also be reduced potency of liver and muscle glucose uptake because of insufficient or inadequate liver glucose output inhibition. Diabetes and obesity have been reported to be chronic medical issues that are continuously increasing all over the world. Body mass index has a major link with insulin tolerance and diabetes (Anyanwu et al. 2020). Being obese in the early phases of life has been correlated with the occurrence of DM. Furthermore, the latest methods for the treatment of DM and its prevention in individuals who are overweight and obese require further studies (Pugazhenthi et al. 2017).

2 Obesity and Body Mass Index

There exists different methods used for the measurement and classification of body fats. One of such methods is the Dual-energy X-ray absorptiometry (DXA). DXA is arguably the most accurate method but its expensive and requires large sampling. Also, body fat thickness is also a common method used for the assessment of fat in the body. Although these methods are accurate, they are also expensive. The WHO and National Heart, Lung, and Blood Institute (NHBLI) have defined the body mass index (BMI) cut off values for obesity as BMI greater than 30 (BMI > 30). BMI is a method that is used for the measurement of body weight. It is defined as weight in Kg divided by height in m^2 (Levian et al. 2014; Peng et al. 2014; Anyanwu et al. 2020).

3 Pathophysiology of Obesity

3.1 Development of Fat Cells

Excess energy in the body accumulates in the adipose tissues. This often occurs when energy intake is more than the energy expenditure. In the body, fat can be measured as both cell number and cell size. A fat cell may continue to increase during positive energy balance. Obese people have a large number of fat cells that are also big in size. As fat in the form of triglycerides starts to accumulate in adipose tissues, their sizes begin to expand which then lead to an increase in cell number, and subsequent development of obesity (McCarthy et al. 2007).

When energy expenditure increases more than energy intake, fat cells start to shrink but their number does not necessarily change, often leading to noticeable reduction in overall body size. This is usually known as wasting in underweight or malnourished individuals. People who lose weight intentionally or unintentionally, can regain weight when this physiological process is reversed (i.e., energy intake being more than energy expenditure), and fat cells regain their increase in size. Excess fat first fills natural storage sites, such as adipose tissues, and then starts to accumulate, around the organs, including the heart, liver, kidney, etc., which play a major role in the development of CVDs, renal failure, insulin resistance, and DM (Nelson-Dooley et al. 2005; Angulo 2007).

3.2 Fat Cell Metabolism

In fat cell metabolism, an important enzyme known as lipoprotein lipase (LPL) is involved. The function of LPL is to remove extra fat from the blood for storage in the muscle cells and adipose tissues. Studies have shown that overweight people have more LPL enzyme activity in adipose tissues compared to normal individuals. This explains why even a moderate amount of fat intake has a significant impact on obese people compared to individuals with lean body mass (Holland et al. 2007). The activity of LPL is also affected by gender. LPL has more activity in women compared to men. In women, LPL activity is more at hips, thighs, and breasts while in men at the abdomen region. That explains the reason why men mostly develop pear-shaped obesity while women develop apple-shaped obesity. Fat breakdown is

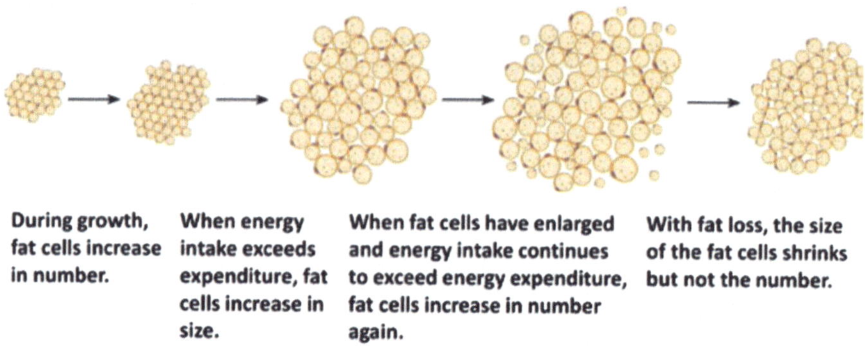

Fig. 1 Fat cell development

less active in the thigh and hip regions; this is one of the reasons why women have more difficulty in losing fat compared to men (Votruba and Jensen 2007).

When a person losses weight, the activity of LPL enzymes increases. Weight reduction gives signal to gene expression of LPL enzymes. This is the reason why people who lose weight are likely to regain some weight due to the enzyme activity. When fat oxidation occurs, it negatively co-relate with body fatness. Fat oxidation leads to excess production of free radicals and contributes to further metabolic impairment (Fig. 1) (Patalay et al. 2005).

3.3 Oxidative Stress

When reactive oxygen species (ROS) increases in the body more than antioxidants, it causes oxidative stress. Increase in the amount of accumulated fat contributes to oxidative stress, which is a major pathological mechanism in obesity and other metabolic disorders. In cultured cells of adipocytes, increased levels of triglycerides stimulates oxidative stress by activation of NADPH oxidase, which leads to impaired production of adipocytokines that are fat deprived hormones (Westerterp et al. 2008; Major et al. 2007).

3.4 Weight Stigma

Weight stigma is a stressor that can enhance the risk of cardiovascular diseases. Weight stigma, or weight bias internalization (WBI), is broadly viewed as a major stressor nowadays, which can stimulate biochemical stress via alterations in the nervous system (autonomic nervous system), hypothalamic axis, free radicals, and systemic inflammation (Waterland 2008; Pearl et al. 2017). Recent studies indicated that individuals with weight stigma may likely develop high blood pressure and increased levels of cortisol (Waterland 2008). Increased craving for food resulting from biochemical alterations as well as emotional stimulations also play role in increasing the consumption of calories, leading to eating disorder, such as binge eating and recurrent food craving. People who are weight stigmatized in fitness settings have more likelihood of avoiding preventive care, which pose risk for the progression of related complication and disorders (Pearl et al. 2017).

4 Role of Different Factors in the Pathophysiology of Obesity

4.1 Pathophysiology of Obesity and the Role of Autonomic Nervous System

Brain and obesity progression have a two-way relationship. Any change in the functions of the brain may contribute to obesity by effecting different pathways. On the other hand, obesity also contributes to alterations in brain fuction which may cause metabolic impairment and lead to CVDs. The autonomic nervous system (ANS) is a major food regulator involved in regulating the signals of fullness and hunger by stimulating various hormones (Chrysohoou et al. 2007; Amasyali et al. 2008). This system works with the coexistence of a network of signals that transmit information from the peripheral system to the brain. It may act in a short term, such as gastric hormone and intestinal hormones which help in the digestion of food and regulate the satiety level, or in the long term, such as the release of hormones, including leptin and insulin, that cause dysfunctioning in the homeostasis and contribute to weight gain (Asferg et al. 2010; Marinou et al. 2010).

4.2 Osteopontin and Obesity

Osteopontin (Opn) is a phosphoprotein involved in the inflammation of adipose tissues and insulin resistance. It is also identified as a pro-inflammatory cytokine required for macrophages infiltration and proliferation within adipose tissues. Opn expression in adipose tissue-associated macrophages and T-lymphocytes are inducible during inflammation that contributes to adipose tissues and systemic inflammation, respectively, which are hallmarks of obesity (Fitter and Zannettino 2018; Wardle et al. 2008).

4.3 Obesity and Renal Disease

Recent studies indicate that obesity can be linked to kidney diseases. Excess fat accumulates within the body organs such as the heart, kidney, and liver. Evidence support that an increase in glomerular filtration rate (GFR) is a mechanism underlying obesity-related renal disease and dyslipidemia. A study was conducted to estimate the GFR in 12 non-diabetic individuals with grade III obesity and compared with renal hemodynamic of nine healthy individuals (Bayliss et al. 2012). GFR and renal plasma blood flow in obese people were found higher compared to controls, with both having 52% and 32%, respectively. Albumin excretion was also increased. Oral glucose tolerance test indicated an increase in insulin resistance in obese individuals compared to control group. Many mechanisms are involved in the pathophysiology of obesity. Fetuin-A and adiponectin are inversely related to each other. Increase in the level of fetuin-A is directly linked to high BMI and fatty acids, and decreased level of adiponectin (Table 1) (Laughlin et al. 2007; Baumann et al. 2009).

Macrophages are present in adipose tissues and infiltrate the tissues in obese individuals, causing insulin resistance. Studies showed that adipose tissue macrophages (ATMs) present in both animals and humans with overweight directly correlate with insulin sensitivity. When a person loses weight, reduction is seen both in inflammatory cytokines and ATMs. ATMs are components of proinflammatory cytokines, including tumor necrosis factor-α (TNF-α) and interleukin-6, that affect insulin action in adipocyte (McLaughlin et al. 2017). There is a strong connection between low-grade inflammation and insulin sensitivity. When weight increases or a person has BMI > 30, the consequence of obesity affects different organ systems in the body. At the tissue level, it plays role in the regulation of nutrients in patient and contributes to insulin resistance and consequent development of DM. Obesity has link with other diseases such as cancer and pulmonary disease. Recent studies showed that obesity-mediated diseases have a strong relationship with excess intake of nutrients caused by impairment of the cellular and molecular mediators of the immune system and inflammation (Rankinen and Bouchard 2006). Low-grade inflammation is often a response of the immune system to the foreign substances, which can reverse in normal body functions. The inflammatory response that occurs due to obesity stimulates C-reactive protein (CRP), which also activate tissue leucocytes and the stimulation of reparative tissues. Obesity induced low-grade inflammation is different and unique compared to other inflammatory processes (autoimmune diseases, infections). Obesity can stimulate innate immune response that affects the homeostasis of metabolic states (Lumeng et al. 2007; Lumeng and Saltiel 2011).

Table 1 Role of hormone in obesity and renal dysfunctioning

Hormones	Secretion sites	Effects on kidney
Adiponectin	Adipocytes	Energy, Regulate filtration barrier
Fetuin	Liver	Increased fetuin level is associated with decreased adiponectin
Leptin	Adipocytes	Increased blood pressure directly through increasing sympathetic activity
Resistin	Macrophages	Increased level of resistin is associated with decreased GFR

5 Maternal Obesity

Obesity in the reproductive age and during pregnancy contributes to complications for both the mother and the infant, as well as across the infant life span. Institute of Medicine (IOM) recommends the guidelines for expectant mother weight gain and stated that excess weight gains is associated with many complications during gestation and delivery. Different studies have explored low gestational weight gain (GWG) for obese women. Studies showed that low GWG has health benefits for both mother and the child. About 50% population in the US within reproductive age has BMI > 30; different interventions have been developed to regulate the gestational weight gain and its associated health risks, including DM, obesity, and CVDs (ACOG 2013). Pregnant women are the most vulnerable group of any community or country and are important target for different interventions especially for those who have a higher incidence of gestational weight gain and its relationship with postpartum overweight both in infant and mother (Peaceman et al. 2018).

According to the classification of obesity, morbid obesity (also called grade III obesity) is defined as BMI >40 which causes complications in perinatal (Cedergren 2004). A study reported that during antenatal care, primary outcomes are abruption of placenta, pre-eclampsia, placenta previa, and stillbirths, which occur mostly after 28th weeks of the pregnancy (Stothard et al. 2009). Children born from mothers who are obese in the first trimester have 24% of them obese compared to 0.9% in children born by mothers with healthy weight (Whitaker 2004).

6 Potential Role of Gut Inflammation in Disease Development

The intestine is an important part of gastrointestinal tract (GIT) and is prone to complications in obesity-related insulin resistance. This causes changes in gut flora, "dysbiosis", which has a significant impact on the body fat composition, inflammation, and insulin resistance. Gut flora changes cause low-grade systemic inflammation, which can cause leakage of bacterial products; lipopolysaccharides and local inflammation that occur in the colon and bowel. Systemically, this products accumulate and cause inflammation in the adipose tissues, such as visceral adipose tissues. In a study, diet-induced obese mice indicated that obesity causes a release of pro-inflammatory adipokines in lamina propria immune cells (McLaughlin et al. 2017). Repression of gut immune modulation has been linked to a decrease in insulin resistance. Treatment with anti-inflammatory compounds such as 5-aminosalicylic acid reverses inflammation associated with bowel with improved metabolic health. These effects were dependent on adaptive and gut immune response, linked with decrease permeability and endotoxemia, as well as reduce VAT associated inflammation. Gut flora and body weight have been correlated (Rastelli et al. 2018).

7 Childhood Obesity

Childhood obesity is common nowadays and can affect a child throughout his or her lifespan. Children with obesity are at increased risk of kidney failure and heart diseases. Current recommendations by the American college of obstetricians and gynecologists stipulate 15 kg gestational weight gain for healthy women, while women who have BMI >30 should gain low weight during pregnancy. This is because there is a strong association between maternal weight gain and childhood adiposity, CVDs, and renal failure at the age of 5-9 years. Also, women with greater gestational weight gain have children with high BMI, Interleukin-6, C-reactive protein, leptin level, hypertension, as well as a high amount of low-density lipoproteins (Wander et al. 2014; Gademan et al. 2014).

8 Pathophysiology of Diabetes

Primary pathophysiological characteristics in T2DM represents diabetic occurrence, decreased insulin secretion, and enhanced insulin intolerance. Deficiency in pancreatic cell activity indicates improvement over time (Tilg et al. 2017).

Diabetes and its associated chronic health complications may impair the processing of fat, protein, and glucose. It is activated due to the loss of secretion in insulin owing either to a gradual failure of pancreatic β-Langerhans islet cells to generate insulin and deficiency in the absorption of insulin in the peripheral tissue (Dhatariya et al. 2020).

8.1 Etiology of Diabetes

T2DM is often characterized by disruption of pancreatic cells induced by environmental factors, lifestyle, and/or infectious agents. Nearly 80% of T1DM patients have islet cell antibodies; most of them also present anti-insulin antibodies before insulin therapy (Umpierrez 2019). For T2DM, there are so many determinants such as liftestyle and genetic factors, including impaired insulin secretion. Environmental and lifestyle factors: overweight, lack of physical activity, overeating, and aging can influence T2DM development. The increasing urbanization and changes in lifestyle have been associated with the recent increase in T2DM.

8.2 Role of Genetics in Development of T2DM

Type 2 diabetes melitus is strongly linked to environmental and genetic factors, as well as lifestyle. The pathogenesis has been considered to involve genetic abnormalities at the molecular level related to the glucose metabolism regulatory system

(Hotamisligil 2017). The analysis of candidate genes beset at glucose-stimulated insulin secretion of pancreatic cells and molecules comprising the molecular mechanism for insulin action have identified genetic abnormalities which are independent causes of the pathogenesis, including glucokinase genes, mitochondrial genes, and insulin receptor genes (Heindel et al. 2017). Genetic abnormalities have been reported and explained 30% of genetic factors for diabetics; understanding the determinants of genetic material are likely to be practically complete in future (Fuchsberger et al. 2016).

8.3 *Role of Environmental Determinants in the Development of T2DM*

Age, obesity, loss of nutrition, drug intake, and smoking are risk factors of T2DM pathogenesis. Obesity/overweight often result from lack of physical activities (Sanyal et al. 2018). Variations in dietary energy sources, increase in fat intake, decrease in starch with increase in the consumption of simple sugars, and low intake of dietary fiber are among the causes of obesity and deterioration of glucose tolerance (Gittelsohn and Trude 2017) as shown in Fig. 2.

8.4 *Insulin Resistance*

Insulin resistance is a complication that manifest when insulin in the body does not show satisfactory activity. When insulin response is insignificant to maintain proper functions of most organs, such as the liver and muscles, it usually results in the

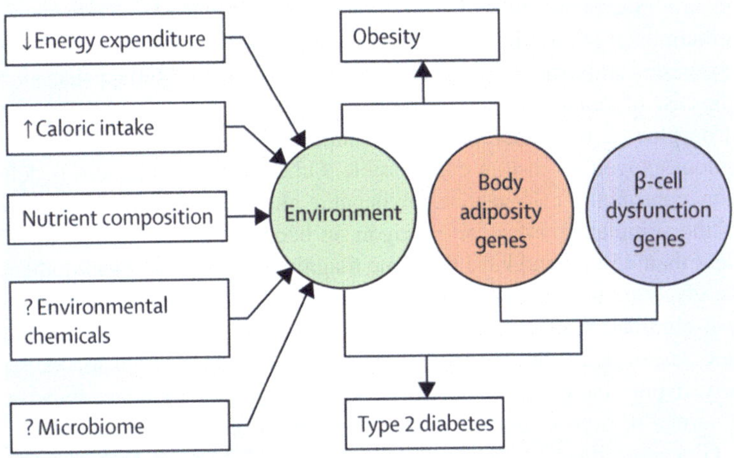

Fig. 2 Risk factors of type 2 diabetes mellitus

pathophysiological changes leading to T2DM. Insulin resistance occurs and intensifies prior to the onset of T2DM. It has been observed that the molecular mechanism for insulin activity has been associated with gene hereditary and factors such as hyperglycemia, free unsaturated fats, fiery system, etc.

Recent considerations have considered the inclusion of adipocyte-inferred biosubstances like adipokines in insulin resistance. While "TNF alpha, leptin, resistin, adiponectin, and free fatty acids may act to force resistance, medical tests of the level of insulin resistance can include homeostasis model assessment for insulin resistance (HOMA-IR), insulin sensitivity test (loading test), the steady-state plasma glucose (SSPG), the minimal model analysis, and insulin clamp technique" (Al-Shoumer and Al-Essa 2015). Matsuda study is currently acknowledged as a basic methodology that can assess insulin obstruction in the liver and muscles (Perry et al. 2014). An increasingly helpful approach to evaluate the level of resistane is to check high fasting blood insulin, instinctive stoutness, and hypertriglyceridemia (Ye 2013).

8.5 *The Role of Glucagon*

Metabolic changes consistently linked with DM have been reported not as a consequence of insulin dysfunction, but as a consequence of bi-hormonal disturbance of α-cell and β- cell activities. The role of glucagon in the production of "diabetic syndrome" is indicated by the observation that glucose loss in glucagon in diabetes is reduced and protein-stimulated glucagon secretion is increased (Ashcroft et al. 2017).

The presence of hyperglucagonemia does not in itself prove glycogenis primary factor in the "abnormal fuel homeostasis" in diabetic patients. Recent studies have shown that glucagon contribution to DM only under the situation of absolute insulin deficiency. An important physiological role of glucagon is to stop "hypoglycemia" through non-glucose stimulated insulin production. "Increase in glucagon level after protein meal allows the complimentary increase in insulin secretion to facilitate the ingested amino acids without risk of hypoglycemia" (Mariappan et al. 2014).

In the case of diabetes pathogenesis, plasma glucagon increases have very little impact on glucose resistance in healthy humans and diabetic patients as long as endogenous/exogenous insulin is accessible. Otherwise, when insulin is deficient, glucagon contributes endogenous on hepatic glucose production is temporary. While "the stimulatory effect of glucagon on hepatic glucose production is transient, less than 45 min and have the same magnitude in a normal and diabetic person, the glycemic response in a diabetic is excessive". Insulin deficiency prevents the rapid elimination of glucose released from the liver. Therefore, the increased glycemic response to glucagon in diabetes is dependent on insulin deficiency. Similarly, hyper-glucagonemia is not very much necessary to induce hyperketonemia in normal persons or diabetic patients, during insulin availability (Wang et al. 2013). Glucagon, therefore, may increase "the production of ketosis in circumstances of total lack of insulin. However, deficiency of insulin is the vital factor that is very important for the changes in fuel utilization which depict the diabetic state (Theodoropoulou and Stalla 2013).

8.6 Somatostatin

Different metabolic effects of infusion exogenous somatostatin and other hormonal activity have been studied. Studies on insulin and glucagon secretion reported that somatostatin associates with the digestion of sugar, reduces gastric motility, and stops production of gastrin. Recent studies revealed that glucose stimulates the release of somatostatin from perfused islet cells. These changes may occur *in vivo* and the extent of the reaction in circulating somatostatin levels is adequate to modify gastrointestinal or pancreatic α and β-cells. Clarification of the physiological function of somatostatin in the metabolism of the body fuel include the consistent assessment of the circulating concentration of peptides and observation of metabolic response to physiologic doses (Zheng et al. 2018).

9 Conclusion

Obesity is a complex disease which plays a significant role in the development of many diseases such as DM. It is a major contributor to the development of DM, CVDs, and cancer. Diabetes is a condition with altered glucose metabolism and hyperglycemia. Obesity and DM have negative impacts on body organs such as the heart, liver, and kidney. Obesity leads to adipose tissue inflammation that contributes to immune system dysfunction. Diabetic retinopathy occurs due to the injury of the veins of the retina, following long term high levels of glucose. Diabetic foot is exceptionally common due to vein injury resulting from prolonged poor management of DM. Some lipolytic processes are impaired by metabolic disorders: obesity and DM. Processed carbohydrates, sugary drinks, and saturated fat contribute to the development of obesity and DM, posing major challenge for public health to develop preventive techniques for the management of obesity as well as DM, and their associated complications. Lifestyle modifications may be effective with pharmacological therapies for the treatment of obesity and DM.

References

Al-Shoumer, K. A., & Al-Essa, T. M. (2015). Is there a relationship between vitamin D with insulin resistance and diabetes mellitus? *World Journal of Diabetes, 6*(8), 1057.

Amasyali, B., Kose, S., Kursaklioglu, H., Kilic, A., & Isik, E. (2008). Leptin in acute coronary syndromes: Has the time come for its use in risk stratification? *International Journal of Cardiology, 130*(2), 264–265.

American College of Obstetricians and Gynecologists. (2013). ACOG Committee opinion No. 548: Weight gain during pregnancy. *Obstetrics & Gynecology, 121*, 210–212.

Angulo, P. (2007). Obesity and nonalcoholic fatty liver disease. *Nutrition Reviews, 65*, S57–S63.

Anyanwu, G. O., Kolb, A. F., & Bermano, G. (2020). Antiobesity functional leads and targets for drug development. In Egbuna et al. (Eds.), *Phytochemicals as lead compounds for new drug discovery* (pp. 143–160). https://doi.org/10.1016/b978-0-12-817890-4.00009-3.

Asferg, C., Møgelvang, R., Flyvbjerg, A., Frystyk, J., Jensen, J. S., Marott, J. L., et al. (2010). Leptin, not adiponectin, predicts hypertension in the Copenhagen City Heart Study. *American Journal of Hypertension, 23*(3), 327–333.

Ashcroft, F. M., Rohm, M., Clark, A., & Brereton, M. F. (2017). Is type 2 diabetes a glycogen storage disease of pancreatic β cells? *Cell Metabolism, 26*(1), 17–23.

Baumann, M., Von Eynatten, M., Dan, L., Richart, T., Kouznetsova, T., Heemann, U., & Staessen, J. A. (2009). Altered molecular weight forms of adiponectin in hypertension. *The Journal of Clinical Hypertension, 11*(1), 11–16.

Bayliss, G., Weinrauch, L. A., & D'Elia, J. A. (2012). Pathophysiology of obesity-related renal dysfunction contributes to diabetic nephropathy. *Current Diabetes Reports, 12*(4), 440–446.

Bhupathiraju, S. N., & Hu, F. B. (2016). Epidemiology of obesity and diabetes and their cardiovascular complications. *Circulation Research, 118*(11), 1723–1735.

Boulton, A. J., Vinik, A. I., Arezzo, J. C., Bril, V., Feldman, E. L., Freeman, R., & Ziegler, D. (2005). Diabetic neuropathies: A statement by the American Diabetes Association. *Diabetes Care, 28*(4), 956–962.

Care, D. (2019). Standards of medical care in diabetes 2019. *Diabetes Care, 42*, S81.

Cedergren, M. I. (2004). Maternal morbid obesity and the risk of adverse pregnancy outcome. *Obstetrics & Gynecology, 103*(2), 219–224.

Chrysohoou, C., Panagiotakos, D. B., Pitsavos, C., Skoumas, I., Papademetriou, L., Economou, M., & Stefanadis, C. (2007). The implication of obesity on total antioxidant capacity in apparently healthy men and women: the ATTICA study. *Nutrition, Metabolism and Cardiovascular Diseases, 17*(8), 590–597.

Czech, M. P. (2017). Insulin action and resistance in obesity and type 2 diabetes. *Nature Medicine, 23*(7), 804–814.

Dhatariya, K. K., Glaser, N. S., Codner, E., & Umpierrez, G. E. (2020). Diabetic ketoacidosis. *Nature Reviews Disease Primers, 6*(1), 1–20.

Egbuna, C., Palai, S., Ebhohimen, I. E., Mtewa, A. G., Ifemeje, J. C., Tupas, G. D., & Kryeziu, T. L. (2019). *Screening of natural antidiabetic agents. Phytochemistry: An in-silico and in-vitro update* (pp. 203–235). https://doi.org/10.1007/978-981-13-6920-9_11.

Fitter, S., & Zannettino, A. C. (2018). Osteopontin in the pathophysiology of obesity: Is Opn a fat cell foe?.

Fuchsberger, C., Flannick, J., Teslovich, T. M., Mahajan, A., Agarwala, V., & Gaulton, K. J. (2016). The genetic architecture of type 2 diabetes. *Nature [Internet]., 536*(7614), 41–47.

Gademan, M. G., Vermeulen, M., Oostvogels, A. J., Roseboom, T. J., Visscher, T. L., van Eijsden, M., et al. (2014). Maternal prepregancy BMI and lipid profile during early pregnancy are independently associated with offspring's body composition at age 5–6 years: the ABCD study. *PLoS One, 9*(4), e94594.

Gittelsohn, J., & Trude, A. (2017). Diabetes and obesity prevention: Changing the food environment in low-income settings. *Nutrition Reviews, 75*(Suppl. 1), 62–69.

Heindel, J. J., Blumberg, B., Cave, M., Machtinger, R., Mantovani, A., Mendez, M. A., & Vandenberg, L. N. (2017). Metabolism disrupting chemicals and metabolic disorders. *Reproductive Toxicology, 68*, 3–33.

Holland, W. L., Knotts, T. A., Chavez, J. A., Wang, L. P., Hoehn, K. L., & Summers, S. A. (2007). Lipid mediators of insulin resistance. *Nutrition Reviews, 65*(Suppl. 1), S39–S46.

Hotamisligil, G. S. (2017). Inflammation, metaflammation and immunometabolic disorders. *Nature, 542*(7640), 177–185.

Katsarou, A., Gudbjörnsdottir, S., Rawshani, A., Dabelea, D., Bonifacio, E., Anderson, B. J., & Lernmark, Å. (2017). Type 1 diabetes mellitus. *Nature Reviews Disease Primers, 3*(1), 1–17.

Laughlin, G. A., Barrett-Connor, E., May, S., & Langenberg, C. (2007). Association of adiponectin with coronary heart disease and mortality: The Rancho Bernardo study. *American Journal of Epidemiology, 165*(2), 164–174.

Levian, C., Ruiz, E., & Yang, X. (2014). The pathogenesis of obesity from a genomic and systems biology perspective. *The Yale Journal of Biology and Medicine, 87*(2), 113.

Lumeng, C. N., Bodzin, J. L., & Saltiel, A. R. (2007). Obesity induces a phenotypic switch in adipose tissue macrophage polarization. *The Journal of Clinical Investigation, 117*(1), 175–184.

Lumeng, C. N., & Saltiel, A. R. (2011). Inflammatory links between obesity and metabolic disease. *The Journal of Clinical Investigation, 121*(6), 2111–2117.

Major, G. C., Doucet, E., Trayhurn, P., Astrup, A., & Tremblay, A. (2007). Clinical significance of adaptive thermogenesis. *International Journal of Obesity, 31*(2), 204–212.

Mariappan, M. M., Prasad, S., D'Silva, K., Cedillo, E., Sataranatarajan, K., Barnes, J. L., & Kasinath, B. S. (2014). Activation of glycogen synthase kinase 3β ameliorates diabetes-induced kidney injury. *Journal of Biological Chemistry, 289*(51), 35363–35375.

Marinou, K., Tousoulis, D., Antonopoulos, A. S., Stefanadi, E., & Stefanadis, C. (2010). Obesity and cardiovascular disease: From pathophysiology to risk stratification. *International Journal of Cardiology, 138*(1), 3–8.

McCarthy, A., Hughes, R., Tilling, K., Davies, D., Davey Smith, G., & Ben-Shlomo, Y. (2007). Birth weight; postnatal, infant, and childhood growth; and obesity in young adulthood: Evidence from the Barry Caerphilly Growth Study. *The American Journal of Clinical Nutrition, 86*(4), 907–913.

McLaughlin, T., Ackerman, S. E., Shen, L., & Engleman, E. (2017). Role of innate and adaptive immunity in obesity-associated metabolic disease. *The Journal of Clinical Investigation, 127*(1), 5–13.

Nelson-Dooley, C., Della-Fera, M. A., Hamrick, M., & Baile, C. A. (2005). Novel treatments for obesity and osteoporosis: Targeting apoptotic pathways in adipocytes. *Current Medicinal Chemistry, 12*(19), 2215–2225.

Patalay, M., Lofgren, I. E., Freake, H. C., Koo, S. I., & Fernandez, M. L. (2005). The lowering of plasma lipids following a weight reduction program is related to increased expression of the LDL receptor and lipoprotein lipase. *The Journal of Nutrition, 135*(4), 735–739.

Peaceman, A. M., Clifton, R. G., Phelan, S., Gallagher, D., Evans, M., Redman, L. M., et al. (2018). Lifestyle interventions limit gestational weight gain in women with overweight or obesity: LIFE-Moms prospective Meta-Analysis. *Obesity, 26*(9), 1396–1404.

Pearl, R. L., Wadden, T. A., Hopkins, C. M., Shaw, J. A., Hayes, M. R., Bakizada, Z. M., et al. (2017). Association between weight bias internalization and metabolic syndrome among treatment-seeking individuals with obesity. *Obesity, 25*(2), 317–322.

Peng, Y., Yu, S., Li, H., Xiang, H., Peng, J., & Jiang, S. (2014). MicroRNAs: Emerging roles in adipogenesis and obesity. *Cellular Signalling, 26*(9), 1888–1896.

Perry, R. J., Samuel, V. T., Petersen, K. F., & Shulman, G. I. (2014). The role of hepatic lipids in hepatic insulin resistance and type 2 diabetes. *Nature, 510*(7503), 84–91.

Pugazhenthi, S., Qin, L., & Reddy, P. H. (2017). Common neurodegenerative pathways in obesity, diabetes, and Alzheimer's disease. *Biochimicaetbiophysicaacta (BBA)-Molecular Basis of Disease, 1863*(5), 1037–1045.

Rankinen, T., & Bouchard, C. (2006). Genetics of food intake and eating behavior phenotypes in humans. *Annual Review of Nutrition, 26*, 413–434.

Rastelli, M., Knauf, C., & Cani, P. D. (2018). Gut microbes and health: A focus on the mechanisms linking microbes, obesity, and related disorders. *Obesity, 26*(5), 792–800.

Sanyal, S., Grimalt, J. O., Horvat, M., Johnstone, E., Maio, S., Polanska, K., & Annesi-Maesano, I. (2018). The environment-wide approach for the assessment of the effect of environmental stressors on overweight, obesity and diabetes: A Study on Singletons for the Heals Project. In ISEE Conference Abstracts (Vol. 2018, No. 1).

Saravanan, K., Egbuna, C., Averal, H. I., Kannan, S., Elavarasi, S., & Bahadur, B. (Eds.). (2020). *Drug Development for Cancer and Diabetes: A Path to 2030* (1st ed.). Apple Academic Press. https://doi.org/10.1201/9780429330490.

Stothard, K. J., Tennant, P. W., Bell, R., & Rankin, J. (2009). Maternal overweight and obesity and the risk of congenital anomalies: A systematic review and meta-analysis. *Journal of the American Medical Association, 301*(6), 636–650.

Theodoropoulou, M., & Stalla, G. K. (2013). Somatostatin receptors: From signaling to clinical practice. *Frontiers in Neuroendocrinology, 34*(3), 228–252.

Tilg, H., Moschen, A. R., & Roden, M. (2017). NAFLD and diabetes mellitus. *Nature Reviews Gastroenterology & Hepatology, 14*(1), 32–42.

Tupas, G. D., Otero, M. C. B., Ebhohimen, I. E., Egbuna, C., & Aslam, M. (2020). Antidiabetic lead compounds and targets for drug development. *Phytochemicals as Lead Compounds for New Drug Discovery*, 127–141. https://doi.org/10.1016/b978-0-12-817890-4.00008-1.

Umpierrez, G. E. (2019). Diabetic ketoacidosis. In *The diabetes textbook* (pp. 619–627). Cham: Springer.

Votruba, S. B., & Jensen, M. D. (2007). Sex differences in abdominal, gluteal, and thigh LPL activity. *American Journal of Physiology-Endocrinology and Metabolism, 292*(6), E1823–E1828.

Wander, P. L., Hochner, H., Sitlani, C. M., Enquobahrie, D. A., Lumley, T., Lawrence, G. M., et al. (2014). Maternal genetic variation accounts in part for the associations of maternal size during pregnancy with offspring cardiometabolic risk in adulthood. *PLoS One, 9*(3), e91835.

Wang, Q., Liang, X., & Wang, S. (2013). Intra-islet glucagon secretion and action in the regulation of glucose homeostasis. *Frontiers in Physiology, 3*, 485.

Wardle, J., Carnell, S., Haworth, C. M., & Plomin, R. (2008). Evidence for a strong genetic influence on childhood adiposity despite the force of the obesogenic environment. *The American Journal of Clinical Nutrition, 87*(2), 398–404.

Waterland, R. A. (2008). Epigenetic epidemiology of obesity: application of epigenomic technology. *Nutrition Reviews, 66*(Suppl. 1), S21–S23.

Westerterp, K. R., Smeets, A., Lejeune, M. P., Wouters-Adriaens, M. P., & Westerterp-Plantenga, M. S. (2008). Dietary fat oxidation as a function of body fat. *The American Journal of Clinical Nutrition, 87*(1), 132–135.

Whitaker, R. C. (2004). Predicting preschooler obesity at birth: the role of maternal obesity in early pregnancy. *Pediatrics, 114*(1), e29–e36.

Ye, J. (2013). Mechanisms of insulin resistance in obesity. *Frontiers of Medicine, 7*(1), 14–24.

Zheng, Y., Ley, S. H., & Hu, F. B. (2018). Global aetiology and epidemiology of type 2 diabetes mellitus and its complications. *Nature Reviews Endocrinology, 14*(2), 88.

Pathophysiology of Obesity and Cancer: Drugs and Signaling Targets

Ahood Khalid, Hira Khalid, Neelam Faiza, Anees Ahmed Khalil, Kiran Shahbaz, Ayesha Aslam, Quratul Ain Shahid, Surajudeen Abiola Abdulrahman, Chukwuebuka Egbuna, and Kingsley C. Patrick-Iwuanyanwu

1 Introduction

Obesity is defined as BMI (body mass index) that is greater than or equal to 30 kg/m^2 typically resulting from a difference between the intake and the expenditure of energy by the human body. Obesity is of increasing public health concern in both developed and developing countries around the world. In the past decade, the global

A. Khalid (✉) · N. Faiza · A. A. Khalil · A. Aslam · Q. A. Shahid
University Institute of Diet and Nutritional Sciences, Faculty of Allied Health Sciences, University of Lahore, Lahore, Pakistan

H. Khalid
Sharif Medical and Dental College, Raiwind Road, Jati Umra, Lahore, Pakistan

K. Shahbaz
Institute of Biotechnology and Biochemistry, University of Veterinary and Animal Sciences, Lahore, Pakistan

S. A. Abdulrahman
Health Education England, East Midlands Deanery, Leicester, UK

C. Egbuna
World Bank Africa Centre of Excellence in Public Health and Toxicological Research (ACE-PUTOR), University of Port Harcourt, Port Harcourt, Nigeria

Department of Biochemistry, Toxicology Unit, Faculty of Science, University of Port Harcourt, Port Harcourt, Rivers State, Nigeria

Department of Biochemistry, Faculty of Natural Sciences, Chukwuemeka Odumegwu Ojukwu University, Uli-Campus, Anambra State, Nigeria

K. C. Patrick-Iwuanyanwu
World Bank Africa Centre of Excellence in Public Health and Toxicological Research (ACE-PUTOR), University of Port Harcourt, Port Harcourt, Nigeria

Department of Biochemistry, Toxicology Unit, Faculty of Science, University of Port Harcourt, Port Harcourt, Rivers State, Nigeria

C. Egbuna, S. Hassan (eds.), *Dietary Phytochemicals*,
https://doi.org/10.1007/978-3-030-72999-8_3

prevalence of obesity has tripled among adults, affecting an estimated 30–35% of the adult population of United States of America compared to 25% in the United Kingdom (National Audit Office 2001). In 2005, the global estimate of the obese and overweight adults was reported to be 400 million and 1.6 billion, respectively. The rising prevalence of overweight and obesity among children and adolescents is of increasing public health concern due to its short and long-term consequences. In 2016, an estimated 40 million children under the age of 5 years and more than 330 million children and adolescents aged 5–19 years were overweight or obese (Development Initiatives 2018).

Obesity is associated with a higher incidence of chronic non-communicable diseases (NCDs) including diabetes mellitus, hypertension, coronary heart disease, metabolic syndrome, cancers, osteoarthritis etc. (Flegal et al. 2007), as well as increased morbidity and mortality from these diseases. The economic burden of obesity includes costs associated with long-term care and management of the resultant NCDs as well as diminished quality of life of those affected (Kopelman et al. 2010; Haslam and James 2005).

Historically, public health approaches to prevention and control of obesity have been focused on improving awareness and health education, encouraging physical activity and healthy diet and providing treatment where lifestyle changes alone prove inadequate (Dansinger et al. 2005; LeBlanc et al. 2011). More recently, surgical treatment in form of Bariatric surgery has been used successfully and effectively to treat obesity and improve health outcomes (Kral and Naslund 2007; Sjostrom et al. 2007). The use of therapeutic agents or drugs remains an alternative treatment option for management of obesity (Sargent and Moore 2009), albeit with notable side effects profile. Balancing safety profile of some anti-obesity drugs with their clinical impact especially at population level remains a challenge precluding licensure of such medications for wider use (Rodgers et al. 2012). Nonetheless, recent advancement in understanding of the neurobiological link between hunger and the homeostasis of energy intake has helped in identifying the exact targets for the development of anti-obesity drugs for clinical use (Wilding 2007; Heal et al. 2012; Halford et al. 2010).

Cancer, one of the most life-threatening diseases, has risen sharply in global prevalence in recent years. The underlying pathophysiology of cancer is chronic, progressive inflammation which may include a number of different physical, biological and chemical determinants in its promotion and progression (Bartsch and Nair 2006). Such underlying chronic inflammation linked with the development and progression of cancers highlights the role of drugs with anti-inflammatory property in the treatment of cancers (Gonda et al. 2009). Different stages of the inflammation observed in patients with cancer include, invasion, proliferation, angiogenesis and metastasis, respectively (Coussens and Werb 2002; Mantovani 2005; Chung et al. 2018). In addition to chronic inflammation, the stress caused by oxidation in the body contributes towards the initiation, promotion and progression of cancer and is being targeted for treatment (Vallée and Lecarpentier 2018). Inflammation increases oxidative stress by elevating the production of reactive oxygen radicals (Yang and Kim 2019), hence contributing towards the different stages involved in the promotion and progression of a cancer (Storz 2005). The inflammation and oxidative stress that is observed in cancer happens through an aberrant WNT/β-catenin pathway (Vallée and Lecarpentier 2018; Onyido et al. 2016). Recent studies have demonstrated that NSAIDs i.e. non-steroidal anti-inflammatory drugs have shown both preventative and therapeutic properties against cancers (Rothwell et al. 2012), particularly lung cancer, breast cancer and gliomas (Vallée and Lecarpentier 2018). The inflammatory factors such as IL (interleukins), TNF (tumor necrosis factor), NF-κB (nuclear factor kappa B) and production of ROS, all of these lead to the damage caused to the DNA and thus propagating the process inducing cancer in the body (Sgarbi et al. 2018).

The current focus is to develop such drugs that have therapeutic properties to mitigate the epidemics of obesity and cancer worldwide. The treatments for both obesity and cancer through the use of these therapeutic drugs is well documented and required for prevention and control of these disorders.

2 Pathophysiology of Obesity and Cancer

Obesity is caused by an underlying chronic positive energy balance often stored in the tissues (specifically adipose) in the form of triacylglycerol. This conversion then results in the inflammation of the fat cells (also known as the adipocytes) from deposition of the triacylglycerols, thus contributing towards the dysfunctional metabolic syndromes in obesity individuals. The effect of these adipocyte compaction may lead to adversity with the endocrine functions. The enzyme named leptin is released by the fat cells in relation to the energy stores (Antel and Hebebrand 2012) and is also involved in signaling of satiety in the hypothalamus which regulates the energy homeostasis (Kirkham 2005). Consequently, this results in high levels of leptin circulating in obese people. Studies have shown a positive association between high levels of leptin and incidence of cancer, specifically cancers of colon (Cota et al. 2003) and breast (Hrabovszky et al. 2012). Another hormone that is released by the adipose tissues is the adiponectin, however, the relationship between adiponectin and adiposity is inverse i.e. this hormone is in a relatively lower amount in obese people as compared to leptin (Kunos and Tam 2011). Nonetheless, lower levels of adiponectin is equally a risk factor for cancers of colon (Christensen et al. 2007) and breast (Tam et al. 2012; Meye et al. 2013) as is high levels of leptin. Compared to lean individuals, the adipose tissues in obese individuals is believed to favor the influx of a significant number of macrophages and other immune related cells into them (Chronaiou et al. 2012; Cummings et al. 2002), making it a fertile environment for the chronic inflammatory changes that underpin initiation and progress of cancers. The stromal-vascular fraction of the adipose tissues consists of the immune cells which are demonstrated to complement the natural expression of the cytokines by the adipocytes. The proinflammatory cytokines that include the TNF-α, IL-6, IL-1β are found to be in higher circulating levels in individuals that are overweight or obese (Cummings et al. 2002). Studies show that there has been a positive association between the higher circulating levels of the proinflammatory cytokines and the risk of cancer (Sumithran et al. 2011; Delporte 2012).

3 History of Anti-obesity Drugs

In the early nineteenth century, different preparations consisting of thyroid hormone were utilized in treatment of obesity (Derosa and Maffioli 2012). These preparations were used because they helped in increasing the basal metabolic rate thus elevating the expenditure of energy in the body. The main drawback of these preparations was side effects such as hyperthyroidism related symptoms such as sleeping problems and restlessness along with the higher risk of heart failure (Valentino et al. 2010). Another drug that improves energy expenditure is dinitrophenol, which was used previously used to aid weight loss centuries ago. Within the mitochondria, this drug helps in separating the oxidative phosphorylation from ATP generation, so that heat is generated instead of the ATPs. While this drug showed reasonable efficacy in enhancing weight loss, it was also associated with a number of deaths due to the risk of overheating from the heat generation (Valentino et al. 2010; Derosa and Maffioli 2012).

Drugs that increase the locomotory action and decreases feeding activity are the amphetamines. These amphetamines were very efficient for weight loss since 1930s, but their use was limited by side effects on the heart and potential for addiction. However, phentermine, a similar drug, is being prescribed till date by weight loss programs for first few weeks of weight loss commencement (Rodgers et al. 2012). Of all the amphetamines, phentermine was found to be the safest and equally efficacious one as it resulted in three to four kg weight loss when compared to the control in a previous study (Li et al. 2005). About twenty years ago a combination of fenfluramine and phentermine named as FenPhen was popular for its effectiveness in reducing weight gain and the results showed that the combination was 10% more efficient than both drugs showed separately (Derosa and Maffioli 2012). This combination was later discontinued due to the incidence of pulmonary hypertension in the patients.

Later in 1990s, another drug introduced for weight loss was sibutramine which inhibited the reuptake of serotonin in the brain (James et al. 2000; Nelson and Gehlert 2006). The results for this drug seemed to have shown positive effect of up to four to five kg weight loss over the year (Li et al. 2005), but it was abandoned in 2010 because of some of the harmful side effects related to heart functions (Derosa and Maffioli 2012). One of the drugs that was approved by the EMA and was brought into the market in 2006 in Europe was the Rimonabant. This drug worked as an antagonist at the CB1 receptor and its use was reported in a previous study to have resulted in weight loss of up to five kgs over one year compared to the control (Padwal and Majumdar 2007). Rimonabant was also abandoned later on in 2008 because of its effect of worsening depression and anxiety in patients.

Quite remarkably, the only drug that has remained in use till date for treating obesity is Orlistat. In a 2005 study conducted by Li and colleagues, Orlistat was reported to have resulted in loss in body weight of about 2–3 kgs on average among users (Li et al. 2005). Orlistat, a gastric and pancreatic lipase inhibitor, got its approval in 1998 (Padwal and Majumdar 2007) and has shown consistent efficacy in management of obesity. Its side effects include oily stools and fecal incontinence, however, the side effects have not limited its use in clinical practice to date.

Other additions to the already available drug variety in management of obesity are Qsymia and Lorcaserin. The latter resulted in a 3–4% greater reduction in weight over one year duration among users when compared with the control (O'neil et al. 2012; Wong et al. 2012). Compared to earlier drugs, topiramate showed significantly better results (i.e. 6–7% more loss) when used singly (Kopelman et al. 2010) or in combination with phentermine (i.e. up to 9–10% more weight loss) (Wong et al. 2012). Bupropion showed similar results to the earlier drugs and resulted in a loss of 2–3 kgs (Li et al. 2005) among users, increasing up to 4–5% when given in combination with naltrexone (combined form is called Contrave). However, nausea was reported as the main side effect of this drug (Greenway et al. 2010).

4 How to Treat Obesity?

Drugs used for the purpose of weight reduction should be able to decrease the intake of energy and/or enhance the expense of energy without affecting the body adversely. Energy expenditure is usually carried out by the BMR (basal metabolic rate) (Ramirez-Zea 2005). There are drugs that target the basal metabolic rate such as thyroid hormone and dinitrophenol which reduce weight efficiently but also cause some side effects as well. Throughout the day, energy expenditure remains constant except for the periods of strenuous physical activity. Contrastingly, the intake of energy varies throughout the day. So, the drugs that are responsible for targeting the energy intake (sibutramine and phentermine) have a greater role to play in the energy balance system. Energy intake may depend on a number of factors including habits, triggers and social pressure. The amount of energy intake depends upon the size of the portion consumed and the satiety level of an individual. The anti-obesity drugs help in reducing hunger by delaying initiation of meals, they affect satiety by decreasing the portion of meal or may affect the choice of meal, thus efficiently lowering the intake of energy which in turn lowers the body weight. The brain houses the neurophysiological centers and signaling pathways responsible for the different phases of energy intake including initiation, termination and also the choice of meal. There are various neurotransmitters and peptides which play specific roles in the energy balance through the endocrine system and the peripheral nerve cells and tissues (Lenard and Berthoud 2008; Williams and Elmquist 2012). The more specific these signals are, the better the chances of developing specific anti-obesity drugs that can directly attack the target cells in order to reduce weight. The anti-obesity drugs mostly affect the monoaminergic systems in the brain however, the effect of monoamines (dopamine, serotonin, etc.) on the energy balance remains poorly understood. The targets that are affected by the anti-obesity medications are important to be studied as the earlier drugs approved by FDA such as Qsymia and Lorcaserin and also some of the new drugs like Tesofensine were found to also target the monoaminergic systems of the brain. There are also other anti-obesity drugs being developed that are targeting other signaling pathways and molecules to take care of the energy balance (Ramirez-Zea 2005; Valentino et al. 2010; Derosa and Maffioli 2012; Dietrich and Horvath 2012; Heal et al. 2012; Witkamp 2011).

5 Different Targets in Brain to Treat Obesity

5.1 Serotoninergic System

Serotonin is a neurotransmitter in the brain which controls satiety (feeling of fullness) (Halford et al. 2010) through three different types of nerve cells (neurons) which assist in the expression of the serotonin receptors in the brain. The first serotonergic drug named as the meta-chlorophenyl piperazine (mCPP) was shown to have anorectic effect through serotonin 2C receptor activation on pro-opiomelanocortin nerve cells (POMC) (Sohn et al. 2011). The main target for the leptin hormone as well as other peripheral hormones is these POMC neurons, the activation rate of which inhibits increases substantially after consumption of a meal or whenever there's positive energy balance. There is a release of b-endorphin and melanocortin (MC) as a result of the increasing POMC nerve cells activation rate. The MC4 receptors are then activated in the secondary nerve cells which is an intermediate to the POMC nerve cells affected by serotonin (Lam et al. 2008). One of the major anti-obesity drug targets is the MC4 receptor due to the genetic variants that are present in vicinity of the MC4 receptor accounting for differences in the BMI of different individuals (Loos et al. 2008). A mutation in the MC4 receptor was shown to have induced obesity in human and mice subjects in a study carried out earlier (Yeo and Heisler 2012). Lowered MC receptor signaling in different parts of the brain including paraventricular nucleus (PVN), dorsal vagal complex (DVC), ventromedial hypothalamus (VMH) and lateral hypothalamus (LH) causes a greater body weight and size of meal (De Backer et al. 2011; Skibicka and Grill 2009). A study was carried out to test the efficacy of one of the MC4 receptor agonist (MK0493) in weight reduction and was found to have displayed a beneficial result in reduction of weight in the clinical trial (Krishna et al. 2009). The MC receptor agonist showed visible difference in weight reduction but was also found to have elevated the blood pressure of participants (Greenfield 2011).

Secondly, serotonin's satiating effect was facilitated by 5-HT1b receptors which are expressed on the arcuate neuropeptide Y/Agouti-related protein (NPY/AgRP) nerve cells. Both of these are orexigenic neuropeptides that play a role as an agonist on the neuropeptide Y receptors and as an antagonist on the MC receptors, respectively. The orexigenic effect of the neuropeptide Y is because of its receptor subtypes Y1 and Y5 (Mashiko et al. 2009). One of the important antagonists on MC receptor, MK-0557 was shown to have significantly lowered the human body weight (Erondu et al. 2006). The NPY/AgRP nerve cells consist of another very important neurotransmitter known as GABA (gamma -Aminobutyric acid). Through 5-HT1b receptors, the serotonin in the brain hyperpolarizes the nerve cells. One of the projection sites is to the arcuate POMC nerve cells, while the other projection site is to the paraventricular nucleus (PVN). The connection with food intake is importantly displayed by the Optogenetic stimulation of arcuate neuropeptide Y/Agouti-related protein GABAergic nerve cells that projected onto the paraventricular nucleus (Atasoy et al. 2012), and a reduction in the performance of these nerve cells was noted due to lower levels of serotonin following decreased food intake.

Arising from the medial raphe nucleus, the nerve cells of serotonergic nature in the brain stem which are the raphe obscurus and the raphe magnus respectively, are also effective in the uptake of food. These neurons are projected onto the Tractus solitarius' (NTS) nucleus and then synapse onto the 5-HT3-receptor -expressing the glutamatergic nerve cells. These then consequently project onto the parabrachial nucleus (Wu et al. 2012). The activation of the parabrachial nucleus nerve cells is linked to the aversion of taste and also to the decreased intake of food (Yamamoto 2007). The release of serotonin helps in the reduction of the active Tractus solitarius nucleus glutamatergic nerve cells thus projecting towards the parabrachial nucleus and resultantly lowering the intake of food. Hence, targeting an increase in the activity of serotonergic system as a way of treating obesity has been quite fruitful. The drug named Fenfluramine has been established to increase the release of serotonin by its action on the vesicular monoamine transporter (VMAT2) and also works in the brain as a selective serotonin reuptake inhibitor (SSRI). This drug assists in decreasing the appetite of an individual and also reduce excessive body weight. Unlike other SSRIs (such as paroxetine) which generally cause weight gain, Fenfluramine notably causes weight loss due to its interference with the VMAT2 (Marks et al. 2008). One of the disadvantages of the serotonergic activating drugs is that they activate the 5-HT2b receptors thereby increasing the likelihood of developing a heart valvulopathy (Elangbam 2010). In contrast, the receptor 5-HT2c is not expressed in the heart, therefore such side effect does not occur in the case of 5-HT2c receptor agonist. The drug named Lorcaserin is a selective 5-HT2c agonist, and has shown quite beneficial effect in weight reduction (O'neil et al. 2012).

5.2 *Noradrenergic System*

Another system in the brain targeted for weight loss is the noradrenergic system. An increase in the signaling of noradrenalin helps in reduction of body weight. Noradrenalin is the primary neurotransmitter in the postganglionergic sympathetic nerve cells and the drugs that imitate the action of noradrenalin in the brain are termed as Sympatico-mimetics. The nerve cells of the noradrenergic system in the brain stem projects to the hypothalamus (Wellman 2000). Small lesions in the sympathetic pathway results in the development of obesity and weakening of the anorectic effect of drugs such as amphetamine. The a1-adrenoreceptor and the b2-adrenoreceptor when activated helps in the reduction of food intake (Wellman 2000). Present within the paraventricular nucleus, the a1-adenoreceptors are of excitatory nature and effective in suppressing the intake of food by an individual (Wellman 2000). The paraventricular nucleus a1-adrenoreceptor is one of the sites where the sympatico-mimetics drugs act thereby causing a reduction in food intake. These nerve cells of the paraventricular nucleus are of unknown identity, but the oxytocin nerve cells of the same nucleus have also been shown to decrease food intake (Valassi et al. 2008). The noradrenergic nerve cells that are present in the brain stem also supply nerves to the serotonergic nerve cells. The nerve cells of the serotonergic system in the medial raphe nucleus are activated by the a1-adrenoreceptor agonist that initiates the physiological response (Mansur et al. 2011) and the noradrenalin released might help in mediating some of the anorectic effects through this pathway. Thus, the sympathetico-mimetics drugs increases the noradrenergic signaling in the brain stem tissue where the serotonergic nerve cells deliver the satiating effects to the brain. The nerve cells of the postganglionic sympathetic system secrete noradrenalin in the target organs. The process of lipolysis in the adipose tissues is stirred by the stimulation of b3-adrenoreceptors through this pathway, as these adrenoreceptors are specifically expressed on the adipocytes (Larsen et al. 2002). Some of the anti-obesity effect of the amphetamine drug is carried out through this pathway. The b3-adrenoceptor agonists have been used as weight reducing drugs, although in practice they have been unable to demonstrate any efficacy in achieving considerable amount of weight loss in the body (Frühbeck et al. 2009). In a previous clinical trial, b3 agonist was found to have increased the basal metabolic rate during the initial weeks but yet again it did not show promising results in the long-term (Redman et al. 2007). An amphetamine-like drug known as the Phentermine also helps with the release of noradrenalin however, its side effects include addiction and cardiovascular problems due to its higher concentrations (Derosa and Maffioli 2012). It is often used as a combined drug called Fen-Phen formulated to treat obesity (Kang et al. 2010).

5.3 Dopaminergic System

The pleasure derived from food consumption is due of the release of dopamine in the striatum inside the brain (Small et al. 2003). The D1 receptor agonists in the brain (in striatum) on stimulation with dopamine increases the palatability of the food consumed, while, activation of the D2 receptor on the other hand causes reduction in food intake (Small et al. 2003). The release of dopamine in the brain majorly affects the activity of the nerve cells of the striatum. The D1 receptor activation causes stimulation, whereas, in contrast, the D2 receptor activation causes the inhibition. There is an inconsistency between the severe effects of the dopamine during the activity of feeding and the results that we get from long-term interference with the signaling of dopamine. In a 2007 study, it was found that D2 antagonists caused an increase in body weight, which therefore suggested that activation of the D2 receptor agonists might suppress the intake of food over the longer period of time (Palmiter 2007). Moreover, a study on mice noted that deficiency of dopamine resulted in failure to eat while restoration of dopamine signaling in those mice led to a re-establishment of feeding in the study subjects (Palmiter 2007). The findings suggest the vital role of dopamine tin activity of feeding, while also highlighting that a higher concentration of dopamine may decrease food intake as well. Dopamine helps in the reduction of food intake at hypothalamus level through lateral and ventromedial hypothalamus and also through the expression of POMC situated in the arcuate nucleus (Tong and Pelletier 1992).

Another group of drugs used are the monoamines, which are known to interact with each other making it difficult to isolate the effect of each monoamine separately. There is an inhibitory effect of serotonin on the dopamine signaling by receptors (5-HT1b and 5-HT2c) in the pathways namely mesolimbic and nigrostriatal, respectively (O'Dell and Parsons 2004). This inhibitory effect is an additional hurdle for the neurotransmitters thus affecting the feeding pathway in an individual. One unique class of anti-obesity drugs is the monoamine reuptake inhibitors. The drug vanoxerine displays an anorectic effect when the reuptake of dopamine is inhibited without the vesicular monoamine transporter (VMAT2) being inhibited (Van der Hoek and Cooper 1994), signifying that the inhibition of dopamine reuptake is well suited to decrease the food intake. Subsequently, the elevated dopamine signaling was found to have increased the locomotory activity which helped in reduction of body weight due to higher energy expenditure. Another important drug that works as a noradrenalin and dopamine reuptake inhibitor is Bupropion which helps in weight loss. This drug is responsible for greater excitability of the nerve cells of the arcuate POMC (Billes and Greenway 2011). The inhibition of both adrenalin and dopamine reuptake was noted to have greater effect on weight reduction as compared to the drugs which targeted only a single transmitter (Billes and Greenway 2011; Plodkowski et al. 2009). Similarly, Tesofensine, a drug that targets noradrenalin, dopamine and serotonin reuptake and inhibits them, has been examined for its weight loss potential. Its original purpose was to treat depression but when it was tested in a clinical trial, a considerable weight loss was observed as well. A 6 months study of Tesofensine showed that at its highest dosage, it resulted in weight reduction up to 10% as compared to the control group (Astrup et al. 2008).

6 Link Between Obesity and Cancer

The dysfunction of the adipose tissues that is linked with obesity is carried out by the signaling pathways that disturb the normal mechanism of energy storage in the body, thus promoting the growth of different tumors. In addition to its effect on satiety, the protein leptin is also responsible for a number of metabolic reactions and effects on other tissues of the body. Leptin is responsible for the activation of the PI3K/Akt/mTOR and JAK/STAT (Janus kinase/signal transducer and activator of transcription) pathways which is signaled by its receptor (LEP-R) (Fazolini et al. 2015). The PI3K/Akt/mTOR pathway severely changes the internal metabolic pathways whereas the JAK/STAT affects the metabolic pathways that are vital for proliferation and cell growth. For example, the activation of JAK is due to the binding of the leptin protein with LEP-R resulting in phosphorylation of STAT3 and its dimerization. The dimerized STAT3 acts as a transcription factor due to its translocation to the nucleus, thereby upregulating the genes that are responsible for the process of glycolysis (Camporeale et al. 2014; Niu et al. 2008), with proof of this activity being HIF-1α-dependent (Demaria et al. 2010).

On the other hand, the hormone adiponectin works in a contrasting manner to leptin in signaling pathways. This hormone helps in the activation of AMPK through the receptors (Adipo R1 & 2) in numerous tissues where it conserves the expense of energy by halting the processes of anabolic nature (Nigro et al. 2014). Additionally, adiponectin has an anti-proliferative property which has been demonstrated by a number of studies. The metabolic dysregulation that is related to obesity helps in filling the adipose tissues with immune cells. In obese subjects, it was observed that the monocyte chemo-attractant protein-1 expression was higher in the fat tissue (adipose tissues), instigating a huge concentration of monocytes towards differentiation into macrophages (Sartipy and Loskutoff 2003). Consequently, the flow of cytokine secretion, typically tumor necrosis factor alpha, is identified to be the reason for obesity-linked insulin resistance in the fat tissues by endorsing serine phosphorylation of IR substrate-1, thereby halting the spreading of downstream signaling of insulin (Ishizuka et al. 2007). Moreover, resistance to insulin in the adipose cells allows persistent basal lipid breakdown (Morigny et al. 2016). it is this lipolysis followed by the release of free fatty acids, which further increase the growth of cancer cells through uptake of exogenous free fatty acids (Balaban et al. 2015). These released free fatty acids serve as ligands for TLR4 (toll-like receptor 4), activating the NF-κB signaling pathway in macrophages to prolong the activity of inflammation and production of cytokine (Suganami et al. 2007; Fessler et al. 2009).

In addition, the cytokines have also been found to have a number of direct effects on the cancer cells. Interleukin-6 binds to the IL-6 receptor and likely to leptin, in addition to its role in activating Janus kinase/signal transducer and transcription 3 signaling (Kumari et al. 2016). Interleukin-6 has been shown to increase the process of glycolysis by the up-regulation of the hexokinase-2 and PFKFB3 which was STAT3 dependent (Ando et al. 2010). The activation of nuclear factor kappa B is carried out by the binding of interleukin-6, tumor necrosis factor alpha, interleukin-1 beta and the free fatty acids to their specific receptors (Suganami et al. 2007; Kumari et al. 2016; Ben-Neriah and Karin 2011). Nuclear factor kappa B is known as the strong mediator of the signaling by working as transcriptional factor targeting a number of genes which might have pleiotropic carcinogenic effects (Bassères et al. 2014). It also helps in reestablishing the metabolic pathways in order to provide strength and support to the cell growth as well as the initiation of aerobic glycolysis (Kawauchi et al. 2008) and angiogenesis (Tabruyn and Griffioen 2008). The up-regulation of the transcription of these cytokines i.e. IL-6 and TNF-α results in propagation of proinflammatory signals leading to activation of the pathways that encourage synthesis (Grivennikov and Karin 2010). Furthermore, it was noted that the cells of the immune system in the adipose tissues release a flood of reactive oxygen species (ROS) for the purpose of destroying the dangerous cells that are foreign to the environment. It was later hypothesized that these ROS might cause damage to the DNA and also mutagenesis in other tissues in the vicinity (Wagner et al. 2015), thus linking the obesity associated inflammation in the tissues to tumorigenesis.

7 Drugs Targeting Cancer Cells

The drugs used for treating cancer are required to block the signaling pathways that are activated by the overexpression of the cancer cells. For instance, in treating lung and breast cancer, the drugs used are EGFR inhibitors and HER2-targeted, respectively (Roca et al. 2018; Slamon et al. 2011). One of the important inventive approaches to tackle cancer is to target the tumor cell glycolysis (Tennant et al. 2010). Yet, developing drugs for cancer is a hurdle due to the challenge in finding out the specific targets for cancer cells that are not expressed on the normal cells. Some of the targets for development of cancer drug include: GLUT1: glucose transporter 1, HKs: hexokinases, PHGDH: phosphoglycerate dehydrogenase and LDHA: lactate dehydrogenase A, respectively, but these were often found to be overexpressed in different types of cancers (Hamanaka and Chandel 2012). The lactate dehydrogenase A helps in converting pyruvate to lactate which a vital part of the anaerobic breakdown of glucose, thereby providing a target for therapeutic purpose. Specific inhibitory effect against human LDH-5 was authorized in a research by several specific small molecules in clinical models (Rani and Kumar 2016), however, these compounds have not been permitted for use in clinical set ups. Likewise, drugs that targeted PHGDH, GLUT1, HKs and other enzymes that catalyze the process are still under evaluation due to their lack of specificity to the cancer cells (Hamanaka and Chandel 2012).

8 Drugs Targeting the Tumor Microenvironment's Cellular and Molecular Components

A class of drug called the antiangiogenic drugs (AADs) have been an important part of cancer therapeutic targeted at the blood vessels of tumor tissue thereby offering a therapeutic option for treatment of cancers (Tennant et al. 2010; Cao et al. 2011). These are often used in combination chemotherapy for treatment of cancers (Hurwitz et al. 2004; Miller et al. 2007; Ueda et al. 2017). Another option of treatment for cancers is immunotherapy using genetically spread tumor antigen-recognizing white blood cells (CAR-T i.e. chimeric antigen receptor T cells), immune checkpoint inhibitors and immune cytokines, respectively (Quintás-Cardama 2018; Abramson et al. 2017; Brahmer et al. 2012; Ribas 2012; Robert et al. 2015). The inflammation within tumors and the fibrotic portions are also viable targets for treating different types of cancers (Lees et al. 2008). Adipocytes, being nonmalignant cells, make up quite a large proportion of the cancer mass, and the constituents of the tumor are correlated with rapidly growing phenotype. For instance, the fibrotic part of the pancreatic ductal cancers can usually constitute up to 90% of the entire tumor mass, and the destruction of the fibrotic portion is inversely related to the survival chances of the tumor (Kong et al. 2014). In the tumor microenvironment of all solid tumors, there are some noncancer cells which are targeted by certain drugs in cancer treatment, for example, the growth of tumor depends on angiogenesis, and the drug bevacizumab (a neutralizing antibody) targeting the vascular endothelial growth factor, is widely known to be used in cancer treatment (Hurwitz et al. 2004). Emactuzumab, a drug that targets tumor-associated macrophages is a humanized anti-CSF1R neutralizing antibody, which was found to be effective in treating breast and ovarian cancer in clinical trials (Cannarile et al. 2017). Also, a nonsteroidal anti-inflammatory drug known as the Celecoxib showed clinical importance in treating breast cancer and adenomatous polyposis (Farooqui et al. 2007). Some of the immunotherapeutic agents that were approved by the FDA include checkpoint blockades such as antibodies blockers and immune-cytokines, which do not specifically target cancer cells, but helps to replenish the immune activity required in the tumor microenvironment to the kill the cancer cells.

9 Changes in the Pharmacokinetics

The effect of obesity on the pharmacokinetics of the chemotherapy utilized for cancer treatment has been well documented (Duong et al. 2015; Behan et al. 2010; Jain et al. 2011). Drugs that are lipophilic in nature require a greater volume for distribution in obese individuals as compared to those who are lean. The liver detoxifies the body from any sort of drug ingested for treatment, hence, a patient with blood perfusion problems and steatosis may have difficulty in clearing the drug through the liver. Patients with obesity are prone to have steatotic liver which then causes a difference in drug clearance (Buechler and Weiss 2011). Clearance of drug from the kidney is equally important and obesity has been associated with poor glomerular filtration rate (Hanley et al. 2010).

10 Changes in Microbiota

The consumption of food with high fat content and subsequent obesity causes dysbiosis and an imbalance in the microbiota inside the body that may contribute towards the resistance of chemotherapy. For colorectal cancers, the increased concentration of *Fusobacterium nucleatum* increases the chances of recurrence of the cancer even after chemotherapy (Yu et al. 2017).

11 Combination Therapy

Combination chemotherapy has been well evaluated in significant detail in previous studies. Research evidence suggest that bevacizumab and other vascular endothelial growth factor (VEGF) inhibitors were found to change the distribution of chemotherapeutics in tumors (Van der Veldt et al. 2012). A study on obese animal models reported that anti-VEGF treatment on these animals resulted in increased tumor hypoxia (oxygen deprivation) in comparison to the lean animals (Iwamoto et al. 2018; Incio et al. 2018).

The majority of cancer chemotherapeutics target the factors that are responsible for up-regulating the pro-apoptotic aspects or diminish the pro-survival mechanism of the cancer cells. A number of the repurposed drugs have displayed many beneficial effects including apoptosis-inducing properties, and are being evaluated for future clinical use. One such important therapeutic agent known to stimulate apoptosis indirectly is nitroglycerin. This agent is also used for the treatment of angina. The nitric oxide donor in nitroglycerin plays a role in secondarily inducing apoptosis, whereas, it is primarily known for its role in halting angiogenesis and also down-regulating HIF1-α (He et al. 2016; Maeda 2010). Moreover, a trial carried out to examine the combined effect of nitroglycerin and other chemotherapeutic agents like vinorelbine and cisplatin, reported an enhanced survival among patients who had untreated stage 3b/4 non-squamous cell lung cancer (Yasuda et al. 2006). Nitroglycerin was found to attenuate hypoxic (deprivation of oxygen) conditions, and was reported to have enhanced resistance when used in combination with other chemotherapeutics. The study showed positive results towards the use of nitroglycerin as it upgraded the rate of response towards vinorelbine with greater tolerance to toxicity as well.

Another important drug that has shown promising results is clarithromycin (CAM), which is a macrolide antibiotic and was used previously for bacterial invasions of the respiratory system and the skin infections. CAM was used earlier in combination with bortezomib (a proteasome inhibitor) to induce apoptosis which was stress-mediated in the endoplasmic reticulum, and the combination was particularly effective in breast cancer and myeloma cells, respectively (Moriya et al. 2013; Komatsu et al. 2013). One study reported that the histone deacetylase inhibitor HDACi SAHA (vorinistat) when combined with bortezomib and clarithromycin, induced the process of apoptosis in the cells of breast cancer. In combination, these drugs worked synergistically to inhibit the capability of the cancer cells to acclimatize to the conditions of cellular stress and spread further (Komatsu et al. 2013). A number of drugs that induce apoptosis are available in the market and are further being examined in different trials to evaluate their efficacy rate (Odia et al. 2015; Zerp et al. 2015; DiPersio et al. 2015). For example, in a previous study in which a concentration of 10 mg/kg of the drug conatumumab (a monoclonal antibody agonist) was combined with FOLFIRI (folinic acid, fluorouracil and irinotecan), resulting in an increased tolerance to toxicity as well as an increase in the disease-free survival in metastatic colorectal cancer (Cohn et al. 2013). Although the results looked promising, further clinical trial is necessary to validate the findings and extend the knowledge further. Similarly, studies are ongoing to evaluate the potential of different drugs and their combined effects in treatment of cancer and its related physiological changes.

12 Conclusion

Globally, the growing epidemic of obesity and cancer has necessitated an evolution of drug discovery and chemotherapeutics at an unprecedented rate. The trial of many chemotherapeutic agents - with unique mechanisms of action - against cancer and obesity, used either singly or in combination therapy has yielded promising results, albeit with different efficacy and success rates. Future studies should focus on development of more sophisticated drugs that may be of particular effectiveness in addressing the combined scourge of obesity and cancer.

References

Abramson, J. S., McGree, B., Noyes, S., Plummer, S., Wong, C., Chen, Y. B., et al. (2017). Anti-CD19 CAR T cells in CNS diffuse large-B-cell lymphoma. *The New England Journal of Medicine, 377*(8), 783.

Ando, M., Uehara, I., Kogure, K., Asano, Y., Nakajima, W., Abe, Y., et al. (2010). Interleukin 6 enhances glycolysis through expression of the glycolytic enzymes hexokinase 2 and 6-phosphofructo-2-kinase/fructose-2, 6-bisphosphatase-3. *Journal of Nippon Medical School, 77*(2), 97–105.

Antel, J., & Hebebrand, J. (2012). Weight-reducing side effects of the antiepileptic agents topiramate and zonisamide. In *Appetite control* (pp. 433–466). Berlin, Heidelberg: Springer.

Astrup, A., Madsbad, S., Breum, L., Jensen, T. J., Kroustrup, J. P., & Larsen, T. M. (2008). Effect of tesofensine on bodyweight loss, body composition, and quality of life in obese patients: A randomised, double-blind, placebo-controlled trial. *The Lancet, 372*(9653), 1906–1913.

Atasoy, D., Betley, J. N., Su, H. H., & Sternson, S. M. (2012). Deconstruction of a neural circuit for hunger. *Nature, 488*(7410), 172–177.

Balaban, S., Lee, L. S., Schreuder, M., & Hoy, A. J. (2015). Obesity and cancer progression: Is there a role of fatty acid metabolism? *BioMed Research International, 2015*.

Bartsch, H., & Nair, J. (2006). Chronic inflammation and oxidative stress in the genesis and perpetuation of cancer: Role of lipid peroxidation, DNA damage, and repair. *Langenbeck's Archives of Surgery, 391*(5), 499–510.

Bassères, D. S., Ebbs, A., Cogswell, P. C., & Baldwin, A. S. (2014). IKK is a therapeutic target in KRAS-induced lung cancer with disrupted p53 activity. *Genes & Cancer, 5*(1–2), 41.

Behan, J. W., Avramis, V. I., Yun, J. P., Louie, S. G., & Mittelman, S. D. (2010). Diet-induced obesity alters vincristine pharmacokinetics in blood and tissues of mice. *Pharmacological Research, 61*(5), 385–390.

Ben-Neriah, Y., & Karin, M. (2011). Inflammation meets cancer, with NF-κB as the matchmaker. *Nature Immunology, 12*(8), 715–723.
Billes, S. K., & Greenway, F. L. (2011). Combination therapy with naltrexone and bupropion for obesity. *Expert Opinion on Pharmacotherapy, 12*(11), 1813–1826.
Brahmer, J. R., Tykodi, S. S., Chow, L. Q., Hwu, W. J., Topalian, S. L., Hwu, P., et al. (2012). Safety and activity of anti–PD-L1 antibody in patients with advanced cancer. *New England Journal of Medicine, 366*(26), 2455–2465.
Buechler, C., & Weiss, S. T. (2011). Does hepatic steatosis affect drug metabolizing enzymes in the liver? *Current Drug Metabolism, 12*(1), 24–34.
Camporeale, A., Demaria, M., Monteleone, E., Giorgi, C., Wieckowski, M. R., Pinton, P., & Poli, V. (2014). STAT3 activities and energy metabolism: Dangerous liaisons. *Cancers, 6*(3), 1579–1596.
Cannarile, M. A., Weisser, M., Jacob, W., Jegg, A. M., Ries, C. H., & Rüttinger, D. (2017). Colony-stimulating factor 1 receptor (CSF1R) inhibitors in cancer therapy. *Journal for Immunotherapy of Cancer, 5*(1), 53.
Cao, Y., Arbiser, J., D'Amato, R. J., D'Amore, P. A., Ingber, D. E., Kerbel, R., et al. (2011). Forty-year journey of angiogenesis translational research. *Science Translational Medicine, 3*(114), 114rv3.
Christensen, R., Kristensen, P. K., Bartels, E. M., Bliddal, H., & Astrup, A. (2007). Efficacy and safety of the weight-loss drug rimonabant: A meta-analysis of randomised trials. *The Lancet, 370*(9600), 1706–1713.
Chronaiou, A., Tsoli, M., Kehagias, I., Leotsinidis, M., Kalfarentzos, F., & Alexandrides, T. K. (2012). Lower ghrelin levels and exaggerated postprandial peptide-YY, glucagon-like peptide-1, and insulin responses, after gastric fundus resection, in patients undergoing Roux-en-Y gastric bypass: A randomized clinical trial. *Obesity Surgery, 22*(11), 1761–1770.
Chung, P. Y., Lam, P. L., Zhou, Y., Gasparello, J., Finotti, A., Chilin, A., et al. (2018). Targeting DNA binding for NF-κB as an anticancer approach in hepatocellular carcinoma. *Cell, 7*(10), 177.
Cohn, A. L., Tabernero, J., Maurel, J., Nowara, E., Sastre, J., Chuah, B. Y. S., et al. (2013). A randomized, placebo-controlled phase 2 study of ganitumab or conatumumab in combination with FOLFIRI for second-line treatment of mutant KRAS metastatic colorectal cancer. *Annals of Oncology, 24*(7), 1777–1785.
Cota, D., Marsicano, G., Tschöp, M., Grübler, Y., Flachskamm, C., Schubert, M., et al. (2003). The endogenous cannabinoid system affects energy balance via central orexigenic drive and peripheral lipogenesis. *The Journal of Clinical Investigation, 112*(3), 423–431.
Coussens, L. M., & Werb, Z. (2002). Inflammation and cancer. *Nature, 420*(6917), 860–867.
Cummings, D. E., Weigle, D. S., Frayo, R. S., Breen, P. A., Ma, M. K., Dellinger, E. P., & Purnell, J. Q. (2002). Plasma ghrelin levels after diet-induced weight loss or gastric bypass surgery. *New England Journal of Medicine, 346*(21), 1623–1630.
Dansinger, M. L., Gleason, J. A., Griffith, J. L., Selker, H. P., & Schaefer, E. J. (2005). Comparison of the Atkins, Ornish, Weight watchers, and zone diets for weight loss and heart disease reduction: A randomised trial. *The Journal of the American Medical Association, 293*, 43–53.
De Backer, M. W. A., La Fleur, S. E., Brans, M. A. D., Van Rozen, A. J., Luijendijk, M. C. M., Merkestein, M., et al. (2011). Melanocortin receptor-mediated effects on obesity are distributed over specific hypothalamic regions. *International Journal of Obesity, 35*(5), 629–641.
Delporte, C. (2012). Recent advances in potential clinical application of ghrelin in obesity. *Journal of Obesity, 2012*.
Demaria, M., Giorgi, C., Lebiedzinska, M., Esposito, G., D'Angeli, L., Bartoli, A., et al. (2010). A STAT3-mediated metabolic switch is involved in tumour transformation and STAT3 addiction. *Aging (Albany NY), 2*(11), 823.
Derosa, G., & Maffioli, P. (2012). Anti-obesity drugs: A review about their effects and their safety. *Expert Opinion on Drug Safety, 11*(3), 459–471.
Development Initiatives. (2018). *2018 Global Nutrition Report: Shining a Light to Spur Action on Nutrition*. Development Initiatives Poverty Research Ltd: Bristol. https://globalnutritionreport.org/. Accessed 16 Jan 2021.

Dietrich, M. O., & Horvath, T. L. (2012). Limitations in anti-obesity drug development: The critical role of hunger-promoting neurons. *Nature Reviews Drug Discovery, 11*(9), 675–691.

DiPersio, J. F., Erba, H. P., Larson, R. A., Luger, S. M., Tallman, M. S., Brill, J. M., et al. (2015). Oral Debio1143 (AT406), an antagonist of inhibitor of apoptosis proteins, combined with daunorubicin and cytarabine in patients with poor-risk acute myeloid leukemia—Results of a phase I dose-escalation study. *Clinical Lymphoma Myeloma and Leukemia, 15*(7), 443–449.

Duong, M. N., Cleret, A., Matera, E. L., Chettab, K., Mathé, D., Valsesia-Wittmann, S., et al. (2015). Adipose cells promote resistance of breast cancer cells to trastuzumab-mediated antibody-dependent cellular cytotoxicity. *Breast Cancer Research, 17*(1), 1–14.

Elangbam, C. S. (2010). Drug-induced valvulopathy: An update. *Toxicologic Pathology, 38*(6), 837–848.

Erondu, N., Gantz, I., Musser, B., Suryawanshi, S., Mallick, M., Addy, C., et al. (2006). Neuropeptide Y5 receptor antagonism does not induce clinically meaningful weight loss in overweight and obese adults. *Cell Metabolism, 4*(4), 275–282.

Farooqui, M., Li, Y., Rogers, T., Poonawala, T., Griffin, R. J., Song, C. W., & Gupta, K. (2007). COX-2 inhibitor celecoxib prevents chronic morphine-induced promotion of angiogenesis, tumour growth, metastasis and mortality, without compromising analgesia. *British Journal of Cancer, 97*(11), 1523–1531.

Fazolini, N. P., Cruz, A. L., Werneck, M. B., Viola, J. P., Maya-Monteiro, C. M., & Bozza, P. T. (2015). Leptin activation of mTOR pathway in intestinal epithelial cell triggers lipid droplet formation, cytokine production and increased cell proliferation. *Cell Cycle, 14*(16), 2667–2676.

Fessler, M. B., Rudel, L. L., & Brown, M. (2009). Toll-like receptor signaling links dietary fatty acids to the metabolic syndrome. *Current Opinion in Lipidology, 20*(5), 379.

Flegal, K. M., Graubard, B. I., Williamson, D. F., & Gail, M. H. (2007). Cause-specific excess deaths associated with underweight, overweight, and obesity. *The Journal of the American Medical Association, 298*, 2028–2037.

Frühbeck, G., Becerril, S., Sáinz, N., Garrastachu, P., & García-Velloso, M. J. (2009). BAT: A new target for human obesity? *Trends in Pharmacological Sciences, 30*(8), 387–396.

Gonda, T. A., Tu, S., & Wang, T. C. (2009). Chronic inflammation, the tumor microenvironment and carcinogenesis. *Cell Cycle, 8*(13), 2005–2013.

Greenfield, J. R. (2011). Melanocortin signalling and the regulation of blood pressure in human obesity. *Journal of Neuroendocrinology, 23*(2), 186–193.

Greenway, F. L., Fujioka, K., Plodkowski, R. A., Mudaliar, S., Guttadauria, M., Erickson, J., et al. (2010). Effect of naltrexone plus bupropion on weight loss in overweight and obese adults (COR-I): A multicentre, randomised, double-blind, placebo-controlled, phase 3 trial. *The Lancet, 376*(9741), 595–605.

Grivennikov, S. I., & Karin, M. (2010). Dangerous liaisons: STAT3 and NF-κB collaboration and crosstalk in cancer. *Cytokine & Growth Factor Reviews, 21*(1), 11–19.

Halford, J. C. G., Boyland, E. J., Blundell, J. E., Kirkham, T. C., & Harrold, J. A. (2010). Pharmacological management of appetite expression in obesity. *Nature Reviews Endocrinology, 6*, 255–269.

Hamanaka, R. B., & Chandel, N. S. (2012). Targeting glucose metabolism for cancer therapy. *Journal of Experimental Medicine, 209*(2), 211–215.

Hanley, M. J., Abernethy, D. R., & Greenblatt, D. J. (2010). Effect of obesity on the pharmacokinetics of drugs in humans. *Clinical Pharmacokinetics, 49*(2), 71–87.

Haslam, D. W., & James, W. P. T. (2005). Obesity. *Lancet, 366*, 1197–1209.

He, Z., Wu, Y., Sun, Q., Zhang, M., Wu, G., Chen, X., et al. (2016). Efficacy and safety of nitroglycerin combined with chemotherapy in the elderly patients with advanced non-small cell lung cancer complicated with coronary heart disease. *International Journal of Clinical and Experimental Medicine, 9*(7), 13021–13027.

Heal, D. J., Gosden, J., & Smith, S. L. (2012). What is the prognosis for new centrally-acting anti-obesity drugs? *Neuropharmacology, 63*(1), 132–146.

Hrabovszky, E., Wittmann, G., Kalló, I., Füzesi, T., Fekete, C., & Liposits, Z. (2012). Distribution of type 1 cannabinoid receptor-expressing neurons in the septal-hypothalamic region of the mouse: Colocalization with GABAergic and glutamatergic markers. *Journal of Comparative Neurology, 520*(5), 1005–1020.

Hurwitz, H., Fehrenbacher, L., Novotny, W., Cartwright, T., Hainsworth, J., Heim, W., et al. (2004). Bevacizumab plus irinotecan, fluorouracil, and leucovorin for metastatic colorectal cancer. *New England Journal of Medicine, 350*(23), 2335–2342.

Incio, J., Ligibel, J. A., McManus, D. T., Suboj, P., Jung, K., Kawaguchi, K., et al. (2018). Obesity promotes resistance to anti-VEGF therapy in breast cancer by up-regulating IL-6 and potentially FGF-2. *Science Translational Medicine, 10*(432).

Ishizuka, K., Usui, I., Kanatani, Y., Bukhari, A., He, J., Fujisaka, S., et al. (2007). Chronic tumor necrosis factor-α treatment causes insulin resistance via insulin receptor substrate-1 serine phosphorylation and suppressor of cytokine signaling-3 induction in 3T3-L1 adipocytes. *Endocrinology, 148*(6), 2994–3003.

Iwamoto, H., Abe, M., Yang, Y., Cui, D., Seki, T., Nakamura, M., et al. (2018). Cancer lipid metabolism confers antiangiogenic drug resistance. *Cell Metabolism, 28*(1), 104–117.

Jain, R., Chung, S. M., Jain, L., Khurana, M., Lau, S. W. J., Lee, J. E., et al. (2011). Implications of obesity for drug therapy: Limitations and challenges. *Clinical Pharmacology & Therapeutics, 90*(1), 77–89.

James, W. P. T., Astrup, A., Finer, N., Hilsted, J., Kopelman, P., Rössner, S., et al. (2000). Effect of sibutramine on weight maintenance after weight loss: A randomised trial. *The Lancet, 356*(9248), 2119–2125.

Kang, J. G., Park, C. Y., Kang, J. H., Park, Y. W., & Park, S. W. (2010). Randomized controlled trial to investigate the effects of a newly developed formulation of phentermine diffuse-controlled release for obesity. *Diabetes, Obesity and Metabolism, 12*(10), 876–882.

Kawauchi, K., Araki, K., Tobiume, K., & Tanaka, N. (2008). p53 regulates glucose metabolism through an IKK-NF-κB pathway and inhibits cell transformation. *Nature Cell Biology, 10*(5), 611–618.

Kirkham, T. C. (2005). Endocannabinoids in the regulation of appetite and body weight. *Behavioural Pharmacology, 16*(5–6), 297–313.

Komatsu, S., Moriya, S., Che, X. F., Yokoyama, T., Kohno, N., & Miyazawa, K. (2013). Combined treatment with SAHA, bortezomib, and clarithromycin for concomitant targeting of aggresome formation and intracellular proteolytic pathways enhances ER stress-mediated cell death in breast cancer cells. *Biochemical and Biophysical Research Communications, 437*(1), 41–47.

Kong, X., Sun, T., Kong, F., Du, Y., & Li, Z. (2014). Chronic pancreatitis and pancreatic cancer. *Gastrointestinal Tumors, 1*(3), 123–134.

Kopelman, P., de Groot, H. G., Rissanen, A., Rossner, S., Toubro, S., Palmer, R., et al. (2010). Weight loss, HbA1c reduction, and tolerability of cetilistat in a randomized, placebo-controlled phase 2 trial in obese diabetics: Comparison with orlistat (Xenical). *Obesity, 18*(1), 108–115.

Kral, J. G., & Naslund, E. (2007). Surgical treatment of obesity. *Nature Clinical Practice Endocrinology & Metabolism, 3*, 574–583.

Krishna, R., Gumbiner, B., Stevens, C., Musser, B., Mallick, M., Suryawanshi, S., et al. (2009). Potent and selective agonism of the melanocortin receptor 4 with MK-0493 does not induce weight loss in obese human subjects: Energy intake predicts lack of weight loss efficacy. *Clinical Pharmacology & Therapeutics, 86*(6), 659–666.

Kumari, N., Dwarakanath, B. S., Das, A., & Bhatt, A. N. (2016). Role of interleukin-6 in cancer progression and therapeutic resistance. *Tumor Biology, 37*(9), 11553–11572.

Kunos, G., & Tam, J. (2011). The case for peripheral CB1 receptor blockade in the treatment of visceral obesity and its cardiometabolic complications. *British Journal of Pharmacology, 163*(7), 1423–1431.

Lam, D. D., Przydzial, M. J., Ridley, S. H., Yeo, G. S., Rochford, J. J., O'Rahilly, S., & Heisler, L. K. (2008). Serotonin 5-HT2C receptor agonist promotes hypophagia via downstream activation of melanocortin 4 receptors. *Endocrinology, 149*(3), 1323–1328.

Larsen, T. M., Toubro, S., van Baak, M. A., Gottesdiener, K. M., Larson, P., Saris, W. H., & Astrup, A. (2002). Effect of a 28-d treatment with L-796568, a novel β3-adrenergic receptor agonist, on energy expenditure and body composition in obese men. *The American Journal of Clinical Nutrition, 76*(4), 780–788.

LeBlanc, E. S., O'Connor, E., Wtitlock, P. D., Patnode, C. D., & Kapka, T. (2011). Effectiveness of primary care -relevant treatments for obesity in adults: A systematic evidence reviews for the U.S. Preventive Services Task Force. *Annals of Internal Medicine, 155*, 434–447.

Lees, C. W., Ironside, J., Wallace, W. A., & Satsangi, J. (2008). Resolution of non–small-cell lung cancer after withdrawal of anti-TNF therapy. *New England Journal of Medicine, 359*(3).

Lenard, N. R., & Berthoud, H. R. (2008). Central and peripheral regulation of food intake and physical activity: Pathways and genes. *Obesity, 16*(S3), S11–S22.

Li, Z., Maglione, M., Tu, W., Mojica, W., Arterburn, D., Shugarman, L. R., et al. (2005). Meta-analysis: Pharmacologic treatment of obesity. *Annals of Internal Medicine, 142*(7), 532–546.

Loos, R. J., Lindgren, C. M., Li, S., Wheeler, E., Zhao, J. H., Prokopenko, I., et al. (2008). Common variants near MC4R are associated with fat mass, weight and risk of obesity. *Nature Genetics, 40*(6), 768–775.

Maeda, H. (2010). Nitroglycerin enhances vascular blood flow and drug delivery in hypoxic tumor tissues: Analogy between angina pectoris and solid tumors and enhancement of the EPR effect. *Journal of Controlled Release, 142*(3), 296–298.

Mansur, S. S., Terenzi, M. G., Neto, J. M., Faria, M. S., & Paschoalini, M. A. (2011). Alpha1 receptor antagonist in the median raphe nucleus evoked hyperphagia in free-feeding rats. *Appetite, 57*(2), 498–503.

Mantovani, A. (2005). Inflammation by remote control. *Nature, 435*(7043), 752–753.

Marks, D. M., Park, M. H., Ham, B. J., Han, C., Patkar, A. A., Masand, P. S., & Pae, C. U. (2008). Paroxetine: Safety and tolerability issues. *Expert Opinion on Drug Safety, 7*(6), 783–794.

Mashiko, S., Moriya, R., Ishihara, A., Gomori, A., Matsushita, H., Egashira, S., et al. (2009). Synergistic interaction between neuropeptide Y1 and Y5 receptor pathways in regulation of energy homeostasis. *European Journal of Pharmacology, 615*(1–3), 113–117.

Meye, F. J., Trezza, V., Vanderschuren, L. J., Ramakers, G. M. J., & Adan, R. A. H. (2013). Neutral antagonism at the cannabinoid 1 receptor: A safer treatment for obesity. *Molecular Psychiatry, 18*(12), 1294–1301.

Miller, K., Wang, M., Gralow, J., Dickler, M., Cobleigh, M., Perez, E. A., et al. (2007). Paclitaxel plus bevacizumab versus paclitaxel alone for metastatic breast cancer. *New England Journal of Medicine, 357*(26), 2666–2676.

Morigny, P., Houssier, M., Mouisel, E., & Langin, D. (2016). Adipocyte lipolysis and insulin resistance. *Biochimie, 125*, 259–266.

Moriya, S., Che, X. F., Komatsu, S., Abe, A., Kawaguchi, T., Gotoh, A., et al. (2013). Macrolide antibiotics block autophagy flux and sensitize to bortezomib via endoplasmic reticulum stress-mediated CHOP induction in myeloma cells. *International Journal of Oncology, 42*(5), 1541–1550.

National Audit Office. (2001). *Tackling obesity in England.* London: Report by the Comptroller and Auditor General. The Stationery Office.

Nelson, D. L., & Gehlert, D. R. (2006). Central nervous system biogenic amine targets for control of appetite and energy expenditure. *Endocrine, 29*(1), 49–60.

Nigro, E., Scudiero, O., Monaco, M. L., Palmieri, A., Mazzarella, G., Costagliola, C., et al. (2014). New insight into adiponectin role in obesity and obesity-related diseases. *BioMed Research International, 2014*.

Niu, G., Briggs, J., Deng, J., Ma, Y., Lee, H., Kortylewski, M., et al. (2008). Signal transducer and activator of transcription 3 is required for hypoxia-inducible factor-1α RNA expression in both tumor cells and tumor-associated myeloid cells. *Molecular Cancer Research, 6*(7), 1099–1105.

O'Dell, L. E., & Parsons, L. H. (2004). Serotonin1B receptors in the ventral tegmental area modulate cocaine-induced increases in nucleus accumbens dopamine levels. *Journal of Pharmacology and Experimental Therapeutics, 311*(2), 711–719.

Odia, Y., Kreisl, T. N., Aregawi, D., Innis, E. K., & Fine, H. A. (2015). A phase II trial of tamoxifen and bortezomib in patients with recurrent malignant gliomas. *Journal of Neuro-Oncology, 125*(1), 191–195.

O'neil, P. M., Smith, S. R., Weissman, N. J., Fidler, M. C., Sanchez, M., Zhang, J., et al. (2012). Randomized placebo-controlled clinical trial of lorcaserin for weight loss in type 2 diabetes mellitus: The BLOOM-DM study. *Obesity, 20*(7), 1426–1436.

Onyido, E. K., Sweeney, E., & Nateri, A. S. (2016). Wnt-signalling pathways and microRNAs network in carcinogenesis: Experimental and bioinformatics approaches. *Molecular Cancer, 15*(1), 1–17.

Padwal, R. S., & Majumdar, S. R. (2007). Drug treatments for obesity: Orlistat, sibutramine, and rimonabant. *The Lancet, 369*(9555), 71–77.

Palmiter, R. D. (2007). Is dopamine a physiologically relevant mediator of feeding behavior? *Trends in Neurosciences, 30*(8), 375–381.

Plodkowski, R. A., Nguyen, Q., Sundaram, U., Nguyen, L., Chau, D. L., & St Jeor, S. (2009). Bupropion and naltrexone: A review of their use individually and in combination for the treatment of obesity. *Expert Opinion on Pharmacotherapy, 10*(6), 1069–1081.

Quintás-Cardama, A. (2018). CAR T-cell therapy in large B-cell lymphoma. *The New England Journal of Medicine, 378*(11), 1065.

Ramirez-Zea, M. (2005). Validation of three predictive equations for basal metabolic rate in adults. *Public Health Nutrition, 8*(7a), 1213–1228.

Rani, R., & Kumar, V. (2016). Recent update on human lactate dehydrogenase enzyme 5 (h LDH5) inhibitors: A promising approach for cancer chemotherapy: Miniperspective. *Journal of Medicinal Chemistry, 59*(2), 487–496.

Redman, L. M., de Jonge, L., Fang, X., Gamlin, B., Recker, D., Greenway, F. L., et al. (2007). Lack of an effect of a novel β3-adrenoceptor agonist, TAK-677, on energy metabolism in obese individuals: A double-blind, placebo-controlled randomized study. *The Journal of Clinical Endocrinology & Metabolism, 92*(2), 527–531.

Ribas, A. (2012). Tumor immunotherapy directed at PD-1. *The New England Journal of Medicine, 366*(26), 2517–2519.

Robert, C., Long, G. V., Brady, B., Dutriaux, C., Maio, M., Mortier, L., et al. (2015). Nivolumab in previously untreated melanoma without BRAF mutation. *New England Journal of Medicine, 372*(4), 320–330.

Roca, E., Amoroso, V., & Berruti, A. (2018). Osimertinib in EGFR mutation–positive advanced NSCLC. *The New England Journal of Medicine, 378*(13), 1261–1262.

Rodgers, R. J., Tschöp, M. H., & Wilding, J. P. (2012). Anti-obesity drugs: Past, present and future. *Disease Models & Mechanisms, 5*(5), 621–626.

Rothwell, P. M., Wilson, M., Price, J. F., Belch, J. F., Meade, T. W., & Mehta, Z. (2012). Effect of daily aspirin on risk of cancer metastasis: A study of incident cancers during randomised controlled trials. *The Lancet, 379*(9826), 1591–1601.

Sargent, B. J., & Moore, N. A. (2009). New central targets for the treatment of obesity. *British Journal of Clinical Pharmacology, 68*, 852–860.

Sartipy, P., & Loskutoff, D. J. (2003). Monocyte chemoattractant protein 1 in obesity and insulin resistance. *Proceedings of the National Academy of Sciences, 100*(12), 7265–7270.

Sgarbi, G., Gorini, G., Liuzzi, F., Solaini, G., & Baracca, A. (2018). Hypoxia and IF1 expression promote ROS decrease in cancer cells. *Cell, 7*(7), 64.

Sjostrom, L., Narbro, K., Sjostrom, C. D., Karason, K., Larsson, B., Wedel, H., Lystig, T., Sullivan, M., Bouchard, C., Carlsson, B., et al. (2007). Swedish Obese Subjects Study Effects of bariatric surgery on mortality in Swedish obese subjects. *New England Journal of Medicine, 357*, 741–752.

Skibicka, K. P., & Grill, H. J. (2009). Hypothalamic and hindbrain melanocortin receptors contribute to the feeding, thermogenic, and cardiovascular action of melanocortins. *Endocrinology, 150*(12), 5351–5361.

Slamon, D., Eiermann, W., Robert, N., Pienkowski, T., Martin, M., Press, M., et al. (2011). Adjuvant trastuzumab in HER2-positive breast cancer. *New England Journal of Medicine, 365*(14), 1273–1283.

Small, D. M., Jones-Gotman, M., & Dagher, A. (2003). Feeding-induced dopamine release in dorsal striatum correlates with meal pleasantness ratings in healthy human volunteers. *NeuroImage, 19*(4), 1709–1715.

Sohn, J. W., Xu, Y., Jones, J. E., Wickman, K., Williams, K. W., & Elmquist, J. K. (2011). Serotonin 2C receptor activates a distinct population of arcuate pro-opiomelanocortin neurons via TRPC channels. *Neuron, 71*(3), 488–497.

Storz, P. (2005). Reactive oxygen species in tumor progression. *Frontiers in Bioscience, 10*(1–3), 1881–1896.

Suganami, T., Tanimoto-Koyama, K., Nishida, J., Itoh, M., Yuan, X., Mizuarai, S., et al. (2007). Role of the Toll-like receptor 4/NF-κB pathway in saturated fatty acid–induced inflammatory changes in the interaction between adipocytes and macrophages. *Arteriosclerosis, Thrombosis, and Vascular Biology, 27*(1), 84–91.

Sumithran, P., Prendergast, L. A., Delbridge, E., Purcell, K., Shulkes, A., Kriketos, A., & Proietto, J. (2011). Long-term persistence of hormonal adaptations to weight loss. *New England Journal of Medicine, 365*(17), 1597–1604.

Tabruyn, S. P., & Griffioen, A. W. (2008). NF-κB: A new player in angiostatic therapy. *Angiogenesis, 11*(1), 101–106.

Tam, J., Cinar, R., Liu, J., Godlewski, G., Wesley, D., Jourdan, T., et al. (2012). Peripheral cannabinoid-1 receptor inverse agonism reduces obesity by reversing leptin resistance. *Cell Metabolism, 16*(2), 167–179.

Tennant, D. A., Durán, R. V., & Gottlieb, E. (2010). Targeting metabolic transformation for cancer therapy. *Nature Reviews Cancer, 10*(4), 267–277.

Tong, Y., & Pelletier, G. (1992). Role of dopamine in the regulation of proopiomelanocortin (POMC) mRNA levels in the arcuate nucleus and pituitary gland of the female rat as studied by in situ hybridization. *Molecular Brain Research, 15*(1–2), 27–32.

Ueda, S., Saeki, T., Osaki, A., Yamane, T., & Kuji, I. (2017). Bevacizumab induces acute hypoxia and cancer progression in patients with refractory breast cancer: Multimodal functional imaging and multiplex cytokine analysis. *Clinical Cancer Research, 23*(19), 5769–5778.

Valassi, E., Scacchi, M., & Cavagnini, F. (2008). Neuroendocrine control of food intake. *Nutrition, Metabolism and Cardiovascular Diseases, 18*(2), 158–168.

Valentino, M. A., Lin, J. E., & Waldman, S. A. (2010). Central and peripheral molecular targets for antiobesity pharmacotherapy. *Clinical Pharmacology & Therapeutics, 87*(6), 652–662.

Vallée, A., & Lecarpentier, Y. (2018). Crosstalk between peroxisome proliferator-activated receptor gamma and the canonical WNT/β-catenin pathway in chronic inflammation and oxidative stress during carcinogenesis. *Frontiers in Immunology, 9*, 745. (28)13.

Van der Hoek, G. A., & Cooper, S. J. (1994). The selective dopamine uptake inhibitor GBR 12909: Its effects on the microstructure of feeding in rats. *Pharmacology Biochemistry and Behavior, 48*(1), 135–140.

Van der Veldt, A. A., Lubberink, M., Bahce, I., Walraven, M., de Boer, M. P., Greuter, H. N., et al. (2012). Rapid decrease in delivery of chemotherapy to tumors after anti-VEGF therapy: Implications for scheduling of anti-angiogenic drugs. *Cancer Cell, 21*(1), 82–91.

Wagner, M., Samdal Steinskog, E. S., & Wiig, H. (2015). Adipose tissue macrophages: The inflammatory link between obesity and cancer? *Expert Opinion on Therapeutic Targets, 19*(4), 527–538.

Wellman, P. J. (2000). Norepinephrine and the control of food intake. *Nutrition, 16*(10), 837–842.

Wilding, J. (2007). Clinical investigations of antiobesity drugs. In T. C. Kirkham & S. J. Cooper (Eds.), *Appetite and body weight* (pp. 337–355). London: Academic.

Williams, K. W., & Elmquist, J. K. (2012). From neuroanatomy to behavior: Central integration of peripheral signals regulating feeding behavior. *Nature Neuroscience, 15*(10), 1350–1355.

Witkamp, R. F. (2011). Current and future drug targets in weight management. *Pharmaceutical Research, 28*(8), 1792–1818.

Wong, D., Sullivan, K., & Heap, G. (2012). The pharmaceutical market for obesity therapies.

Wu, Q., et al. (2012). Deciphering a neuronal circuit that mediates appetite. *Nature, 483*, 594–597.

Yamamoto, T. (2007). Brain regions responsible for the expression of conditioned taste aversion in rats. *Chemical Senses, 32*(1), 105–109.

Yang, D., & Kim, J. (2019). Mitochondrial retrograde signalling and metabolic alterations in the tumour microenvironment. *Cell, 8*(3), 275.

Yasuda, H., Yamaya, M., Nakayama, K., Sasaki, T., Ebihara, S., Kanda, A., et al. (2006). Randomized phase II trial comparing nitroglycerin plus vinorelbine and cisplatin with vinorelbine and cisplatin alone in previously untreated stage IIIB/IV non–small-cell lung cancer. *Journal of Clinical Oncology, 24*(4), 688–694.

Yeo, G. S., & Heisler, L. K. (2012). Unraveling the brain regulation of appetite: Lessons from genetics. *Nature Neuroscience, 15*(10), 1343–1349.

Yu, T., Guo, F., Yu, Y., Sun, T., Ma, D., Han, J., et al. (2017). Fusobacterium nucleatum promotes chemoresistance to colorectal cancer by modulating autophagy. *Cell, 170*(3), 548–563.

Zerp, S. F., Stoter, T. R., Hoebers, F. J., van den Brekel, M. W., Dubbelman, R., Kuipers, G. K., et al. (2015). Targeting anti-apoptotic Bcl-2 by AT-101 to increase radiation efficacy: Data from in vitro and clinical pharmacokinetic studies in head and neck cancer. *Radiation Oncology, 10*(1), 1–9.

Peptides Involved in Body Weight Regulation

Lisbeth Vallecilla-Yepez

1 Introduction

Peptides are protein-based fragments that act as a source of nutrients, which have positive influences on the physiological functions in the body, and therefore may affect human health (Kaur et al. 2020). In a database known as 'Biopep,' over 1500 variations of peptides have been reported (Singh et al. 2014). Proteins and peptides are both organic substances, but peptides are composed of amino acids linked through covalent bonds known as peptide bonds, while proteins are polypeptides having a high molecular weight (Gupta 2019). Animal and plant-based proteins are sources of different types of peptides depending on the structure of the organism. Exogenous (attained through the diet) and endogenous (synthesized within an organism from amino acids) sources of protein also influence the peptides present in an organism (Bhat et al. 2015).

Peptides and proteins exhibit drug or hormone-like functions, which can be classified based on the mode of action like anti-microbial, anti-hypertensive, anti-oxidative, anti-thrombotic, immune-modulatory, mineral binding, and opioid activity (Sánchez and Vázquez 2017). During gastrointestinal digestion, enzymatic proteolysis releases the peptides. Peptides can also be produced in an in vitro process through food-grade proteolytic enzymes, which release peptides from food through protein hydrolysis. During food processing like cooking, fermentation, and ripening, peptides may be found in hydrolysates (Abdel-Hamid et al. 2017). As peptides are released from precursor proteins, their activity depends on the composition and sequence of amino acids. The composition, interaction, and amino acid

L. Vallecilla-Yepez (✉)
Biological Systems Engineering Department, College of Engineering, University of Nebraska-Lincoln, Lincoln, NE, USA
e-mail: lvallecillay@huskers.unl.edu

C. Egbuna, S. Hassan (eds.), *Dietary Phytochemicals*,
https://doi.org/10.1007/978-3-030-72999-8_4

sequence can regulate the different physiological activities occurring in the body (Chakrabarti et al. 2018).

Humans all over the world face various health issues, including blood pressure, cardiovascular problems, cancer, diabetes, and obesity. The main approach taken to treating these health issues has mainly been through pharmaceuticals, which can result in drug toxicity (Girgih et al. 2014). Recently, both functional foods (Chakrabarti et al. 2018) and nutraceuticals (Moldes et al. 2017) have attained major consideration because of their positive influences on human health. Peptides have the potential to cure different medical issues. Therefore, significant attention has been given to the production and characteristics of peptides (Lemes et al. 2016; Przybylski et al. 2016).

In recent years one of the major issues in the developing world has been the devastating increase in obesity and overweight people, leading to an increased risk of multiple other health problems (Chooi et al. 2019). Also, an increased rate of mortality, morbidity, distressing economic, personal, and social outcomes has been reported (Abdelaal et al. 2017). Moderate weight loss can be attained through the regulation of food intake and physical activity while gastrointestinal surgery results in sustainable weight loss (Hall and Kahan 2018). Gastrointestinal surgery is linked to modifications in the functioning of different gut hormones and their appetite-regulating abilities (Dimitriadis et al. 2017). Globally, the overwhelming epidemic of obesity has led to an increased understanding of the complicated interaction between the central nervous system and gut hormones, along with food intake regulation by appetite modulation (Sorrentino and Ragozzino 2017). Researches have revealed various circulating appetite modulators impacting appetite (cholecystokinin, ghrelin, glucagon-like peptide, peptide YY, pancreatic polypeptide, and oxyntomodulin) (Suzuki et al. 2011).

Consequently, there has been an increasing trend to convert this evidence-based information into practical anti-obesity intervention. Newly emerging research studies are concentrating on developing nutraceuticals and functional foods by using different peptides to capitalize on their particular physiological health benefits (Cheung et al. 2015). The long-term use of any medicine may have negative aftereffects leading to increased health care costs. Natural foods composed of peptides can considerably decrease the dependency on medication, help prevent diseases, and reduce healthcare costs, hence leading to an increased interest in healthy lifestyle choices (Girgih et al. 2014). This chapter reviews some of the new insights gained from studies of different endogenous peptides and biopeptides and how their activities are linked with body weight regulation.

2 Appetite, Food Intake, and Obesity

Appetite can be defined as the desire to eat food or drink regardless of the presence or absence of hunger and serves to maintain the metabolic needs by the ingestion of adequate energy intake (Airaodion et al. 2019; Gregersen et al. 2011). Appetite can be present when hunger is absent if it is stimulated by novel foods (Airaodion et al. 2019). It is controlled by the interplay of many mechanisms principally in the gut, brain, and adipose tissue. External factors such as stress and exercise routines can increase or decrease appetite, which affects food intake. Also, appetite is highly related to age and sex (Gregersen et al. 2011; Wynne et al. 2005a, b).

Khalid et al. (2014) defined food intake as "the ingestion of any substance consisting of carbohydrates, proteins, fats, vitamins, and minerals." Food ingestion plays a fundamental role in human existence because it boosts growth, promotes energy, and maintains life. Multiplex interactions between nutrients, neuropeptides, hormones, and numerous different brain areas influence food intake (Airaodion et al. 2019). Understanding food intake regulation is crucial to understand body weight and obesity.

According to the WHO (World Health Organization), obesity or being overweight is the condition of having excessive and abnormally accumulated fat in the body leading to health risks (WHO 2014). This condition is complex and includes many factors, such as a diet with excessive calories, sedentary life, genetic factors, and metabolic disorders (Crespo et al. 2014). Some authors claim that obesity is an association of many factors, such as metabolic and eating disorders. These eating disorders include the omission of breakfast and overeating at night (Jakubowicz and Froy 2013). The obesity pandemic is a global emerging issue. Its prevalence has almost tripled since 1975 (Kumar 2019), which is an alarming sign and represents a significant concern in human health. Being obese and its consequences constitute an important source of morbidity and a substantial increase in overall mortality (Abdelaal et al. 2017). Obesity raises the likelihood of developing numerous diseases, such as systemic hypertension, heart disease, hyperlipidemia, sleep apnea, diabetes mellitus, some various types of cancer, left ventricular hypertrophy, and osteoarthritis, among others (Airaodion et al. 2019; Kumar 2019; Madamanchi et al. 2014).

3 Peptides and Body Weight

Endogenous peptides, as well as food-protein-derived functional peptides (exogenous peptides), play an essential role in body fat and lipid metabolism and have anti-obesity properties. Endogenous peptides possess a fundamental role in food intake, energy consumption, and body weight regulation by different mechanisms (Kumar 2019; Morley 1989). Current strategies targeting appetite and body weight regulation involve understanding the roles of the central nervous system, the control of food intake through appetite modulation, the control of the metabolism through actions from the periphery, the absorption of fat, and the modulation of gut peptide receptors (Perry and Wang 2012; Torres-Fuentes et al. 2015).

Food intake behavior is modulated at different levels. The sensory-specific satiety behavior takes place at the level of the forebrain in the hippocampus and occurs when a subject stops eating when the same food is offered but continues when an appealing food is provided (Morley 1989). In the central nervous system, the hypothalamus appears to play an important role both in maintaining the feeding behavior and in modulating the level of satiety by integrating the multiple input peripherical signals (Perry and Wang 2012). The regulation of body weight is a homeostatic process; however, this process contains few mechanisms that encourage weight loss, and strongly tends toward the storage of fat and weight gain (Wilding 2002). The food intake and energy equilibrium are regulated by the arcuate nucleus (ARC) (Crespo et al. 2014), which is located on either side of the third cerebral ventricle at the base of the hypothalamus (Kalra and Kalra 2004). The activity of the ARC neurons can be influenced by gut hormones, hypothalamic signals, and other circulating signals. Gut hormone receptors are localized within the ARC in neuronal populations (Small and Bloom 2004). The receptors in the G-protein, which are activated once the nutrients are ingested, are also in charge of stimulating the release of gut hormones. These hormones, together with the hormones released by the adipose tissue, play a fundamental role in food intake regulation and energy expenditure (Crespo et al. 2014).

In the ARC, neurons can be of two types and possess contrasting effects in the process of food intake and weight regulation. The first group is in charge of appetite stimulation (Parkinson et al. 2008) and expresses agouti-related protein (AgRP) and neuropeptide Y (NPY) (Bewick et al. 2005). The second group contains the appetite-inhibiting neurons that express the alpha-melanocyte-stimulating hormone (α-MSH) derived from cocaine and amphetamine-regulated transcript (CART) and pro-opiomelanocortin (POMC). This group's principal function is to inhibit appetite (Elias et al. 1998). Peripheral signals emerge in changes in the analogous activity of these two neuronal subpopulations and the production of their respective neuropeptides. A balance of the activities of these two types of neurons is necessary to maintain body weight regulation, feeding behavior, and energy storage and expenditure (Crespo et al. 2014). The first and second group of peptides are known as orexigenic peptides and anorexigenic peptides, respectively (Valassi et al. 2008; Williams et al. 2001). In the following sections, this review will seek to guide the reader through the effects of hormonal peptides and some hypothalamic peptides that are involved in the regulation of body weight.

3.1 *Effects of Hormonal Peptides*

3.1.1 Orexigenic Peptides

(i) Ghrelin peptide, the only known orexigenic peptide gut hormone to date, (Perry and Wang 2012). It is known as the "hunger hormone" and consists of 28 amino acids. Ghrelin is secreted mainly in the stomach in response to starvation and circulates in the blood. It serves as a signal from the periphery to the central nervous system to foster feeding (Airaodion et al. 2019; Crespo et al. 2014; Kojima et al. 1999). Ghrelin is also present in the intestine, where its concentration declines continuously from the duodenum to the colon. Ghrelin plays a crucial role in regulating appetite and in the distribution and rate of energy expenditure (Sakata and Sakai 2010). It has been found that the secretion of this peptide is regulated by food intake. Ghrelin concentration decreases after feeding and increases when fasting. In their study, Sakata and Sakai (2010) found that injections of ghrelin in humans and rats increase appetite. Moreover, the levels of plasma ghrelin are high during the night and in underweight subjects, and by contrast, they are low in obese people (Cummings et al. 2002; Dzaja et al. 2004). The factors involved in this mechanism are not well understood, however, some research studies have found that blood glucose influences ghrelin secretion (Shiiya et al. 2002). Glucose administration decreases the concentrations of ghrelin in plasma. When ghrelin is administered, cumulative food intake increases, and energy expenses decrease, resulting in body weight gain. On the contrary, when an individual gains weight, ghrelin concentration decreases, reducing food intake and weight loss. These phenomena indicate that ghrelin is a bodyweight regulator.

3.1.2 Anorexigenic Peptides

1. **Leptin peptide** is known as the "satiety hormone" and consists of a 167 amino acid protein (Airaodion et al. 2019; Crespo et al. 2014). Leptin is secreted mainly by adipose cells (around 95% of the leptin production occurs here) (Rolls 2011); however, it is also found in the stomach and pituitary gland. Leptin possesses opposed properties to ghrelin; it regulates body weight by enhancing the metabolic rate, controlling energy expenditure while inhibiting hunger. Body max index (the weight of an individual divided by the square of their height: BMI = Weight [kg]/Height^2[m^2]) and adipose cells are positively correlated with leptin levels (Airaodion et al. 2019; Crespo et al. 2014). In most overweight and obese people, the sensitivity to leptin levels decreases, failing to detect satiety regardless of the high energy accumulated. Therefore, these individuals generally have higher leptin levels compared to normal-weight people (Brennan and Mantzoros 2006). Furthermore, obesity patients have shown resistance to leptin, similarly to insulin resistance in diabetes type 2. This phenomenon results in an inability to control hunger and balanced weight. In their study, Spiegelman and Flier (1996) reported that two strains of mutant rodents [diabetes (db/db) and obesity (ob/ob)], which are distinguished by an early onset of severe obesity were unable to respond to (db/db) or to produce (ob/ob) leptin. When leptin was administered to ob/ob mice, they ate less and their metabolic rates increased. They became more active and lost weight considerably. Nowadays, it is understood that obese people may have a deficient sensitivity to the leptin but not an inability to produce it.
2. **Cholecystokinin**, known as an appetite-suppressing hormone is a peptide hormone produced as a response to stimulate the digestion of protein digestion byproducts; these byproducts include amino acids, small peptides, long-chain fatty acids, and saturated fat (Liddle et al. 1985; Merritt and Julliand 2013). After

food consumption, cholecystokinin is secreted mainly in the small intestine, from the duodenum to the jejunum. It flows rapidly into the circulation and acts as a hunger suppressant. Cholecystokinin's receptors have been identified in tissues in the periphery such as the vagal afferent nerve in the gut, the pancreas, the gallbladder, as well as areas within the central nervous system involved in food intake regulation (Moran et al. 1986; Moran and Kinzig 2004). Clinical human studies found that intravenous administration of cholecystokinin resulted in a similar reduction of food consumption and boosting of the perception of satiety in underweight and obese people (Lieverse et al. 1995).

3. **Insulin** is a peptide hormone that consists of 51 amino acids. It is produced by β-cells of the pancreatic islets (Chevenne et al. 1999), and it is considered the principal hormone of the anabolic system. Insulin enters the brain from circulation in response to decreased energy consumption (Crespo et al. 2014); its presence is an indication of satiety and obesity (Schwartz and Science 2005). Insulin's central action enhances anorexia since it stimulates POMC and inhibits NPY/AgRP expressions. Studies in rodents showed that the central injection of insulin causes a decrease in food ingestion and body weight (Air et al. 2002). Moreover, a decrease or deletion of insulin receptors causes obesity, and an extreme desire to eat food, a phenomenon known as hyperphagia (Kuliczkowska-Plaksej et al. 2012).
4. **Glucagon-like peptides** (GLP-1and-2) are yielded In parallel after food intake when the proteolytic cleavage in the L-cells of the distal gut and the nucleus of the tractus solitarius occurs (Vrang et al. 2003). GLP-1 is a 30 amino acid hormone peptide secreted after food intake, and well known as a gut-derived incretin hormone. Some of the agonist receptors of glucagon-like peptide 1 (GLP-1RA) such as exenatide and liraglutide enhance glycaemic control and stimulate satiety and anorexia, causing a decrease in food ingestion and body weight (van Bloemendaal et al. 2014). Gastric emptying and food intake declined in a dose-dependent manner when intravenous GLP-1 was injected into normal and obese individuals (Nauck et al. 1997; Verdich et al. 2001). Since neurons within the brainstem in rats are activated by peripheral administration of GLP-1, it is suggested that GLP-1's effects are mediated through vagal and brainstem pathways. GLP-2 contributes to the regulation of energy absorption, function, and integrity of the intestine. Even though its role in human appetite regulation is not well understood (after peripheral GLP-2 administration in humans, food intake seems to not decrease), recent studies suggest that intraperitoneal administration of GLP-2 decreases food ingestion in rodents (Baldassano et al. 2012).
5. **Peptide tyrosine tyrosine (PYY)** is a full-length peptide that consists of 36 amino acids, with tyrosine residues. This peptide needs C-terminal amidation, which gives its biological activity property. PPY is synthesized and released from the L-cells of the gastrointestinal tract. However, the 34-amino acid PYY_{3-36} constitutes the most common PYY form in circulation, which has been truncated at the N-terminus (Grandt et al. 1994). PYY levels have been detected in the small and large intestines. Major PPY concentrations can be found in the cells in the colon and rectum, in addition to minor levels in the stomach in the

enteroendocrine cells (Adrian et al. 1985). Levels of PYY increase after 30 min of food intake in humans and stay elevated for up to 6 h. Its anorectic effects on appetite control and satiety are mostly regulated via neuronal population in the ARC (Small and Bloom 2004). Batterham et al. (2003a, b) found that PYY has anorexigenic effects in both healthy weight individuals and obese people. In their clinical trials, intravenous administration of PYY caused about 30% restriction in caloric intake and a decrease in appetite in both groups. Also, exaggerated postprandial PPY levels after gastrointestinal surgery suggest a possible contribution to the initial and long-term conservation of weight loss of an individual undergoing the surgery process (Korner et al. 2005).

6. **Pancreatic polypeptide (PP)** is released by pancreatic islet polypeptide cells and consists of 36 amino acids with an amide. PP's secretions depend on and rise proportionally to caloric intake, and are low during the fasting period (Track et al. 1980). The effects of the pancreatic polypeptide appear to be mediated in areas central to the control of appetite, reducing appetite and food intake in humans, which seems to be independent of gastric motility changes (Batterham et al. 2003a, b). In a research study, a reduction in food consumption was found after peripherally-administered PP in both mice and humans (Batterham et al. 2003a, b; Malaisse-Lagae et al. 1977).
7. **Oxyntomodulin** is a 37 amino acid peptide. Production occurs from L-cells and is directly proportional to caloric intake. It was found that oxyntomodulin administration reduces feeding in normal-weight individuals (Cohen et al. 2003). Moreover, there is evidence that repeated injections of this peptide may increase energy consumption in people, leading to a reduction of body weight in obese and overweight individuals (Katie Wynne et al. 2005a, b). Oxyntomodulin has a similar precursor molecule as GLP-1, therefore it is secreted following meal intake and is proportional to caloric content. It was found that food consumption in rodents was reduced after oxyntomodulin administered centrally and peripherally. However, energy expenditure increased, leading to reductions in body weight (Dakin et al. 2004).
8. **Obestatin** is a recently discovered 23-amino acid peptide produced by additional proteolytic cleavage of preproghrelin (Crespo et al. 2014). Obestatin possesses anorexigenic effects, impedes contraction of the jejunum, suppresses gastric emptying, and represses body weight gain (Stanley et al. 2005). Tang and Zhang (2013) observed that after obestatin administration, there was a decrease in jejunum muscle contraction in vitro and suppression of gastric emptying in vivo.

3.1.3 Effects of Hypothalamic Peptides

1. **Neuropeptide Y** is a 36 amino acid peptide hormone with tyrosine (Y) residues at the beginning and end of the molecule (Spiegelman and Flier 1996). Neuropeptide Y is mostly expressed in the ARC within the neurons in the hypothalamus. Shimada et al. (1998) showed an increase of neuropeptide Y levels in

response to starvation in mice. In a study conducted by Schwartz et al. (2000), it was found that the repeated administration of neuropeptide Y into the hypothalamus resulted in mice becoming overweight. Besides, some studies have found that agents such as insulin and leptin, and stimulatory signals such as glucocorticoids, influence neuropeptide Y secretion. It seems that the action of leptin to inhibit the neuropeptide Y production explains, in part, the ability of leptin to promote weight loss and hypophagia (Crespo et al. 2014). On the other hand, it has been found that insulin seems to hinder neuropeptide Y synthesis and secretion in the paraventricular nucleus.
2. **Pro-opiomelanocortin (POMC)** possesses complex and rich biology. Minor fragments derived from the inert POMC precursor play an important role in the integration of physiological functions (Airaodion et al. 2019). Some research has investigated the involvement of POMC-derived peptides in the control of food consumption and body weight regulation. Yoo et al. (2014) found that high pro-opiomelanocortin methylation in cord blood was linked with underweight newborns. Also, these children showed higher insulin levels in the blood, as well as higher triacylglycerides concentrations.

3.2 *Bioactive Peptides and Body Weight*

Bioactive peptides are food-derived peptides, which are organic compounds formed by amino acids joined by covalent bonds (Sánchez and Vázquez 2017). Bioactive peptides have been defined as food derived (protein-derived) substances (nutraceuticals), which usually consist of 3–20 amino acids and have the potential of providing health benefits to the host beyond nutritional values (Manikkam et al. 2016). The reduction of the risk of many chronic diseases coupled with changes in the human diet has led to a growing interest in bioactive peptides in the scientific community and the food industry. Therefore, the development of foods that promote health and wellness based on bioactive peptides has been sought after.

At present, a considerable number of bioactive peptides have been identified and isolated. Bioactive peptide properties have been investigated for their fundamental role in the prevention of certain diseases and their potential impact on human health. Research began in the 1950s when Mellander proposed that vitamin D-independent bone calcification in rachitic children could be enhanced by the ingestion of phosphorylated casein-derived peptides (Manikkam et al. 2016). Studies have found that bioactive peptides have some functional properties; they are antimicrobial, antioxidative, antiobesity, antihypertensive, immunomodulatory, and antithrombotic (Bhandari et al. 2020; Jakubowicz and Froy 2013; Kumar 2019; Torres-Fuentes et al. 2015). In the human body, these functionalities are determined principally by the amino acid composition and sequence once they have been released from their parent protein. In natural sources, the release mechanism occurs mainly by enzymatic processes; however, some bioactive peptides are also synthesized chemically (Manikkam et al. 2016).

Extensive research on functional foods that contain bioactive peptides has been performed for products currently on the market. 2,5-Diketopiperazines (DKPs) are cyclic dipeptides that can be found in a variety of food such as cocoa, roasted coffee, malt, chicken essence, whey protein hydrolysates, beer, aged sake among others (Sánchez and Vázquez 2017). The N-terminal amino acid residues of a linear protein or peptide can be used to form DKPs. They have received significant attention as bioactive compounds. Yamamoto et al. (2016) observed food ingestion and body weight reduction in rats after the DKPs administration.

Animal studies have shown that soy protein can be used to treat metabolic syndrome (Sánchez and Vázquez 2017). It has been found that soy protein may have several health benefits, including the prevention of cardiovascular diseases and fat and weight loss. In addition, the enhancement of lipid profile, glucose, and insulin homeostasis has been demonstrated by Velasquez and Bhathena (2007). On the other hand, after whey protein was consumed in a ration of 54 g/day, overweight/obese subjects showed that starvation plasma insulin amounts decreased by 11%, proposing a long-term enhancement of insulin sensitivity (Pal et al. 2010). This finding is critical since insulin secretion promoted by whey intake may affect food ingestion that can reduce appetite and therefore body weight. Baer et al. (2011) found that supplementation with whey protein alters body weight by causing an increase of satiety compared to supplementation with carbohydrates or soy protein.

Bioactive peptides from marine sources have been studied amply. A marine-derived compound known as Nutripeptin has been found to be effective in incrementing weight loss response and satiety (Manikkam et al. 2016). In addition, bioactivities of fish collagen peptide were fed to obese mice with a high-fat diet (HFD), and it was found that the administration of biopeptides from marine sources substantially reduced the HFD-induced body weight increase without a substantial difference in food consumption (Lee et al. 2017). Additionally, the fish collagen peptide consumption reduced serum levels of triglyceride, total cholesterol, and low-density lipoprotein, while the serum high-density lipoprotein was increased. These observations suggest adipogenesis and body weight loss in in vivo models of obesity.

Moreover, the renin-angiotensin system has been identified as a regulatory system that is responsible for several physiological and pathophysiological functions. In this system, the angiotensin-converting enzyme (ACE) transforms angiotensin-I (ANG I), a peptide hormone, to angiotensin-II (ANG II), resulting in vasoconstriction. Preventing the formation of ANG II and inhibiting the activity of ACE could result in a promising and therapeutic aspect of treating obesity and hypertension. Manikkam et al. (2016) reported that fish-derived proteins and peptides could potentially be excellent sources of ACE inhibitors. Their paper gives an overview of the synthesis and characterization of marine-derived peptides, their ACE inhibitory activity, and bioavailability. The effect of ANG II on body weight regulation has been studied in vivo. Porter et al. (2003) found that when ANG II was administered into young rats, their food intake and body weight gain decreased compared to the control group. Moreover, Brink et al. (1996) observed that male Sprague-Dawley rats lost between 18 and 26% of body weight after having been treated for 1 week with ANG II infusion. However, it was shown that ANG II increased the expression of fatty acid synthase in human adipose cells, promoting adipocyte hypertrophy, as well as an increase in triglyceride content (Jones et al. 1997).

4 Conclusion

Peptides from different sources have been identified for having positive activities, but insufficient research work considering the in vivo study has been conducted. For several decades, the occurrence of obesity, overweight, and other related diseases has increased significantly. Now obesity and overweight are commonly known as a major worldwide nutritional challenge. At present gastrointestinal surgery is considered as the most effective approach among anti-obesity therapies leading to significant weight loss, but it has some limitations like mortality, high cost, and changes in functions and circulation of gut hormones. The most important aspects of controlling body weight are the regulation of food intake and proper physical activity. The brain changes the appetite pattern to regulate food intake by having interactions between the central nervous system, gut hormones, and peptides. Therefore, efforts have now increased to understand the interaction of these three factors in the regulation of food intake. Besides, peptides can be used to develop anti-obesity drug formulations. However, it needs more research and efforts for the development of peptide-based therapeutic drugs. Additionally, research work concerning the obesity mediators, the structural framework of peptides, their mechanism, and functions is required, which could be helpful to develop novel peptides.

References

Abdelaal, M., le Roux, C., & Docherty, N. (2017). Morbidity and mortality associated with obesity. *Annals of Translational Medicine, 5*(7), 161. https://www.ncbi.nlm.nih.gov/pmc/articles/PMC5401682/.

Abdel-Hamid, M., Otte, J., De Gobba, C., Osman, A., & Hamad, E. (2017). Angiotensin I-converting enzyme inhibitory activity and antioxidant capacity of bioactive peptides derived from enzymatic hydrolysis of buffalo milk proteins. *International Dairy Journal, 66*, 91–98. https://doi.org/10.1016/j.idairyj.2016.11.006.

Adrian, T. E., Ferri, G.-L., Bacarese-Hamilton, A. J., Fuessl, H. S., Polak, J. M., & Bloom, S. R. (1985). Human distribution and release of a putative' new gut hormone, peptide YY. *Gastroenterology, 89*(5), 1070–1077. https://www.gastrojournal.org/article/0016-5085(85)90211-2/fulltext.

Air, E., Benoit, S., Smith, K., Clegg, D., & Woods, S. (2002). Acute third ventricular administration of insulin decreases food intake in two paradigms. *Pharmacology Biochemistry and Behavior, 72*(1–2), 423–429. https://www.sciencedirect.com/science/article/pii/S0091305701007808.

Airaodion, A. I., Ogbuagu, U., Oloruntoba, A. P., Paul, A., Airaodion, E. O., Mokelu, I. P., & Ekeh, S. C. (2019). Biochemical mechanisms involved in the regulation of appetite and weight. *Review, 06*, 397–409.

Baer, D., Stote, K., Paul, D., Harris, G., Rumpler, W. V., & Clevidence, B. A. (2011). Whey protein but not soy protein supplementation alters body weight and composition in free-living overweight and obese adults. *The Journal of Nutrition, 141*(8), 1489–1494. https://academic.oup.com/jn/article-abstract/141/8/1489/4630526.

Baldassano, S., Bellanca, A. L., Serio, R., & Mulè, F. (2012). Food intake in lean and obese mice after peripheral administration of glucagon-like peptide 2. *Journal of Endocrinology, 213*, 277–284. https://doi.org/10.1530/JOE-12-0092.

Batterham, R., Le Roux, C., Cohen, M., Park, A., Ellis, S., Patterson, M., & Bloom, S. (2003b). Pancreatic polypeptide reduces appetite and food intake in humans. *The Journal of Clinical Endocrinology & Metabolism, 88*(8), 3989–3992. https://academic.oup.com/jcem/article-abstract/88/8/3989/2845640.

Batterham, R. L., Cohen, M. A., Ellis, S. M., Le Roux, C. W., Withers, D. J., Frost, G. S., Ghatei, M. A., & Bloom, S. R. (2003a). Inhibition of food intake in obese subjects by peptide YY3-36. *New England Journal of Medicine, 349*(10), 941–948. https://doi.org/10.1056/NEJMoa030204.

Bewick, G., Dhillo, W., Darch, S., Murphy, K., Gardiner, J., Jethwa, P., & Bloom, S. (2005). Hypothalamic Cocaine- and Amphetamine-Regulated Transcript (CART) and Agouti-Related Protein (AgRP) neurons coexpress the NOP1 receptor and nociceptin. *Endocrinology, 146*(8) https://academic.oup.com/endo/article-abstract/146/8/3526/2500521.

Bhandari, D., Rafiq, S., Gat, Y., Gat, P., Waghmare, R., & Kumar, V. (2020). A review on bioactive peptides: Physiological functions, bioavailability and safety. *International Journal of Peptide Research and Therapeutics, 26*(1), 139–150. https://doi.org/10.1007/s10989-019-09823-5.

Bhat, Z. F., Kumar, S., & Bhat, H. F. (2015). Bioactive peptides from egg: A review. *Nutrition & Food Science, 45*(2), 190–212. https://doi.org/10.1108/NFS-10-2014-0088.

Brennan, A., & Mantzoros, C. M. (2006). Drug insight: The role of leptin in human physiology and pathophysiology—Emerging clinical applications. *Nature Clinical Practice Endocrinology*. https://www.nature.com/articles/ncpendmet0196.

Brink, M., Wellen, J., & Delafontaine, P. (1996). Angiotensin II and insulin-like growth factor I angiotensin II causes weight loss and decreases circulating insulin-like Growth factor I in rats through a pressor-independent mechanism key words: Congestive heart fail-ure • hypertension • IGF binding proteins • metabolism • re-ceptors, growth factor. *International Journal of Clinical Investigation, 97*(11) https://www.jci.org/articles/view/118698.

Chakrabarti, S., Guha, S., & Majumder, K. (2018). Food-derived bioactive peptides in human health: Challenges and opportunities. *Nutrients, 10*(11), 1738. https://doi.org/10.3390/nu10111738.

Cheung, R. C. F., Ng, T. B., & Wong, J. H. (2015). Marine peptides: Bioactivities and applications. *Marine Drugs, 13*(7), 4006–4043. https://doi.org/10.3390/md13074006.

Chevenne, D., Trivin, F., & Porquet, D.-. (1999). Insulin assays and reference values. *Diabetes & Metabolism*. https://europepmc.org/abstract/med/10633871.

Chooi, Y., Ding, C., & Magkos, F. (2019). The epidemiology of obesity. *Metabolism, 92*, 6–10. https://www.sciencedirect.com/science/article/pii/S002604951830194X.

Cohen, M., Ellis, S., Le Roux, C., Batterham, R. L., Park, A., Patterson, M., & Bloom, S. R. (2003). Oxyntomodulin suppresses appetite and reduces food intake in humans. *The Journal of Clinical Endocrinology & Metabolism, 88*(10), 4696–4701. https://academic.oup.com/jcem/article-abstract/88/10/4696/2845751.

Crespo, C. S., Cachero, A. P., Jiménez, L. P., Barrios, V., & Ferreiro, E. A. (2014). Peptides and food intake. *Frontiers in Endocrinology, 5*, 1–13. https://doi.org/10.3389/fendo.2014.00058.

Cummings, D. E., Weigle, D. S., Scott Frayo, R., Breen, P. A., Ma, M. K., Patchen Dellinger, E., & Purnell, J. Q. (2002). Plasma ghrelin levels after diet-induced weight loss or gastric bypass surgery. *New England Journal of Medicine, 346*(21), 1623–1630. https://doi.org/10.1056/NEJMoa012908.

Dakin, C., Small, C., Batterham, R., Neary, N., Cohen, M. A., Patterson, M., & Bloom, S. R. (2004). Peripheral oxyntomodulin reduces food intake and body weight gain in rats. *Endocrinology, 145*(6), 2687–2695. https://academic.oup.com/endo/article-abstract/145/6/2687/2878151.

Dimitriadis, G. K., Randeva, M. S., & Miras, A. D. (2017). Potential hormone mechanisms of bariatric surgery. *Current Obesity Reports, 6*(3), 253–265. https://doi.org/10.1007/s13679-017-0276-5.

Dzaja, A., Dalal, M. A., Himmerich, H., Uhr, M., Pollmächer, T., & Schuld, A. (2004). Sleep enhances nocturnal plasma ghrelin levels in healthy subjects. *American Journal of Physiology - Endocrinology and Metabolism, 286*(6), 49–56. https://doi.org/10.1152/ajpendo.00527.2003.

Elias, C., Lee, C., Kelly, J., Aschkenasi, C., & Ahima, R. (1998). Leptin activates hypothalamic CART neurons projecting to the spinal cord. *Neuron, 21*(6), 1375–1385. https://www.sciencedirect.com/science/article/pii/S089662730080656X.

Girgih, A. T., He, R., Malomo, S., Offengenden, M., Wu, J., & Aluko, R. E. (2014). Structural and functional characterization of hemp seed (Cannabis sativa L.) Protein-derived antioxidant and antihypertensive peptides. *Journal of Functional Foods, 6*, 384–394. https://www.sciencedirect.com/science/article/pii/S1756464613002594.

Grandt, D., Schimiczek, M., Beglinger, C., Layer, P., Goebell, H., Eysselein, V., & Reeve, J., Jr. (1994). Two molecular forms of peptide YY (PYY) are abundant in human blood: Characterization of a radioimmunoassay recognizing PYY 1–36 and PYY 3–36. *Regulatory Peptides, 51*(2) https://www.sciencedirect.com/science/article/pii/0167011594902046.

Gregersen, N., Møller, B., Raben, A., Kristensen, S., Holm, L., Flint, A., Astrup, A., Gregersen, N. T., Møller, B. K., & Kristensen, S. T. (2011). *Food & Nutrition Research Determinants of appetite ratings: The role of age, gender, BMI, physical activity, smoking habits, and diet/weight concern* (p. 55). Taylor & Francis. https://doi.org/10.3402/fnr.v55i0.7028.

Gupta, A. (2019). Protein and amino acids. In *Comprehensive biochemistry for dentistry* (pp. 35–65). Springer Singapore. https://doi.org/10.1007/978-981-13-1035-5_3.

Hall, K., & Kahan, D. (2018). Maintenance of lost weight and long-term management of obesity. *Medical Clinics, 102*(1), 183–197. https://www.medical.theclinics.com/article/S0025-7125(17)30136-0/abstract.

Jakubowicz, D., & Froy, O. (2013). Biochemical and metabolic mechanisms by which dietary whey protein may combat obesity and Type 2 diabetes. *Journal of Nutritional Biochemistry, 24*(1), 1–5. https://doi.org/10.1016/j.jnutbio.2012.07.008.

Jones, B., Standridge, M., & Moustaid, N. (1997). Angiotensin II increases lipogenesis in 3T3-L1 and human adipose cells. *Endocrinology, 138*(4), 1512–1519. https://academic.oup.com/endo/article-abstract/138/4/1512/2987396.

Kalra, S., & Kalra, P. (2004). Hypothalamic regulation of appetite and obesity.

Kaur, J., Kumar, V., Sharma, K., Kaur, S., Gat, Y., Goyal, A., & Tanwar, B. (2020). Opioid peptides: An overview of functional significance. *International Journal of Peptide Research and Therapeutics, 26*(1), 33–41. Springer. https://doi.org/10.1007/s10989-019-09813-7.

Khalid, M., Khan, M., Kausar, N., Yousaf, K., & Khalid, S. (2014). Actual intake verses recommended intake amongst female adolescents. *European Scientific Journal, 10*(36).

Kojima, M., Hosoda, H., Date, Y., Nakazato, M., Matsuo, H., & Kangawa, K. (1999). Ghrelin is a growth-hormone-releasing acylated peptide from stomach. *Nature, 402*(6762), 656–660. https://idp.nature.com/authorize/casa?redirect_uri=https://www.nature.com/articles/45230&casa_token=Ru4oCMyCGpQAAAAA:miJE8MNuxE7gbLGIXCRcimsyi9MYwjC9E23Rvd7TxbcccaS1hRYgmhLcj0TXmEESz8zHrxO97EbPVsWn.

Korner, J., Bessler, M., Cirilo, L., Conwell, I., Daud, A., Restuccia, N., & Wardlaw, S. (2005). Effects of Roux-en-Y gastric bypass surgery on fasting and postprandial concentrations of plasma ghrelin, peptide YY, and insulin. *The Journal of Clinical Endocrinology & Metabolism, 90*(1), 359–365. https://academic.oup.com/jcem/article-abstract/90/1/359/2835647.

Kuliczkowska-Plaksej, J., Milewicz, A., & Jakubowska, J. (2012). Neuroendocrine control of metabolism. *Gynecological Endocrinology, 28*(Suppl.1), 27–32. https://doi.org/10.3109/09513590.2012.651930.

Kumar, M. S. (2019). Peptides and peptidomimetics as potential antiobesity agents: Overview of current status. *Frontiers in Nutrition, 6*, 1–12. https://doi.org/10.3389/fnut.2019.00011.

Lee, E. J., Hur, J., Ham, S. A., Jo, Y., Lee, S. Y., Choi, M. J., & Seo, H. G. (2017). Fish collagen peptide inhibits the adipogenic differentiation of preadipocytes and ameliorates obesity in high fat diet-fed mice. *International Journal of Biological Macromolecules, 104*, 281–286. https://doi.org/10.1016/j.ijbiomac.2017.05.151.

Lemes, A. C., Sala, L., Da Costa Ores, J., Cavalcante Braga, A. R., Buranelo Egea, M., & Fernandes, K. F. (2016). A review of the latest advances in encrypted bioactive peptides from protein-rich waste. *Journal of Molecular Sciences, 17*(6), 950. https://doi.org/10.3390/ijms17060950.

Liddle, R. A., Goldfine, I. D., Rosen, M. S., Taplitz, R. A., & Williams, J. A. (1985). Cholecystokinin bioactivity in human plasma molecular forms, responses to feeding, and relationship to gallbladder contraction. *The Journal of Clinical Investigation, 75*(4), 1144–1152. https://www.jci.org/articles/view/111809.

Lieverse, R. J., Jansen, J. B. M. J., Masclee, A. A. M., Lamers, H. W., & Masclee, C. B. (1995). Satiety effects of a physiological dose of cholecystokinin in humans. *Gut, 36*(2), 176–179. https://doi.org/10.1136/gut.36.2.176.

Madamanchi, C., Alhosaini, H., Sumida, A., & Runge, M. S. (2014). Obesity and natriuretic peptides, BNP and NT-proBNP: Mechanisms and diagnostic implications for heart failure. *International Journal of Cardiology, 176*(3), 611–617. https://doi.org/10.1016/j.ijcard.2014.08.007.

Malaisse-Lagae, F., Carpentier, J. L., Patel, Y. C., Malaisse, W. J., & Orci, L. (1977). Pancreatic polypeptide: A possible role in the regulation of food intake in the mouse. *Hypothesis Experientia, 33*(7), 915–917. https://doi.org/10.1007/BF01951279.

Manikkam, V., Vasiljevic, T., Donkor, O. N., & Mathai, M. L. (2016). A review of potential marine-derived hypotensive and anti-obesity peptides. *Critical Reviews in Food Science and Nutrition, 56*(1), 92–112. https://doi.org/10.1080/10408398.2012.753866.

Merritt, A., & Julliand, V. (2013). Gastrointestinal physiology. In *Equine applied and clinical nutrition e-book: Health, welfare and performance*. https://books.google.com/books?hl=es&lr=&id=rlBfYgLiqtwC&oi=fnd&pg=PA3&dq=Johnson,+L.+R.+(2013).+Gastrointestinal+Physiology&ots=ShlGQBTP_m&sig=ALlPgFpy897W_37yaeTLiAeb97c.

Moldes, A., Vecino, X., & Cruz, J. (2017). Nutraceuticals and food additives. In *Current developments in biotechnology and bioengineering* (pp. 143–164) https://www.sciencedirect.com/science/article/pii/B9780444636669000066.

Moran, T., Robinson, P., Goldrich, M., & McHugh, P. (1986). Two brain cholecystokinin receptors: Implications for behavioral actions. *Brain Research, 362*(1), 175–179. https://www.sciencedirect.com/science/article/pii/0006899386914137.

Moran, T. H., & Kinzig, K. P. (2004). Gastrointestinal satiety signals. II. Cholecystokinin. *American Journal of Physiology: Gastrointestinal and Liver Physiology, 286*(2), 49–52. https://doi.org/10.1152/ajpgi.00434.2003.

Morley, J. E. (1989). Appetite regulation: The role of peptides and hormones. *Journal of Endocrinological Investigation, 12*(2), 135–147. https://doi.org/10.1007/BF03349944.

Nauck, M. A., Niedereichholz, U., Ettler, R., Holst, J. J., Ørskov, C., Ritzel, R., & Schmiegel, W. H. (1997). Glucagon-like peptide 1 inhibition of gastric emptying outweighs its insulinotropic effects in healthy humans. *American Journal of Physiology - Endocrinology and Metabolism, 273*(5), E981–E988. https://doi.org/10.1152/ajpendo.1997.273.5.e981.

Pal, S., Ellis, V., & Dhaliwal, S. (2010). Effects of whey protein isolate on body composition, lipids, insulin and glucose in overweight and obese individuals. *British Journal of Nutrition, 104*(5), 716–723. https://doi.org/10.1017/S0007114510000991.

Parkinson, J. R. C., Dhillo, W. S., Small, C. J., Chaudhri, O. B., Bewick, G. A., Pritchard, I., Moore, S., Ghatei, M. A., & Bloom, S. R. (2008). PYY3-36 injection in mice produces an acute anorexigenic effect followed by a delayed orexigenic effect not observed with other anorexigenic gut hormones. *American Journal of Physiology - Endocrinology and Metabolism, 294*(4). https://doi.org/10.1152/ajpendo.00405.2007.

Perry, B., & Wang, Y. (2012). Appetite regulation and weight control: The role of gut hormones. *Nutrition and Diabetes, 2*, e26–e27. https://doi.org/10.1038/nutd.2011.21.

Porter, J., Anderson, J., Robison, R., & Phillips, A. (2003). Effect of central angiotensin II on body weight gain in young rats. *Brain Research, 959*(1), 20–28. https://www.sciencedirect.com/science/article/pii/S0006899302036764.

Przybylski, R., Firdaous, L., Châtaigné, G., & Dhulster, P. (2016). Production of an antimicrobial peptide derived from slaughterhouse by-product and its potential application on meat as preservative. *Food Chemistry, 211*, 306–313. https://www.sciencedirect.com/science/article/pii/S0308814616307543.

Rolls, E. (2011). Taste, olfactory and food texture reward processing in the brain and obesity. *International Journal of Obesity.*. https://idp.nature.com/authorize/casa?redirect_uri=https://www.nature.com/articles/ijo2010155&casa_token=rqCiH7_0G5gAAAAA:HuJSktrQVZqDk7-VJ4G-ZlFbQsPvpVThMIXJ8vt0rZQEvqFYNcA-njmT7uNjJJEXaQHfqwiQZMo1Gky0.

Sakata, I., & Sakai, T. (2010). Ghrelin cells in the gastrointestinal tract. *International Journal of Peptides, 2010*, 945056. *Hindawi.Com* https://www.hindawi.com/journals/ijpep/2010/945056/abs/.

Sánchez, A., & Vázquez, A. (2017). Bioactive peptides: A review. *Food Quality and Safety, 1*(1), 29–46. https://doi.org/10.1093/fqs/fyx006.

Schwartz, M., & Science, D. P. (2005). Diabetes, obesity, and the brain. *Science, 307*(5708), 375–379. Sciencemag.Org. https://doi.org/10.1126/science.1104342.

Schwartz, M., Woods, S., Porte, D., Seeley, R., & Baskin, D. (2000). Central nervous system control of food intake. *Nature, 404*(6778), 661–671. https://idp.nature.com/authorize/casa?redirect_uri=https://www.nature.com/articles/35007534&casa_token=9Eo_nnRzrAgAAAAA:POhSWXUVkik7iO8qolLVbZMtIeCiWsIXfnSit5smE1xCGryzu8MOE9XDqafzhfKwohmhtAHa2ISZ_0N_.

Shiiya, T., Nakazato, M., Mizuta, M., et al. (2002). *Plasma ghrelin levels in lean and obese humans and the effect of glucose on ghrelin secretion. Academic.Oup.Com.* https://academic.oup.com/jcem/article-abstract/87/1/240/2847073

Shimada, M., Tritos, N., Lowell, B., & Flier, J. (1998). Mice lacking melanin-concentrating hormone are hypophagic and lean. *Nature, 396*(6712), 670–674. www.nature.com.

Singh, B., Vij, S., & Hati, S. (2014). Functional significance of bioactive peptides derived from soybean. *Peptides, 54*, 171–179. https://www.sciencedirect.com/science/article/pii/S0196978114000369.

Small, C. J., & Bloom, S. R. (2004). Gut hormones and the control of appetite. *Trends in Endocrinology & Metabolism, 15*(6), 259–263. https://doi.org/10.1016/j.tem.2004.06.002.

Sorrentino, M., & Ragozzino, G. (2017). The regulation of food intake: The brain-endocrine network. *International of Journal of Clinical Endocrinology & Metabolism, 1*(1), 41–48.

Spiegelman, B., & Flier, J. (1996). Adipogenesis and obesity: Rounding out the big picture. *Cell, 87*(3), 377–389. https://www.cell.com/fulltext/S0092-8674(00)81359-8?scriptOff=true&code=cell-site.

Stanley, S., Wynne, K., McGowan, B., & Bloom, S. (2005). Hormonal regulation of food intake. *Physiological Reviews, 85*(4), 1131–1158. https://doi.org/10.1152/physrev.00015.2004.

Suzuki, K., Jayasena, C., & Bloom, S. (2011). The gut hormones in appetite regulation. *Journal of Obesity, 2011*. http://downloads.hindawi.com/journals/jobes/2011/528401.pdf.

Tang, H., & Zhang, W. (2013). The diversity of ghrelin gene in its products and functions. *Sheng Li Ke Xue Jin Zhan (Progress in Physiology), 44*(3), 169–176. https://europepmc.org/article/med/24027822.

Torres-Fuentes, C., Schellekens, H., Dinan, T. G., & Cryan, J. F. (2015). A natural solution for obesity: Bioactives for the prevention and treatment of weight gain. *A review. Nutritional Neuroscience, 18*(2), 49–65. https://doi.org/10.1179/1476830513Y.0000000099.

Track, N. S., McLeod, R. S., & Mee, A. V. (1980). Human pancreatic polypeptide: Studies of fasting and postprandial plasma concentrations. *Canadian Journal of Physiology and Pharmacology, 58*(12), 1484–1489. https://doi.org/10.1139/y80-223.

Valassi, E., Scacchi, M., & Cavagnini, F. (2008). Neuroendocrine control of food intake. *Nutrition, Metabolism and Cardiovascular Diseases, 18*(2), 158–168. https://doi.org/10.1016/j.numecd.2007.06.004.

van Bloemendaal, L., ten Kulve, J. S., La Fleur, S. E., Ijzerman, R. G., & Diamant, M. (2014). Effects of glucagon-like peptide 1 on appetite and body weight: Focus on the CNS. *Journal of Endocrinology, 221*(1). https://doi.org/10.1530/JOE-13-0414.

Velasquez, M., & Bhathena, S. (2007). Role of dietary soy protein in obesity. *International Journal of Medical Sciences, 4*(2), 72. https://www.ncbi.nlm.nih.gov/pmc/articles/pmc1838825/.

Verdich, C., Flint, A., & Gutzwiller, J. (2001). A meta-analysis of the effect of glucagon-like peptide-1 (7–36) amide on Ad libitum energy intake in humans. *The Journal of Clinical Endocrinology & Metabolism, 86*(9), 4382–4389. https://academic.oup.com/jcem/article-abstract/86/9/4382/2848979.

Vrang, N., Phifer, C. B., Corkern, M. M., & Berthoud, H. R. (2003). Gastric distension induces c-Fos in medullary GLP-1/2-containing neurons. *American Journal of Physiology: Regulatory Integrative and Comparative Physiology, 285*(2), 54–62. https://doi.org/10.1152/ajpregu.00732.2002.

WHO. (2014). *Obesity*. World Health Organization.

Wilding, J. P. H. (2002). Neuropeptides and appetite control. *Diabetic Medicine, 19*(8), 619–627. https://doi.org/10.1046/j.1464-5491.2002.00790.x.

Williams, G., Bing, C., Cai, X., Harrold, J., & King, P. (2001). The hypothalamus and the control of energy homeostasis: Different circuits, different purposes. *Physiology & Behavior, 74*(4–5), 683–701. https://www.sciencedirect.com/science/article/pii/S0031938401006126.

Wynne, K., Stanley, S., & McGowan, B. (2005b). Appetite control. *Journal of Endocrinology, 184*(2), 291–318. https://joe.bioscientifica.com/view/journals/joe/184/2/1840291.xml.

Wynne, K., Park, A. J., Small, C. J., Patterson, M., Ellis, S. M., Murphy, K. G., Wren, A. M., Frost, G. S., Meeran, K., Ghatei, M. A., & Bloom, S. R. (2005a). Subcutaneous oxyntomodulin reduces body weight in overweight and obese subjects a double-blind, randomized, controlled trial. *Diabetes, 54*(8), 2390–2395. https://diabetes.diabetesjournals.org/content/54/8/2390.short.

Yamamoto, K., Hayashi, M., Murakami, Y., Araki, Y., Otsuka, Y., Kashiwagi, T., Shimamura, T., & Ukeda, H. (2016). Bioscience, biotechnology, and biochemistry development of LC-MS/MS analysis of cyclic dipeptides and its application to tea extract development of LC-MS/MS analysis of cyclic dipeptides and its application to tea extract. *Bioscience, Biotechnology, and Biochemistry, 80*(1), 172–177. https://doi.org/10.1080/09168451.2015.1075865.

Yoo, J. Y., Lee, S., Lee, H. A., Park, H., Park, Y. J., Ha, E. H., & Kim, Y. J. (2014). Can proopiomelanocortin methylation be used as an early predictor of metabolic syndrome? *Diabetes Care, 37*(1), 734–739. https://doi.org/10.2337/dc13-1012.

Insulin Resistance: A Link Between Obesity and Cancer

Saira Sattar, Muhammad Faisal Nisar, and Onyeka Kingsley Nwosu

1 Introduction

Insulin resistance is a condition in which the body tissues (adipose, skeletal, liver, etc) do not respond properly to insulin action. Environmental and genetic factors are also influencers to the susceptibility of insulin resistance. For example, a sedentary lifestyle and overeating are major environmental factors that leads to obesity thereby promoting insulin resistance. A genetic defect in the insulin receptor signaling pathway is associated with a hindrance in the regulation of metabolic responses and growth to insulin in body tissues (Melvin et al. 2018).

According to different epidemiological studies, the increased risk of cancers are associated with obesity and insulin resistance (type 2 diabetes mellitus (T2DM)) (Gallagher et al. 2010). Some studies have also revealed that T2DM is an independent factor which causes various types of cancer including liver, pancreas, colorectum, and breast cancer (Jalving et al. 2010). Cancer is the second most lethal disease that causes death worldwide, whereas diabetes is ranked at twelve (Lopez et al. 2006). In the nineteenth Century, an English Physician reported the first case of diabetes related cancer and by further studies by epidemiologists, it was deduced that the carcinoma of the pancreas is associated with hyperglycemia and glycosuria (Bell 1957).

S. Sattar (✉) · M. F. Nisar
Institute of Home and Food Sciences, Government College University Faisalabad, Faisalabad, Punjab, Pakistan

O. K. Nwosu
National Biosafety Management Agency, Abuja, Nigeria

Department of Biochemistry, School of Life Sciences, Federal University of Technology, Minna, Nigeria

C. Egbuna, S. Hassan (eds.), *Dietary Phytochemicals*,
https://doi.org/10.1007/978-3-030-72999-8_5

In 2005 the International Obesity Taskforce reported that up to 312 million people are obese worldwide; i.e. having Body Mass Index (BMI) 25–30 kg/m^2 which increase the likelihood of developing T2DM. The body also reported that by the year 2030, it is estimated to extend its prevalence to affect nearly 366 million more people (Gahagan et al. 2005). In recent years, an increasing body of evidence is supporting the hypothesis that T2DM and obesity are responsible for inducing carcinoma in cells and tissues. Several epidemiological studies have reported that T2DM and obesity cause 6% annual cancer cases in the United States (Ogden et al. 2006), 7.7% in Canada (Pan et al. 2004), 5% in Europe (Bergström et al. 2001) and highest cases up to 12% in Asia-Pacific countries including Japan (Kuriyama et al. 2005).

The effect of insulin resistance such as increased obesity and cancer has significant consequences in terms of their impact on economic cost and public health. To control the increasing diabetes and cancer diseases, preventative recommendations for a healthy lifestyle and clinical care must be adopted (Giovannucci et al. 2010).

2 Obesity and Cancer

The studies since the 1990s reveal that obesity and cancer are the most prevalent epidemics of the century (Vucenik and Stains 2012). According to "The American Cancer Society" the yearly cancer cases reporting in the United States are around 1.5 million with 550,000 deaths (Jemal et al. 2011). According to the World Health Organization (WHO), a BMI of 25 kg/m^2 and above is considered as overweight. In the United States over the past 40 years, the prevalence of obesity is rising constantly, 66% of people are obese having BMI $\geq$ 30 kg/m^2 respectively (Flegal et al. 2012).

Epidemiological studies have shown the link between obesity and the risk of cancer that includes breast, pancreas, kidney, endometrium, liver, gall bladder, esophagus, hematological malignancy, and colon cancer (Fund and Research 2007; Lichtman 2010) (Table 1). Cancer Prevention Study II (CPS-II) which examined a large number of patients in United States of America from 1982 to 1996 became an important reference study that demonstrated the link between the high mortality rate of cancer with obesity. The authors proposed that 40–80% of obese men and women with BMI greater than 35 kg/m^2 were at highest risk of cancer mortality as compared to normal-weight people having BMI 18.5–24.9 kg/m^2 (Calle et al. 2003). In another study, involving 221 datasets from different populations showed that an increasing BMI in men, leads to a high risk of thyroid, leukemia, kidney, esophageal, and colon cancer, whereas in women it leads to a high risk of breast, pancreatic, gallbladder, thyroid, kidney and multiple myeloma type cancer (Renehan et al. 2006). Whereas, no association of obesity was found in people having lung cancer and only 4% of the population reported having prostate cancer were linked with obesity (Calle et al. 2003; Bergström et al. 2001).

Table 1 Type of cancers associated with obesity and T2DM

Cause	Cancer type	References
Obesity	Thyroid	Pappa and Alevizaki (2014)
	Liver	Marengo et al. (2016)
	Esophageal adenocarcinoma	Lagergren (2011)
	Leukemia	Larsson and Wolk (2008)
	Lung	Fujita et al. (2019)
	Pancreas	Rawla et al. (2019)
	Colon	Larsson and Wolk (2007)
	Renal	Aurilio et al. (2019)
	Prostate	Allott et al. (2013)
Type 2 Diabetes	Breast	Boyle et al. (2012)
	Bladder	Gu et al. (2016)
	Endometrium	Friberg et al. (2007)
	Colorectal	Yang et al. (2004)
	Non-Hodgkins Lymphoma	Mitri et al. (2008)
	Prostate	Bansal et al. (2013)
	Pancreatic	Li (2012)
	Hepatocellular	Dyson et al. (2014)

3 Diabetes and Cancer

Diabetes also has an increasing prevalence worldwide along with obesity. Epidemiological studies have projected the association between diabetes and a variety of cancers, including pancreatic, endometrial, colorectal, and breast cancers (Garg et al. 2014) (Table 1). According to multiple prospective cohort studies type 2-diabetes is individually responsible for different types of cancers that include liver, gall bladder, pancreas, and non-Hodgkin lymphoma carcinomas (Kasper and Giovannucci 2006; Johnson and Pollak 2010). The first examination studies that correlated diabetes with cancer was conducted by "The National Health and Nutrition Examination Survey I". According to the study involving 14,407 total population of men and women, diabetic men had a 39% higher risk of developing prostate and colorectal cancer while diabetic women had a 17% increased risk of developing breast cancer (Michels et al. 2003; Sjöström et al. 2009). Also, high mortality of bladder cancer, pharyngeal cancer, and oral cavity was reported in men and endometrial cancer was seen in women (Campbell et al. 2012).

In contrast, type 1 diabetes has less risk of cancer as compared to type 2 diabetes. In type 1 diabetes the autoimmune destruction of pancreatic beta cells causes the deficiency of insulin in the body and it is not highly linked with obesity and metabolic syndrome. Though, hyperglycemia being an important factor promotes cancer in type 2 diabetic patients, the increase in cancer risk and mortality could also be expected in patients with type 1 diabetes as mentioned in some studies. A Swedish group of researchers in one of their studies observed the link between type 1 diabetes and increased risk of endometrial, cervical, and stomach cancer (Zendehdel et al. 2003). In another study by Swedish researchers, it was determined that type 1 diabetic patients were of 17% higher risk in developing cancer, specifically leukemia, skin cancer, and gastric cancer (Shu et al. 2010). The general mechanism that promotes cancer in the body tissue of diabetic persons is due to the alterations caused by hyperglycemia, hyperinsulinemia, and inflammation (Gallagher et al. 2010; Johnson and Pollak 2010). Another cause that promotes cancer in diabetic patients is site-specific mechanisms in a specific organ.

4 Obesity and Diabetes

The prevalence of obesity is the root that potentially leads to the development of various diseases among which the most devastating is type 2 diabetes mellitus (T2DM) (Huang et al. 2018). Type 2 diabetes leads to several comorbidities such as renal failure, cardiovascular disease, and microangiopathy (Holman 2013). The excessive production of insulin in the body decreases insulin sensitivity which is a cause of diabetic hyperglycemia or insulin resistance. The prevalence of diabetes increases four times with the long term obesity as compared to the non-obese population (Bhupathiraju and Hu 2016). In the United States, the increasing number of T2DM (i.e. 1 in every 10 people) has been diagnosed among adolescents and children that are overweight. The yearly health costs that obese and diabetic patients pay is 42% higher than normal-weight individuals (Mozaffarian et al. 2015).

Both diabetes mellitus and obesity are also the root cause of multifactorial diseases such as coronary heart disease, hypertension, cancer, etc. (Li et al. 2014). The American Heart Association in one of their research articles proposed the best body mass index to be <25 kg/m^2 and fasting glucose level < 100 mg/dL for ideal cardiovascular health (Mongraw-Chaffin et al. 2015). The health risks related to obesity and diabetes are mostly preventable if the reasons behind their exceptional growth are known. The association of obesity and T2DM was studied by the Centers for Disease Control and Prevention among the population from 1988–1994 to 2011–2012. It was found that diabetes was increased among the population of all age groups with a period. It was further examined that individuals with greater BMI (obese) showed a greater trend of diabetes mellitus (20%) (Menke et al. 2015).

5 Mechanism Linking Insulin Resistance to Obesity and Cancer

Among numerous biological mechanisms that explain the relationship between obesity and cancer, the best-considered factors are endogenous hormones (insulin, IGF system, and steroid hormones), cytokines, and adipokines (Khandwala et al. 2000). Obesity causes the increment in the formation of free fatty acids, certain cytokines, and leptins, it also causes a decrease in the production of the hormone adiponectin that has antidiabetic properties (Barb et al. 2006). All these abnormalities along with decreased physical activity lead to insulin resistance which is the major metabolic cause that leads to T2DM (Morris et al. 2006). Several studies indicate the high risk of obesity-related cancerous growth associated with greater insulin concentration in the body (Hammarsten and Högstedt 2005; Lawlor et al. 2004). In another review, it was demonstrated that the functionality of bioactive IGF-1 was inhibited by insulin resistance as it reduces the formation and efficiency of IGF-binding protein, thus causing the rise in free IGF-1 levels (McCarty 2004). The study further endorse that these elevated IGF-1 levels could be associated with prostate cancer and other types of malignancies such as lung cancer and breast cancer.

Also, insulin resistance/hyperinsulinemia also causes a reduction in sex-hormone binding globulin, hence resulting in the bioavailability of high levels of estrogen. In obese individuals, the aromatase levels present in adipose tissue increases the estrogen synthesis (LeRoith and Roberts Jr 2003). The experiment of inducing high dose insulin in rodents results in neoplastic abnormalities. High levels of IGF-1 in rodents causes hyperplasia in the skin and mammary cancer, whereas the those subjected to lower IGF-1 levels showed a reduction in cancer growth (Fierz et al. 2010; Yakar et al. 2005). The relationship between insulin resistance with obesity and cancer is demonstrated in Fig. 1 and the detailed mechanism of between obesity, T2DM, and cancer is illustrated in Fig. 2.

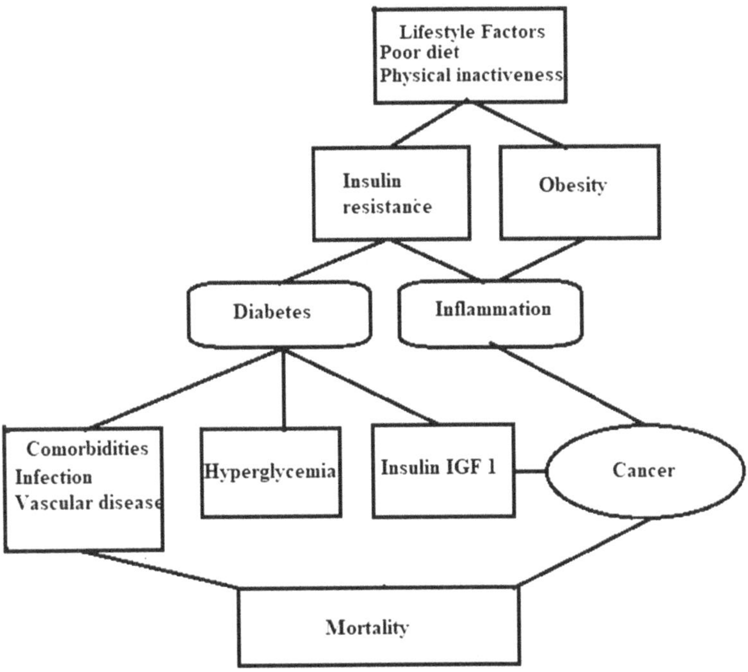

Fig. 1 The interrelationship between diabetes, obesity, and cancer

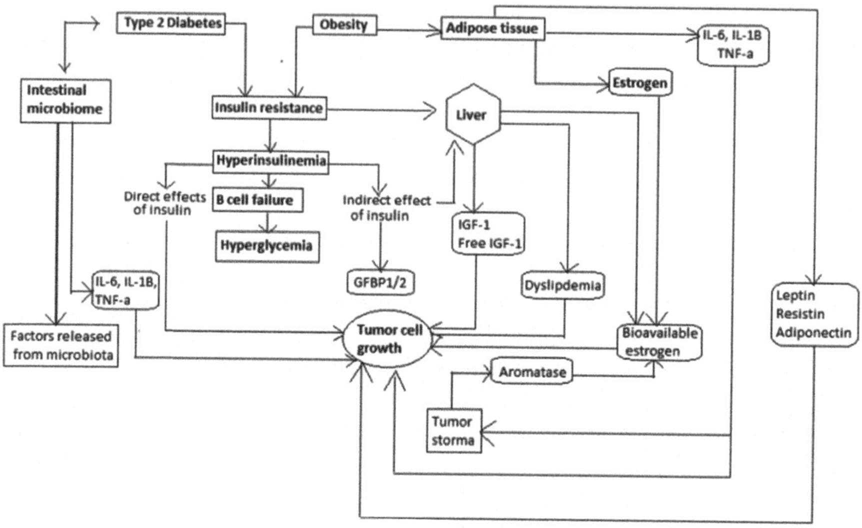

Fig. 2 The mechanism associated with insulin resistance, obesity, and cancer

Diet also play an important role in causing obesity as it is the main determinant of the body composition of an individual (Percik and Stumvoll 2009). Mitogenesis (stimulation of cancer cells) occurs in a person's body when overnutrition leads to the production of reactive oxygen species (Wellen et al. 2010). The consumption of a high amount of food could also enhance the functionality of carcinogenic compounds present in the food. Therefore, the cancer progression is promoted by inflammatory effects (including, production of C-reactive protein, tissue-specific hormone activity, TNF-α, IL-6, IL-1, etc.) on the body due to obesity (Bao et al. 2011). The reduction in these inflammatory markers production is highly based on weight loss, which also decreases the chances of certain carcinoma (Williams et al. 2013).

6 Drugs to Treat Insulin Resistance

6.1 Metformin

The most common therapy to treat individuals of type 2 diabetes mellitus is metformin (Nathan et al. 2009). It may protect against cancer by reducing the insulin levels and also by reducing the effect of AMPK (adenosine monophosphate-activated protein) signaling pathways. In some previous studies of cancer cell lines, it was determined that metformin reduces colony formation and inhibits cell proliferation. In another In vitro study, the effect of metformin in enhancing the efficiency of breast cancer treatment and killing cancer stem cells was also revealed (Jalving et al.

2010). The mechanisms involved in the antineoplastic effect of metformin have been investigated by several In vitro and In vivo studies. The possible mechanisms among different studies include (1) induction of apoptosis (2) reduction in circulating insulin levels (3) activation of liver kinase B1/AMPK pathway that stabilizes the tumor suppressor gene (4) removal of cancer stem cells (5) activation of the immune system (6) inhibition of protein synthesis and (7) inhibition of the unfolded protein response (Kourelis and Siegel 2012).

In a recent study, the link between metformin and the occurrence of breast cancer in T2DM women was investigated (Bosco et al. 2011). It was deduced that a 20% decrease in the incidence of breast cancer was observed in women using metformin as compared to other antidiabetic medication users. In another study, the development of pancreatic cancer was also significantly lower in individuals using metformin (Little et al. 2016).

6.2 Thiazolidinediones

Thiazolidinediones (TZDs) are another major medication used to prevent and treat type 2 diabetes. Currently, TZDs are being used in the prescribed drugs pioglitazone and rosiglitazone in the United States. The main role of TZDs is to activate adipocyte transcription factor (PPAR)-γ peroxisome proliferator-activated receptor which when reacting with retinoid X receptor (RXR) forms a heterodimer that helps in the regulation of genes expression of inflammation, insulin action, lipid metabolism, and adipocyte differentiation. This PPAR-γ activation by TZDs enhances insulin sensitivity leading to a decrease in the release of FFA; causing adipocytes to produce a low amount of resin, prostaglandins, and leptins (Flowers et al. 2015).

The effect of TZDs as an antineoplastic drug was first studied with troglitazone to try and cure prostate cancer. Troglitazone though had an antiproliferative effect in the number of cancer cells but failed to provide the therapeutic benefit on breast cancer and colorectal cancer, thus was withdrawn from clinical studies and was replaced by pioglitazone and rosiglitazone (Burstein et al. 2003; Kulke et al. 2002). Another study revealed that in a clinical trial, patients with thyroid cancer who were given rosiglitazone increased serum thyroglobulin levels and improved radioiodine uptake (Kebebew et al. 2009). The use of rosiglitazone in randomized clinical trials demonstrates that patients treated with rosiglitazone have a lower likelihood of developing a tumor as compared to non-rosiglitazone treated patients (Monami et al. 2014).

The anti-neoplastic effect of pioglitazone is also studied in various In vitro and In vivo studies. Some studies demonstrate that the risk of bladder cancer increases with the use of pioglitazone but various studies do not support this finding. A study with 2-year therapy of pioglitazone in diabetic patients revealed that there was not a high risk to develop cancer in pioglitazone-treated patients (Ferrara et al. 2011).

Another study by United Kingdom Clinical Practice Research Datalink also stated that there is no high risk of developing bladder cancer with the use of pioglitazone respectively (Chen et al. 2013).

6.3 *Insulin Analogues*

Insulin therapy is a keynote treatment to prevent hyperglycemia in both type 1 and type 2 diabetes. In the 1980s, the effort made to replace beef insulin and porcine led to the manufacture of recombinant human insulin produced by DNA technology for clinical use (Galloway 1980). Multiple insulin analogs for the reduction of hyperglycemia were developed. There are two categories of these insulin analogs (i) those with the highest and fastest effect (ii) those with delayed action (Evans et al. 2011; Kang et al. 1991). One of the cohort studies in 2011–2012 deduced that the risk of some cancers, especially breast cancer increases with the use of insulin analogs (Higgins et al. 2012; de Ruiter et al. 2012). In another study, the researchers found that the use of insulin glargine for up to 4 years does not induce any breast cancer in the women population. Moreover, some studies revealed that the use of insulin glargine helps in reducing the risk of various cancers including breast cancer (Ljung et al. 2012).

7 Challenges in Prevention and Treatment

It has been reported that the clinical treatments and procedures of people with obesity have poor outcomes. For instance, in radiotherapy it is difficult to adjust the radiation volume for obese patients, thus it reduces the effectiveness of treatment (Vucenik and Stains 2012). In some cases, during laparoscopic surgery for colorectal cancer, obese people experience longer operating times and also have a greater chance of requiring open surgery (Makino et al. 2012; Whitlock et al. 2012). Several studies also demonstrated that obese people remain under-dosed during chemotherapy as high doses calculated according to their body masses may cause toxicity (Wolin et al. 2010). Hence, the effectiveness of chemotherapy in obese people is less as compared to non-obese patients (Griggs et al. 2012). Moreover, diabetic treatment nowadays may lead to the development of cancer as these medications directly or indirectly affect the level of circulating insulin, insulin resistance, and other mechanisms in the body (Zakikhani et al. 2006). Therefore, there is great concern amongst physicians to lower the risk of cancer while prescribing diabetic medication (Burton et al. 2008).

8 Conclusion

The interrelationship between diabetes, obesity, and cancer as given in various epidemiological studies has opened new understandings of cancer growth and the mechanisms. Uncontrolled obesity and diabetes cause multiple abnormalities in an individual which may increase the risk of cancer. This understanding may help develop strategies to prevent diabetes and obesity and to also control cancer development in diabetic and obese patients. Life style factor is one of the important factors which may prevent a high proportion of obesity, diabetes, and cancer globally. Furthermore, a complete guideline for screening cancer in diabetic and obese patients is required to reduce the rate of cancer mortality.

References

Allott, E. H., Masko, E. M., & Freedland, S. J. (2013). Obesity and prostate cancer: Weighing the evidence. *European Urology, 63*(5), 800–809.

Aurilio, G., Piva, F., Santoni, M., Cimadamore, A., Sorgentoni, G., Lopez-Beltran, A., et al. (2019). The role of obesity in renal cell carcinoma patients: Clinical-pathological implications. *International Journal of Molecular Sciences, 20*(22), 5683.

Bansal, D., Bhansali, A., Kapil, G., Undela, K., & Tiwari, P. (2013). Type 2 diabetes and risk of prostate cancer: A meta-analysis of observational studies. *Prostate Cancer and Prostatic Diseases, 16*(2), 151–158.

Bao, B., Ali, S., Kong, D., Sarkar, S. H., Wang, Z., Banerjee, S., et al. (2011). Anti-tumor activity of a novel compound-CDF is mediated by regulating miR-21, miR-200, and PTEN in pancreatic cancer. *PLoS One, 6*(3), e17850.

Barb, D., Pazaitou-Panayiotou, K., & Mantzoros, C. S. (2006). Adiponectin: A link between obesity and cancer. *Expert Opinion on Investigational Drugs, 15*(8), 917–931.

Bell, E. (1957). Carcinoma of the pancreas: I. A clinical and pathologic study of 609 necropsied cases. II. The relation of carcinoma of the pancreas to diabetes mellitus. *The American Journal of Pathology, 33*(3), 499.

Bergström, A., Pisani, P., Tenet, V., Wolk, A., & Adami, H. O. (2001). Overweight as an avoidable cause of cancer in Europe. *International Journal of Cancer, 91*(3), 421–430.

Bhupathiraju, S. N., & Hu, F. B. (2016). Epidemiology of obesity and diabetes and their cardiovascular complications. *Circulation Research, 118*(11), 1723–1735.

Bosco, J. L. F., Antonsen, S., Sørensen, H. T., Pedersen, L., & Lash, T. L. (2011). Metformin and incident breast cancer among diabetic women: A population-based case–control study in Denmark. *Cancer Epidemiology and Prevention Biomarkers, 20*(1), 101–111.

Boyle, P., Boniol, M., Koechlin, A., Robertson, C., Valentini, F., Coppens, K., et al. (2012). Diabetes and breast cancer risk: A meta-analysis. *British Journal of Cancer, 107*(9), 1608–1617.

Burstein, H. J., Demetri, G. D., Mueller, E., Sarraf, P., Spiegelman, B. M., & Winer, E. P. (2003). Use of the peroxisome proliferator-activated receptor (PPAR)-γ ligand troglitazone as treatment for refractory breast cancer: A phase II study. *Breast Cancer Research and Treatment, 79*(3), 391–397.

Burton, J. B., Priceman, S. J., Sung, J. L., Brakenhielm, E., An, D. S., Pytowski, B., et al. (2008). Suppression of prostate cancer nodal and systemic metastasis by blockade of the lymphangiogenic axis. *Cancer Research, 68*(19), 7828–7837.

Calle, E. E., Rodriguez, C., Walker-Thurmond, K., & Thun, M. J. (2003). Overweight, obesity, and mortality from cancer in a prospectively studied cohort of US adults. *New England Journal of Medicine, 348*(17), 1625–1638.

Campbell, P. T., Newton, C. C., Patel, A. V., Jacobs, E. J., & Gapstur, S. M. (2012). Diabetes and cause-specific mortality in a prospective cohort of one million US adults. *Diabetes Care, 35*(9), 1835–1844.

Chen, S.-W., Tsan, Y.-T., Chen, J.-D., Hsieh, H.-I., Lee, C.-H., Lin, H.-H., et al. (2013). Use of thiazolidinediones and the risk of colorectal cancer in patients with diabetes: A nationwide, population-based, case-control study. *Diabetes Care, 36*(2), 369–375.

de Ruiter, M. B., Reneman, L., Boogerd, W., Veltman, D. J., Caan, M., Douaud, G., et al. (2012). Late effects of high-dose adjuvant chemotherapy on white and gray matter in breast cancer survivors: Converging results from multimodal magnetic resonance imaging. *Human Brain Mapping, 33*(12), 2971–2983.

Dyson, J., Jaques, B., Chattopadyhay, D., Lochan, R., Graham, J., Das, D., et al. (2014). Hepatocellular cancer: The impact of obesity, type 2 diabetes and a multidisciplinary team. *Journal of Hepatology, 60*(1), 110–117.

Evans, M., Schumm-Draeger, P., Vora, J., & King, A. (2011). A review of modern insulin analogue pharmacokinetic and pharmacodynamic profiles in type 2 diabetes: Improvements and limitations. *Diabetes, Obesity and Metabolism, 13*(8), 677–684.

Ferrara, A., Lewis, J. D., Quesenberry, C. P., Peng, T., Strom, B. L., Van Den Eeden, S. K., et al. (2011). Cohort study of pioglitazone and cancer incidence in patients with diabetes. *Diabetes Care, 34*(4), 923–929.

Fierz, Y., Novosyadlyy, R., Vijayakumar, A., Yakar, S., & LeRoith, D. (2010). Insulin-sensitizing therapy attenuates type 2 diabetes–mediated mammary tumor progression. *Diabetes, 59*(3), 686–693.

Flegal, K. M., Carroll, M. D., Kit, B. K., & Ogden, C. L. (2012). Prevalence of obesity and trends in the distribution of body mass index among US adults, 1999–2010. *Journal of the American Medical Association, 307*(5), 491–497.

Flowers, E., Aouizerat, B. E., Abbasi, F., Lamendola, C., Grove, K. M., Fukuoka, Y., & Reaven, G. M. (2015). Circulating microRNA-320a and microRNA-486 predict thiazolidinedione response: Moving towards precision health for diabetes prevention. *Metabolism, 64*(9), 1051–1059.

Friberg, E., Orsini, N., Mantzoros, C., & Wolk, A. (2007). Diabetes mellitus and risk of endometrial cancer: A meta-analysis. *Diabetologia, 50*, 1365–1374. https://doi.org/10.1007/s00125-007-0681-5. Springer.

Fujita, K., Hayashi, T., Matsushita, M., Uemura, M., & Nonomura, N. (2019). Obesity, inflammation, and prostate cancer. *Journal of Clinical Medicine, 8*(2), 201.

Fund, W. C. R., & Research, A. I. F. C. (2007). *Food, nutrition, physical activity, and the prevention of cancer: A global perspective* (Vol. 1). Washington, DC: American Institute for Cancer Research.

Gahagan, S., Haslam, D. W., & James, W. P. (2005). Obesity. *The Lancet, 366*, 1197–1209.

Gallagher, E., Fierz, Y., Ferguson, R., & LeRoith, D. (2010). The pathway from diabetes and obesity to cancer, on the route to targeted therapy. *Endocrine Practice, 16*(5), 864–873.

Galloway, J. A. (1980). Insulin treatment for the early 80s: Facts and questions about old and new insulins and their usage. *Diabetes Care, 3*(5), 615–622.

Garg, S., Maurer, H., Reed, K., & Selagamsetty, R. (2014). Diabetes and cancer: Two diseases with obesity as a common risk factor. *Diabetes, Obesity and Metabolism, 16*(2), 97–110.

Giovannucci, E., Harlan, D. M., Archer, M. C., Bergenstal, R. M., Gapstur, S. M., Habel, L. A., et al. (2010). Diabetes and cancer: A consensus report. *Diabetes Care, 33*(7), 1674–1685.

Griggs, J. J., Mangu, P. B., Temin, S., & Lyman, G. H. (2012). Appropriate chemotherapy dosing for obese adult patients with cancer: American Society of Clinical Oncology clinical practice guideline. *Journal of Oncology Practice, 8*(4), e59.

Gu, J., Yan, S., Wang, B., Shen, F., Cao, H., Fan, J., & Wang, Y. (2016). Type 2 diabetes mellitus and risk of gallbladder cancer: A systematic review and meta-analysis of observational studies. *Diabetes/Metabolism Research and Reviews, 32*(1), 63–72.

Hammarsten, J., & Högstedt, B. (2005). Hyperinsulinaemia: A prospective risk factor for lethal clinical prostate cancer. *European Journal of Cancer, 41*(18), 2887–2895.

Higgins, M. J., Jelovac, D., Barnathan, E., Blair, B., Slater, S., Powers, P., et al. (2012). Detection of tumor PIK3CA status in metastatic breast cancer using peripheral blood. *Clinical Cancer Research, 18*(12), 3462–3469.

Holman, R. R. (2013). Optimal management of T2DM remains elusive. *Nature Reviews Endocrinology, 9*(2), 67–68.

Huang, X., Liu, G., Guo, J., & Su, Z. (2018). The PI3K/AKT pathway in obesity and type 2 diabetes. *International Journal of Biological Sciences, 14*(11), 1483.

Jalving, M., Gietema, J. A., Lefrandt, J. D., de Jong, S., Reyners, A. K., Gans, R. O., & de Vries, E. G. (2010). Metformin: Taking away the candy for cancer? *European Journal of Cancer, 46*(13), 2369–2380.

Jemal, A., Bray, F., Center, M. M., Ferlay, J., Ward, E., & Forman, D. (2011). Global cancer statistics. *CA: A Cancer Journal for Clinicians, 61*(2), 69–90.

Johnson, J., & Pollak, M. (2010). Insulin, glucose and the increased risk of cancer in patients with type 2 diabetes. *Diabetologia, 53*(10), 2086–2088.

Kang, S., Brange, J., Burch, A., Vølund, A., & Owens, D. R. (1991). Subcutaneous insulin absorption explained by insulin's physicochemical properties: Evidence from absorption studies of soluble human insulin and insulin analogues in humans. *Diabetes Care, 14*(11), 942–948.

Kasper, J. S., & Giovannucci, E. (2006). A meta-analysis of diabetes mellitus and the risk of prostate cancer. *Cancer Epidemiology and Prevention Biomarkers, 15*(11), 2056–2062.

Kebebew, E., Lindsay, S., Clark, O. H., Woeber, K. A., Hawkins, R., & Greenspan, F. S. (2009). Results of rosiglitazone therapy in patients with thyroglobulin-positive and radioiodine-negative advanced differentiated thyroid cancer. *Thyroid, 19*(9), 953–956.

Khandwala, H. M., McCutcheon, I. E., Flyvbjerg, A., & Friend, K. E. (2000). The effects of insulin-like growth factors on tumorigenesis and neoplastic growth. *Endocrine Reviews, 21*(3), 215–244.

Kourelis, T. V., & Siegel, R. D. (2012). Metformin and cancer: New applications for an old drug. *Medical Oncology, 29*(2), 1314–1327.

Kulke, M. H., Demetri, G. D., Sharpless, N. E., Ryan, D. P., Shivdasani, R., Clark, J. S., et al. (2002). A phase II study of troglitazone, an activator of the PPARγ receptor, in patients with chemotherapy-resistant metastatic colorectal cancer. *The Cancer Journal, 8*(5), 395–399.

Kuriyama, S., Tsubono, Y., Hozawa, A., Shimazu, T., Suzuki, Y., Koizumi, Y., et al. (2005). Obesity and risk of cancer in Japan. *International Journal of Cancer, 113*(1), 148–157.

Lagergren, J. (2011). Influence of obesity on the risk of esophageal disorders. *Nature Reviews Gastroenterology & Hepatology, 8*(6), 340.

Larsson, S. C., & Wolk, A. (2007). Obesity and colon and rectal cancer risk: A meta-analysis of prospective studies. *The American Journal of Clinical Nutrition, 86*(3), 556–565.

Larsson, S. C., & Wolk, A. (2008). Overweight and obesity and incidence of leukemia: A meta-analysis of cohort studies. *International Journal of Cancer, 122*(6), 1418–1421.

Lawlor, D. A., Smith, G. D., & Ebrahim, S. (2004). Hyperinsulinaemia and increased risk of breast cancer: Findings from the British Women's Heart and Health Study. *Cancer Causes & Control, 15*(3), 267–275.

LeRoith, D., & Roberts, C. T., Jr. (2003). The insulin-like growth factor system and cancer. *Cancer Letters, 195*(2), 127–137.

Li, D. (2012). Diabetes and pancreatic cancer. *Molecular Carcinogenesis, 51*(1), 64–74.

Li, G., Zhang, P., Wang, J., An, Y., Gong, Q., Gregg, E. W., et al. (2014). Cardiovascular mortality, all-cause mortality, and diabetes incidence after lifestyle intervention for people with impaired glucose tolerance in the Da Qing Diabetes Prevention Study: A 23-year follow-up study. *The Lancet Diabetes & Endocrinology, 2*(6), 474–480.

Lichtman, M. A. (2010). Obesity and the risk for a hematological malignancy: Leukemia, lymphoma, or myeloma. *The Oncologist, 15*(10), 1083.
Little, M., Humphries, S., Patel, K., Dodd, W., & Dewey, C. (2016). Factors associated with glucose tolerance, pre-diabetes, and type 2 diabetes in a rural community of south India: A cross-sectional study. *Diabetology & Metabolic Syndrome, 8*(1), 21.
Ljung, R., Talbäck, M., Haglund, B., Jonasson, J. M., Gudbjörnsdòttir, S., & Steineck, G. (2012). Insulin glargine use and short-term incidence of breast cancer—A four-year population-based observation. *Acta Oncologica, 51*(3), 400–402.
Lopez, A. D., Mathers, C. D., Ezzati, M., Jamison, D. T., & Murray, C. J. (2006). Global and regional burden of disease and risk factors, 2001: Systematic analysis of population health data. *The Lancet, 367*(9524), 1747–1757.
Makino, T., Shukla, P. J., Rubino, F., & Milsom, J. W. (2012). The impact of obesity on perioperative outcomes after laparoscopic colorectal resection. *Annals of Surgery, 255*(2), 228–236.
Marengo, A., Rosso, C., & Bugianesi, E. (2016). Liver cancer: Connections with obesity, fatty liver, and cirrhosis. *Annual Review of Medicine, 67*, 103–117.
McCarty, M. F. (2004). Elevated sympathetic activity may promote insulin resistance syndrome by activating alpha-1 adrenergic receptors on adipocytes. *Medical Hypotheses, 62*(5), 830–838.
Melvin, A., O'Rahilly, S., & Savage, D. (2018). Genetic syndromes of severe insulin resistance. *Current Opinion in Genetics & Development, 50*, 60–67.
Menke, A., Casagrande, S., Geiss, L., & Cowie, C. C. (2015). Prevalence of and trends in diabetes among adults in the United States, 1988–2012. *Journal of the American Medical Association, 314*(10), 1021–1029.
Michels, K. B., Solomon, C. G., Hu, F. B., Rosner, B. A., Hankinson, S. E., Colditz, G. A., & Manson, J. E. (2003). Type 2 diabetes and subsequent incidence of breast cancer in the Nurses' Health Study. *Diabetes Care, 26*(6), 1752–1758.
Mitri, J., Castillo, J., & Pittas, A. G. (2008). Diabetes and risk of non-Hodgkin's lymphoma: A meta-analysis of observational studies. *Diabetes Care, 31*(12), 2391–2397.
Monami, M., Dicembrini, I., & Mannucci, E. (2014). Thiazolidinediones and cancer: Results of a meta-analysis of randomized clinical trials. *Acta Diabetologica, 51*(1), 91–101.
Mongraw-Chaffin, M. L., Peters, S. A., Huxley, R. R., & Woodward, M. (2015). The sex-specific association between BMI and coronary heart disease: A systematic review and meta-analysis of 95 cohorts with 1·2 million participants. *The Lancet Diabetes & Endocrinology, 3*(6), 437–449.
Morris, J., George, L., Wu, T., & Wald, N. (2006). Insulin-like growth factors and cancer: No role in screening. Evidence from the BUPA study and meta-analysis of prospective epidemiological studies. *British Journal of Cancer, 95*(1), 112–117.
Mozaffarian, D., Benjamin, E. J., Go, A. S., Arnett, D. K., Blaha, M. J., Cushman, M., et al. (2015). Executive summary: Heart disease and stroke statistics—2015 update: A report from the American Heart Association. *Circulation, 131*(4), 434–441.
Nathan, D. M., Buse, J. B., Davidson, M. B., Ferrannini, E., Holman, R. R., Sherwin, R., & Zinman, B. (2009). Medical management of hyperglycaemia in type 2 diabetes mellitus: A consensus algorithm for the initiation and adjustment of therapy. *Diabetologia, 52*(1), 17.
Ogden, C. L., Carroll, M. D., Curtin, L. R., McDowell, M. A., Tabak, C. J., & Flegal, K. M. (2006). Prevalence of overweight and obesity in the United States, 1999–2004. *Journal of the American Medical Association, 295*(13), 1549–1555.
Pan, S. Y., Johnson, K. C., Ugnat, A.-M., Wen, S. W., & Mao, Y. (2004). Association of obesity and cancer risk in Canada. *American Journal of Epidemiology, 159*(3), 259–268.
Pappa, T., & Alevizaki, M. (2014). Obesity and thyroid cancer: A clinical update. *Thyroid, 24*(2), 190–199.
Percik, R., & Stumvoll, M. (2009). Obesity and cancer. *Experimental and Clinical Endocrinology & Diabetes, 117*(10), 563–566.
Rawla, P., Thandra, K. C., & Sunkara, T. (2019). Pancreatic cancer and obesity: Epidemiology, mechanism, and preventive strategies. *Clinical Journal of Gastroenterology, 12*(4), 285–291.

Renehan, A. G., Frystyk, J., & Flyvbjerg, A. (2006). Obesity and cancer risk: The role of the insulin–IGF axis. *Trends in Endocrinology & Metabolism, 17*(8), 328–336.

Shu, X., Ji, J., Li, X., Sundquist, J., Sundquist, K., & Hemminki, K. (2010). Cancer risk among patients hospitalized for Type 1 diabetes mellitus: A population-based cohort study in Sweden. *Diabetic Medicine, 27*(7), 791–797.

Sjöström, L., Gummesson, A., Sjöström, C. D., Narbro, K., Peltonen, M., Wedel, H., et al. (2009). Effects of bariatric surgery on cancer incidence in obese patients in Sweden (Swedish Obese Subjects Study): A prospective, controlled intervention trial. *The Lancet Oncology, 10*(7), 653–662.

Vucenik, I., & Stains, J. P. (2012). Obesity and cancer risk: Evidence, mechanisms, and recommendations. *Annals of the New York Academy of Sciences, 1271*(1), 37.

Wellen, K. E., Lu, C., Mancuso, A., Lemons, J. M., Ryczko, M., Dennis, J. W., et al. (2010). The hexosamine biosynthetic pathway couples growth factor-induced glutamine uptake to glucose metabolism. *Genes & Development, 24*(24), 2784–2799.

Whitlock, K., Gill, R. S., Birch, D. W., & Karmali, S. (2012). The association between obesity and colorectal cancer. *Gastroenterology Research and Practice, 2012.*

Williams, A. J., Henley, W. E., Williams, C. A., Hurst, A. J., Logan, S., & Wyatt, K. M. (2013). Systematic review and meta-analysis of the association between childhood overweight and obesity and primary school diet and physical activity policies. *International Journal of Behavioral Nutrition and Physical Activity, 10*(1), 1–22.

Wolin, K. Y., Carson, K., & Colditz, G. A. (2010). Obesity and cancer. *The Oncologist, 15*(6), 556.

Yakar, S., LeRoith, D., & Brodt, P. (2005). The role of the growth hormone/insulin-like growth factor axis in tumor growth and progression: Lessons from animal models. *Cytokine & Growth Factor Reviews, 16*(4–5), 407–420.

Yang, Y. X., Hennessy, S., & Lewis, J. D. (2004). Insulin therapy and colorectal cancer risk among type 2 diabetes mellitus patients. *Gastroenterology, 127*(4), 1044–1050.

Zakikhani, M., Dowling, R., Fantus, I. G., Sonenberg, N., & Pollak, M. (2006). Metformin is an AMP kinase–dependent growth inhibitor for breast cancer cells. *Cancer Research, 66*(21), 10269–10273.

Zendehdel, K., Nyrén, O., Östenson, C.-G., Adami, H.-O., Ekbom, A., & Ye, W. (2003). Cancer incidence in patients with type 1 diabetes mellitus: A population-based cohort study in Sweden. *Journal of the National Cancer Institute, 95*(23), 1797–1800.

Role of Cytoskeletal Protein, Actin in Various Diseases

Samridhi Pathak, Avinash Kale, C. M. Santosh Kumar, and Mansoor Sheikh

Abbreviations

ABPs	Actin binding proteins
AD	Alzheimer's disease
ADF	Actin depolymerizing factor
ADP	Adenosine di-phosphate
ALS	Amyotrophic lateral sclerosis
AMPAR	α-amino-3-hydroxy-5-methyl-4-isoxazolepropionic acid receptor
APC	Antigen presenting cells
APP	Amyloid precursor proteins
ASD	Autism spectrum disorder
ATP	Adenosine tri-phosphate
CAP	Cyclase associated protein
CD3	Cluster of differentiation 3
CP-α and β	Capping protein-α and β

All the figures were prepared using MS office (PowerPoint) and are subject for copyright to the authors of the chapter. The figures are made using the in-built tools and software available to us. Figure 5 has a SEM image of paracrystalline actin from pig thigh muscle (*Sus scrofadomesticus*) which was obtained at 200 nm and 50.28× magnification. The SEM analysis was carried out on the ZEISS microscope using the following parameters: EHT: 3 kV, Signal: SE2. It should be noted that all the imaging assay was performed in G-actin buffer at Tata Institute of Fundamental Research (TIFR), Mumbai.

S. Pathak (✉) · A. Kale
School of Chemical Sciences, UM-DAE Centre for Excellence in Basic Sciences, University of Mumbai, Mumbai, Maharashtra, India

C. M. S. Kumar
Institute of Microbiology and Infection (IMI), School of Biosciences, University of Birmingham, Birmingham, UK

M. Sheikh
Advance Centre for Human Genetics, Sher-i-Kashmir Institute of Medical Sciences, Soura, Srinagar, Jammu and Kashmir, India

C. Egbuna, S. Hassan (eds.), *Dietary Phytochemicals*,
https://doi.org/10.1007/978-3-030-72999-8_6

ECM	Extracellular matrix
EGF	Epidermal growth factor
EMT	Epithelial–mesenchymal transition
FGF	Fibroblast growth factor
GAPs	GTPase-activating proteins
GEFs	Guanine nucleotide exchange factors
hnRNPs	Heterogenous nuclear ribonucleoprotein particles
ID	Intellectual disability
IF	Intermediate Filaments
IL-8	Interleukin 8
ITAMs	Immunoreceptor tyrosine-based activation motif
LTD	Long term depression
LTP	Long term potentiation
MEFs	Mouse embryonic fibroblast
MFs	Microfilaments
MHC	Major histocompatibility complex
MTs	Microtubules
NAD	Neutrophil actin dysfunction
NDs	Neurodegenerative disorder
NMDAR	N-methyl-D-aspartate receptor
NM-I	Nuclear myosin I
N-WASP	Neuronal Wiskott–Aldrich Syndrome protein
PD	Parkinson's disease
PH	Periventricular heterotopia
PremRNPs	Precursor messenger ribonucleoproteins
PSD	Post synaptic density
RNPs	Ribonuceloproteins
SCN	Severe congenital neutropenia
TAZ	Transcriptional coactivator with PDZ-binding motif
TCR	T-cell receptor
TNF-α	Tumor necrosis factor-α
VEGF	Vascular endothelial growth factor
WASP	Wiskott–Aldrich Syndrome protein
WHIM syndrome	Warts, Hypogammaglobulinemia, Immunodeficiency, and Myelokathexis syndrome
XLN	X-linked neutropenia
YAP	Yes-associated protein

1 Introduction

The cytoskeletal proteins are important for several functions in the higher eukaryotic cells. The evolution of the eukaryotic cells from the prokaryotic counterparts is quite evident due to the cellular compartmentalization, which gave rise to internally specialized organelles, such as chloroplast, mitochondria, nucleus, endoplasmic reticulum, Golgi apparatus, lysosome, peroxisomes etc. (Zhang et al. 2018). Most of the aforementioned organelle compartments are formed due to the presence of cytoskeletal proteins inside the cell (Fletcher and Mullins 2010; Pessoa-Pureur et al. 2014). The cytoskeletal proteins, which are extensively distributed around the cell, perform a wide variety of cellular functions, such as spatial organization of the cell contents, physical and biochemical connections between the cells and the external environment, providing coordinated effect for cellular motility and changes in the cell shape (Fletcher and Mullins 2010; Klingler-Hoffmann et al. 2019; Logan and Menko 2019; Pessoa-Pureur et al. 2014; Yoon et al. 2002). Further, the cytoskeletal proteins support many phenomena, such as intracellular signaling and transport of essential molecules, cancer related process, such as cell migration and angiogenesis (Paran et al. 1996; Pralhad et al. 2003; Saaristo et al. 2000; Yamazaki et al. 2005).

Three major types of cytoskeletal proteins make up most of the three-dimensional lattice of the cell, viz. (1) actin filaments (2) microtubules and (3) intermediate filaments. The dynamics of these filamentous structures in conjugation with their regulatory proteins, which are in constant flux, is the underlying factor for cell functioning. These three molecules are arranged into large networks that dynamically reorganize to prevent cellular deformations when an external force is applied (Fletcher and Mullins 2010). Actin polymerization dynamics, which is controlled by actin binding proteins (Lambrechts et al. 2004), and microtubule polymerization dynamics control the organization of cellular components. The imperative features that differentiate these three cytoskeletal polymers are their mechanical stiffness, the differential assembly and disassembly dynamics, polarity, the associated regulatory proteins and molecular motors (Fletcher and Mullins 2010). Intermediate filaments (IFs) are a set of closely related proteins that prominently exist in cells that are subject to mechanical stress, as the major function of IFs is to impart physical strength to the cells (Alberts et al. 2002).

The polymerization dynamics, structure and the subsequent regulation of the three proteins differ from each other (Mitchison 1992). Even though a few similarities exist in their polymerization, their lattice structures differ, resulting in differential mechanical strength to their filamentous structure (Mitchison 1992). Actin filaments are usually composed of two protofilaments that form a thin and flexible structure (Balcer et al. 2003; Bugyi and Carlier 2010; Cooper and Schafer 2000; Vavylonis et al. 2005), whereas the microtubules form thicker, hollow, and thirteen stranded assembly, giving rise to much stronger filamentous structure (Logan and Menko 2019; Muñoz-Lasso et al. 2020; Pessoa-Pureur et al. 2014; Yoon et al. 2002). IFs, unlike the other two filaments, are made up of cross section of 32 individual α-helical coils. These helical coils then form a large number of polypeptides which are held together with the strong hydrophobic interactions. Thus, intermediate filaments have a rope like structure and lack overall polarity like in actin filaments and microtubules (Alberts et al. 2002; Lariviere and Julien 2004; Worman and Courvalin 2004; Yoon et al. 2002).

Any deviations in their structures, respective polymerization dynamics and regulation differences, are known to result in various diseases. Here as mentioned in Fig. 1, we discuss such differential aberrations that lead to functional differences of these proteins and consequent multiple disorders.

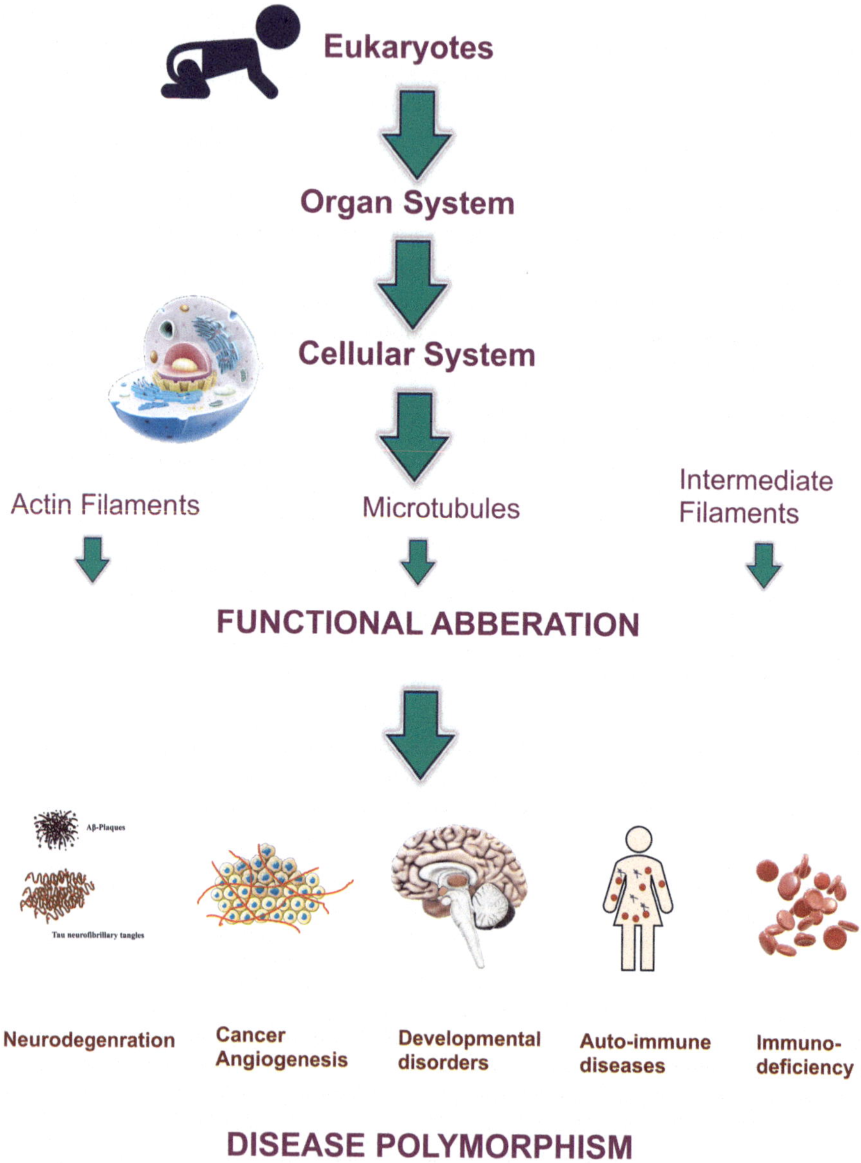

Fig. 1 Overview of the role of cytoskeletal proteins in various disease polymorphisms. The figure explains the different level of cellular organization in eukaryotes and the various disease polymorphism that may manifest due to functional aberrations of the proteins, MFs, MTs and IFs

2 Microfilaments or Actin Filaments

Actin, a 42-kD globular protein, forms one of the major proteins in eukaryotic cytoskeletal niche and is highly conserved (Reisler and Egelman 2007). It exists as a monomeric G-actin as shown in Fig. 2a, which polymerizes to the filamentous F-actin (Reisler and Egelman 2007). The intrinsic property of actin polymerization has been attributed to ATP hydrolysis. ATP binds in the nucleotide-binding cleft of the G-actin monomer (Kudryashov et al. 2005). Structurally, G-actin can be divided into a major and a minor domain. Further, it constitutes four sequential sub-domains, SD-1 to SD-4 (Fig. 2b). The major domain is formed of SD-1 and SD-3 while them in or domain formed of SD-2 and SD-4 (Kudryashov et al. 2005; Carlsson n.d.; Sawaya et al. 2008; Sept et al. 1999). About 150–200 actin binding proteins regulate actin polymerization dynamics, which in turn controls various cellular processes (Papakonstanti and Stournaras 2008; Yamaguchi and Condeelis 2007).

Actin reorganization within a cell is known to be involved in the regulation of several process, such as cellular locomotion, formation of lamellipodia and filipodia, tumor angiogenesis, cancer metastasis, transport of signaling molecules, synaptic plasticity, etc. (Fletcher and Mullins 2010; Papakonstanti and Stournaras 2008; Varland et al. 2019; Yamaguchi and Condeelis 2007). Aberrations in actin and its regulatory proteins might lead to variety of diseased conditions. We, thus, discuss the different cellular process, which are directly or indirectly linked to disease polymorphisms that arise due to the dysregulation of actin polymerization dynamics. All the examples below suggest the need to study the actin polymerization dynamics.

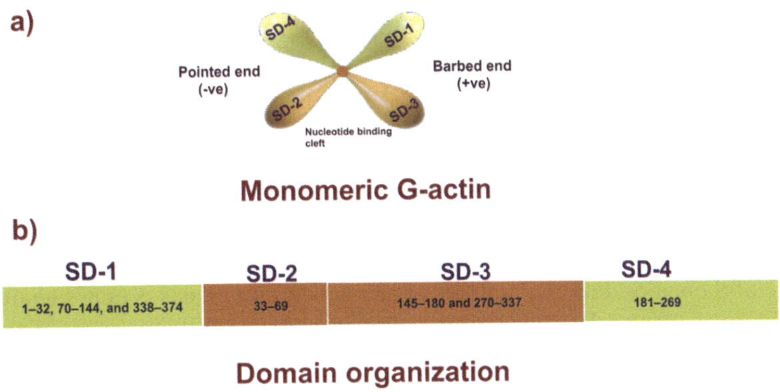

Fig. 2 Cartoon figure represents the domain organization of actin monomer. (**a**) Globular actin monomer with sub-domain (SDs) SD-1 and SD-3 constitute the larger domain, while SD-2 and SD-4 forms the smaller domain. The cleft flanking the SD-1 and SD-3 is known as barbed end or positive end as filament growth occurs at this end. The cleft flanking the SD-2 and SD-4 is known as pointed end or negative end as depolymerization occurs at this end. (**b**) Linear domain organization and residue classification of actin into different sub-domains

2.1 Role in Cancer Metastasis and Tumor Angiogenesis

Cancer metastasis involves various steps, such as cellular migration from the primary tumor, invasion of newer site, intravasation, angiogenesis and anoikic (cell detachment-induced cell death), growth of secondary tumors, evasion, extravasation (Sun et al. 2015). This process requires continuous cellular motility, which is profoundly dependent upon actin polymerization dynamics and bundling (Bayless and Johnson 2011; Lamalice et al. 2007; Sun et al. 2015; Yamazaki et al. 2005). This treadmill formation, involving various key regulators, are involved in the formation of various cellular structures, like lamellipodia, filopodia, invadopodia/podosomes, etc. (Bayless and Johnson 2011; Lamalice et al. 2007; Sun et al. 2015; Yamazaki et al. 2005). In this process, the migrating cell depends on the formation of membrane protrusion, which is generated in response to chemotactic and migratory response. The key element that drives the cellular motility is the localized polymerization of actin within the cell. The actin polymerization increases during cancer invasion and metastasis (Arjonen et al. 2011; Mahadevan and Hart 1990; Prager et al. 2012). There are few key regulatory proteins for actin cytoskeleton that lead to the formation of the membrane protrusion during cancer. These proteins are, WASP family proteins, Arp2/3 complex, LIM-kinase, cofilin, and cortactin (Barron et al. 2017; Bayless and Johnson 2011; Li et al. 2012; Nguyen et al. 2015; Pralhad et al. 2003; Saaristo et al. 2000; Stevenson et al. 2012; Sun et al. 2015; Yamazaki et al. 2005). One of the most important steps for trans-differentiation or lineage reprogramming of epithelial-like cells into motile mesenchymal-like cells, a process known as epithelial-mesenchymal transition (EMT), is the reorganization of actin cytoskeleton (Barron et al. 2017; Bayless and Johnson 2011; Li et al. 2012; Nguyen et al. 2015; Pralhad et al. 2003; Saaristo et al. 2000; Stevenson et al. 2012; Sun et al. 2015; Yamazaki et al. 2005). This is largely involved in the development, wound healing and cancer progression (Lamalice et al. 2007; Li et al. 2012; Paran et al. 1996; Saaristo et al. 2000). However, aggregation of actin might lead to a defect in the reorganization within the cells. Reorganization of actin cytoskeleton involves various regulatory proteins, which are part of different cellular signaling pathways (Papakonstanti and Stournaras 2008). Understanding this reorganization of actin cytoskeleton is a key for controlling cancer invasion.

One of the major defining features of any cancer metastasis is tumor angiogenesis, through which a localized tumor spreads across to different parts of the tissues, thereby affecting the healthy tissue (Barron et al. 2017; Li et al. 2012; Liotta and Stetler-Stevenson 1991; Nguyen et al. 2015; Paran et al. 1996; Stevenson et al. 2012). In case of physiological angiogenesis, which is crucial during the process of wound healing, embryo development, formation of corpus luteum and endometrium, collateral formation of blood vessels for improved organ perfusion actin polymerization dynamics is very tightly regulated by the balance of pro-angiogenetic and anti-angiogenetic factors (Arjonen et al. 2011; Bielenberg and Zetter 2015; Daly et al. 2003; Lugano et al. 2020; Mahadevan and Hart 1990;

Nishida et al. 2006; Prager et al. 2012; Rajabi and Mousa 2017; Sund and Kalluri 2008). However, during the pathological angiogenesis, an aberration in the controlled dynamics of actin polymerization leads to several disease conditions, such as ocular neovascularization, leading to the loss of vision, wet macular degeneration, a rare disease leading to blindness in the elderly, infrequent types of neoplasia, von Hippel–Lindau disease, Lymphangioleiomyomatosis etc. (Rodríguez-Caso et al. 2012).

In growing cancer, endothelial cells become one of the majorly activated cells due to the release of number of regulatory proteins, such as EGF, estrogen, basic and acidic FGF, IL-8, prostaglandin E1 and E2, TNF-α, and VEGF, that can activate endothelial cell growth and motility, when the anti-angiogenic factor production is downregulated (Arjonen et al. 2011; Bielenberg and Zetter 2015; Daly et al. 2003; Lugano et al. 2020; Mahadevan and Hart 1990; Nishida et al. 2006; Prager et al. 2012; Rajabi and Mousa 2017; Sund and Kalluri 2008). These highly activated endothelial cells require oxygen for their metabolism, which is intrinsically provided by the formation of new blood vessels in these tumor cells (Arjonen et al. 2011; Bielenberg and Zetter 2015; Daly et al. 2003; Lugano et al. 2020; Mahadevan and Hart 1990; Nishida et al. 2006; Prager et al. 2012; Rajabi and Mousa 2017; Sund and Kalluri 2008). The process of angiogenesis nourishes the cancer cells and scavenges the waste from the site of tumor formation. Pathological angiogenesis involves a series of sequential steps for tumor nourishment, wherein variety of angiogenic factors bind to the endothelial cell receptors to initiate the process of angiogenesis. The underlying factors for driving this process are the sub-cellular localization and polymerization of actin filaments (Arjonen et al. 2011; Bielenberg and Zetter 2015; Daly et al. 2003; Lugano et al. 2020; Mahadevan and Hart 1990; Nishida et al. 2006; Prager et al. 2012; Rajabi and Mousa 2017; Sund and Kalluri 2008). Protrusive structures, such as filopodia, lamellipodia, and invadopodia/podosomes are formed by migrating and invading cells (Arjonen et al. 2011; Bielenberg and Zetter 2015; Daly et al. 2003; Lugano et al. 2020; Mahadevan and Hart 1990; Nishida et al. 2006; Prager et al. 2012; Rajabi and Mousa 2017; Sund and Kalluri 2008). Upon binding of the angiogenic factors to their respective receptors, the chemoattractant released from the tumor cells stimulate intracellular signaling pathways to regulate the reorganization of actin cytoskeleton as shown in Fig. 3 (Arjonen et al. 2011; Chen et al. 2012; Joo et al. 2019; Nishida et al. 2006; Prager et al. 2012; Rajabi and Mousa 2017; Zuazo-Gaztelu and Casanovas 2018). Lamellipodia, which are flat, sheet-like membrane protrusions are rich in dendritic arrays of actin filaments, and the molecular machinery that controls its polymerization dynamics have been known to play a major role in cell migration, via the attachment to the substrate and generation of force to pull the cell body forward (Arjonen et al. 2011; Prager et al. 2012). Filopodia, which are finger like projections and rich in bundled, crosslinked actin filaments, play a major role at the migrating end of the cell during tumor progression (Arjonen et al. 2011; Prager et al. 2012).

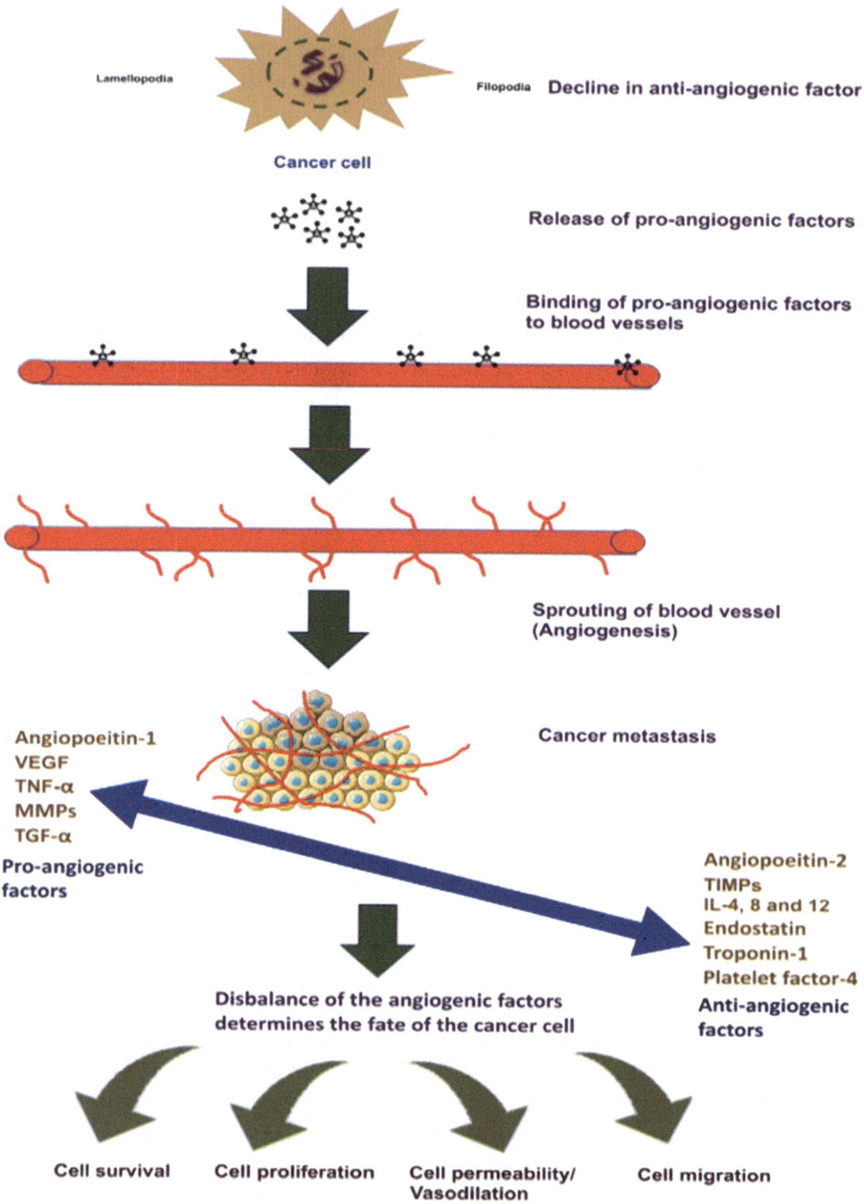

Fig. 3 The cartoon depicts the process of cancer metastasis and tumor angiogenesis. Cancer cells are shown under the influence of different angiogenic factors. Actin reorganization is important for the sprouting of the blood vessels that occurs as a result of interaction with pro-angiogenic factors

2.2 Role in Cytokinesis

Cytokinesis is the process of cell division, wherein one cell is divided into two new daughter cells (Karsenti and Vernos 2001). This involves localized assembly of the contractile ring, which is formed due to the in folding of the plasma membrane between the separating mass of chromosome, thereby forming a crease, known as the cytokinetic furrow (Disanza and Scita 2008). The key feature in this mechanism involves a dynamic actomyosin structure. The stability and organization of the cytokinetic furrow is heavily dependent on localized β-actin filament assembly at the site of cytokinesis (Chen et al. 2017; Mangione and Gould 2019). There are various key regulatory enhancers and activators of this process. β-actin filaments which are assembled directly at the furrow by an anillin-dependent pathway are known to enhance RhoA- dependent activation of the formin DIAPH3, an actin nucleator. Anillin is a specific protein that gets localized at the cytokinetic furrow, thereby interconnecting the actin, myosin, septin, and microtubule cytoskeletons with the plasma membrane (Chen et al. 2017; Mangione and Gould 2019; Miller 2011; Subramanian et al. 2013; Takaine et al. 2015). Cytokinesis should be precisely regulated for temporal and spatial reorganization of actin filament that ensures the position of the cleavage furrow wherein the chromosomes and organelles are bisected. Failure of cytokinesis can result in various types of birth defects and tumor formation (Chen et al. 2017; Mangione and Gould 2019; Miller 2011; Subramanian et al. 2013; Takaine et al. 2015).

Different actin regulators play crucial role during the process of cytokinetic furrow formation. Profilin, which binds to the monomeric G-actin at the barbed ends, sequesters them to provide formin with an ATP-bound G-actin for new rounds of polymerization. Formin then binds to the G-actin provided to it by profilin to carry out the elongation process. Septins are the GTPases that specifically stabilize contractile ring and restrict the spread of cleavage furrow(Chen et al. 2017; Fletcher and Mullins 2010; Mangione and Gould 2019; Miller 2011; Subramanian et al. 2013; Takaine et al. 2015; Varland et al. 2019). The contractile rings in mammalian and yeast cells form dynamic structures, wherein the F-actin filaments and myosin 2 are constantly assembled and turned over (Chen et al. 2017; Fletcher and Mullins 2010; Mangione and Gould 2019; Miller 2011; Subramanian et al. 2013; Takaine et al. 2015; Varland et al. 2019).

Schizosaccharomyces pombe cells were demonstrated to divide by growing through tip extension, which is controlled by actin treadmilling process and actin-conciliated endocytosis (Subramanian et al. 2013; Takaine et al. 2015). *S. pombe* is reported to have actin structures branched actin filaments, cables, and the cytokinetic ring, which are composed of linear and unbranched actin filaments. Approximately 39 mutant alleles with amino acid changes in actin protein led to defects in distinct cellular processes. Seventeen out of the 39 mutants were defective in the process of cytokinesis, while 13 contributed to the defects in the polarized growth of actin filaments. However, these 17 mutants were not defective in actomyosin ring assembly and constriction, but were found to have delocalized actin patches. Also, the 13 mutants which affected the cellular polarity exhibited major interactions between actin cytoskeleton and the Cdc42p GTPase localization(Subramanian et al. 2013; Takaine et al. 2015). The genetic analysis of these mutants revealed synthetic lethal interactions between actin, Cdc8p-tropomyosin, Cdc3p-profilin, Cdc4p-EF hand protein, myosin II heavy chain (Myo2p), and Cdc12p-formin (Subramanian et al. 2013; Takaine et al. 2015). These interactions were quite distinct from each other that the mutant actin could not be translocated by Myosin II and/or Myosin V, to the site of cytokinetic furrow (Subramanian et al. 2013; Takaine et al. 2015).

2.3 Role in Cellular Signaling and Transport

Numerous signaling pathways have been investigated that are known to use actin polymerization and cellular contractility. This helps determine the response to extracellular signals that regulate the movement of the cell, cell secretion, phagocytosis etc. as represented in Fig. 4 (Carpenter 2000). One of the major roles that actin cytoskeleton plays is to provide mechanical cues that are known to control the size of the organs in animals during embryogenesis, such as extracellular matrix (ECM) stiffness, cell-cell adhesion, cell-matrix adhesion, cell density, cell shape and cell polarity in regulating the Hippo signaling pathway (Carlier and Shekhar 2017; Carpenter 2000; Dugina et al. 2019; Kashina 2020; Moujaber and Stochaj 2020; Murrell et al. 2015). This mechano-transduction utilizes YAP and TAZ as its central components. Actin cytoskeleton is involved on the modulation of cell-cell and cell-matrix junction, thereby converting the mechanical forces into the intracellular signals, which lead to the changes in the connection of contractile actin fibers to the junctions (Kashina 2020; Moujaber and Stochaj 2020; Seo and Kim 2018). Various studies have deduced the potential role of Rho GTPases in mediating mechanical stimuli to control the regulation of actin dependent-Hippo-YAP (Kashina 2020; Moujaber and Stochaj 2020; Seo and Kim 2018). Rho works towards the assembly of contractile actin stress fibers, by simulating the downstream effector molecules, such as Rho-associated kinase (ROCK) and mDia1/2 (Kashina 2020; Moujaber and Stochaj 2020; Seo and Kim 2018; Yoon et al. 2002). Other downstream regulators, like Rac and Cdc42, help in promoting the formation of F-actin networks, which lead to the formation of structures like lamellipodia and filopodia (Kashina 2020; Moujaber and Stochaj 2020; Seo and Kim 2018; Yoon et al. 2002). Thus, the most important driving force to regulate the Hippo-YAP signaling is either through the promotion or inhibition of actin polymerization, via Rho GTPases and protein kinase A. However, the mechanistic details of this cascade are unresolved. Developing embryos, which are widely exposed to numerous mechanical stresses, because of morphogenetic movements, cell migration and cell proliferation, are regulated via YAP/TAZ-dependent transcriptional program occurring during the developmental and homeostatic stage (Seo and Kim 2018). This is done by

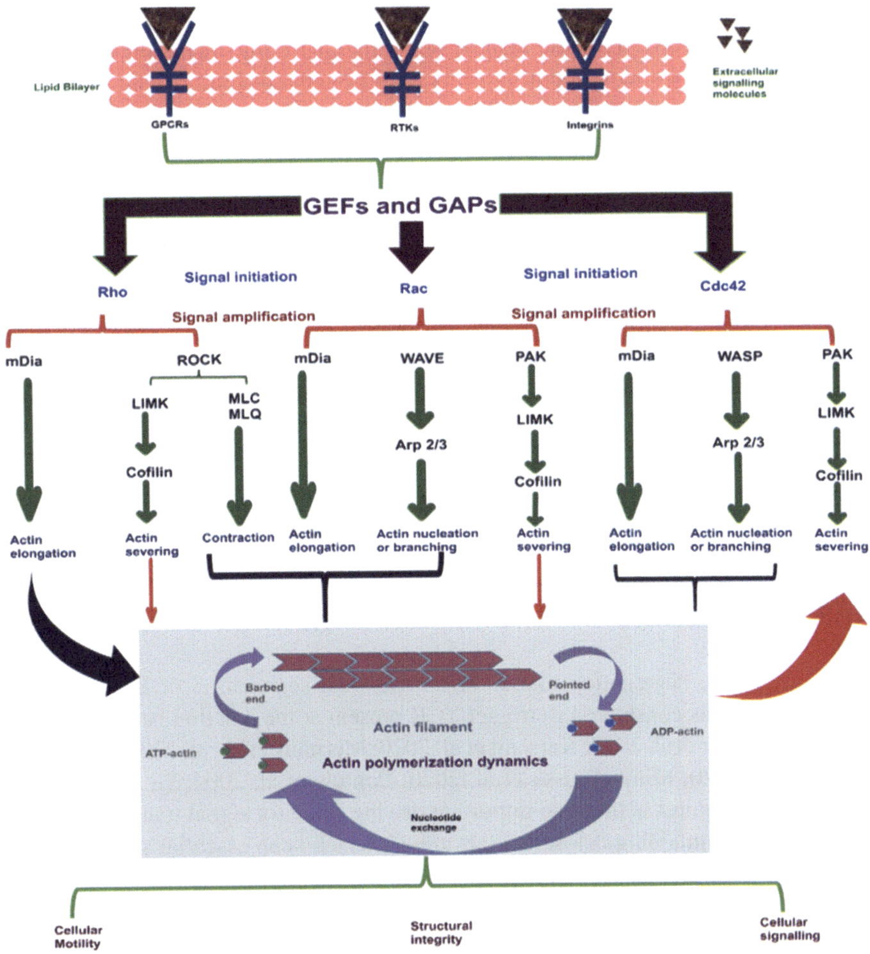

Fig. 4 Cartoon depicting the signal amplification from extracellular signaling molecule that modulates the reorganization and treadmilling of actin filament. Signal initiation occurs upon the activation of GEFs and GAPs which activate Rho, Rac and Cdc42. Signal amplification occurs by the activation of mDia, ROCK, WAVE. WASP and PAK help remodulate the actin filament to perform different cellular functions

organizing the cell junction and the actin filaments, by controlling the expression of actin/focal adhesion-related gene (Kashina 2020; Seo and Kim 2018).

Another pathway where actin cytoskeleton is known to be involved is the process of T-cell activation to mount an immune response during host cell infection by pathogens (Yu et al. 2013). Actin initiatives the cell polarization, thereby maintains cell to cell contact, which is a primary step for T-cell activation (Dugina et al. 2019; Moujaber and Stochaj 2020; Senju and Lappalainen 2019; Yu et al. 2013). The mechanism of TCR triggering requires mechanical forces from actin, which indirectly acts as a scaffold for clustering, translocation and spatial segregation of proteins, which are important for amplification and sustenance of T-cell signaling (Öztekin et al. 2008). It should also be noted that actin polymerization dynamics regulate the spatial and temporal movement of proteins that are essential for both mechanical and biochemical signaling pathways in T cell activation (Burkhardt et al. 2008). There are three possible models that have been deduced for the engagement of TCR in biochemical signaling across the plasma membrane. These processes, which has involvement of actin cytoskeleton, include, (a) conformational change; (b) kinetic segregation; and (c) receptor aggregation (Yu et al. 2013). According to the conformational change model, actin cytoskeleton is considered to bring about necessary conformational changes in the CD3 subunits upon binding of T-cell receptor to the pMHC (Karsenti et al. 2020; Mermelstein et al. 2003; Muñoz-Lasso et al. 2020; Yu et al. 2013). This in turn moves in an actin-dependent manner, which usually occurs as a consequence of attachment of the antigen processing cells (APC). Therefore, tension across the membrane is generated. The tension is subsequently transferred to the TCR signaling cascade, inducing further conformational changes in TCR-CD3 complex thereby making the cytoplasmic immunoreceptor tyrosine-based activation motifs (ITAMs) more accessible to the tyrosine-protein kinase, Lck. Thus, actin might be important for specificity and sensitivity of T-cell triggering, which permits the rapid association and dissociation of TCR-pMHC binding (Karsenti et al. 2020; Mermelstein et al. 2003; Muñoz-Lasso et al. 2020; Yu et al. 2013).

In the kinetic segregation model, size-dependent segregation of kinases from phosphatases was considered to trigger TCR present at the junction between T-cell and APC (Bachir et al. 2017; Karsenti et al. 2020; Mermelstein et al. 2003; Moujaber and Stochaj 2020; Muñoz-Lasso et al. 2020; Pontow et al. 2004; Yu et al. 2013). This process requires actin as an important driving force for signal transduction, by forming and maintaining “close-contact zone”, which is an essential step for TCR signaling. Actin, along with its molecular motor machinery, regulates the tension on the plasma membrane via coupling to them. It also promotes the formation of lipid rafts to modulate TCR triggering at the cluster levels (Bachir et al. 2017; Karsenti et al. 2020; Mermelstein et al. 2003; Moujaber and Stochaj 2020; Muñoz-Lasso et al. 2020; Pontow et al. 2004; Yu et al. 2013). Actin cytoskeleton helps in transporting the TCRs into the kinase-rich lipid rafts and maintaining the environment of lipid raft (Mermelstein et al. 2003; Moujaber and Stochaj 2020; Pontow et al. 2004; Yu et al. 2013). The receptor aggregation model suggests that binding of TCRs and pMHCs promotes aggregation of TCR-CD3 complexes, thereby enhancing the local TCR signaling. This model has several limitations, as it fails to answer how these signaling is triggered at low pMHC concentrations, when the TCR microclusters are small and sparse. However, it considers that actin acts as a scaffold to stabilize and support protein aggregation, thereby contributing to TCR-CD3 aggregation (Mermelstein et al. 2003; Moujaber and Stochaj 2020; Pontow et al. 2004; Yu et al. 2013).

2.4 Role in Immunodeficiency

One of the fundamental features of any immunodeficiency is the failure of actin cytoskeleton to mount immune responses against a particular infection or a foreign antigen (Moulding et al. 2013). Previous studies on the actin-related primary immuno-deficiency unravels the pivotal role played by the actin filament polymerization in innate and adaptive immunity (Keszei and Westerberg 2014; Taratuto 2002). Polymerization dynamics of actin regulates all stages of the immune response, such as haematopoiesis and immune cell development, recruitment, intercellular and intracellular signalling, migration, and activation of both innate and adaptive immune responses (Moulding et al. 2013). Mutations in regulatory proteins of actin leads to major defects in actin polymerization (Wong et al. 2011). There is a dire need to understand the regulation under mutated actin regulators. The better understanding of this regulation is possible with smaller set of model systems especially with the subset from that of the apicomplexan (Blanchoin et al. 2014; Frénal et al. 2017). The structure and dynamics of any cytoskeleton is dependent upon the large number of regulatory proteins and molecules that bind to either monomeric or filamentous actin. Understanding the actin dynamics and control in vitro is important to comprehend actin-ABC interaction (Tangye et al. 2019). Many congenital immunodeficiencies are the result of impaired actin polymerization dynamics. Neutropenia or neutrophil dysfunction are the major causes of primary immuno deficiencies accounting for 10–20% of all known immunodeficiencies (Keszei and Westerberg 2014).

The neutrophil cytoskeleton polarizes highly branched and dynamic actin filaments, forming a leading edge (lamellipodia/pseudopod) towards the signalling molecule. A trailing uropod is formed at the posterior edge of the cell, which are rich in actin-myosin II contractile structures (Keszei and Westerberg 2014; Tangye et al. 2019). These protrusions are formed in response to bacterial or viral infections. The actin cytoskeleton polarity is regulated by Rho GTPases Cdc42, Rac, and Rho etc. β-actin dysfunction is attributed to cause recurrent infections, thrombocytopenia, photosensitivity, and mental retardation. The neutrophils containing the impaired β-actin failed in chemotaxis, superoxide production, and membrane potential response (Keszei and Westerberg 2014; Pontow et al. 2004; Tangye et al. 2019). Neutrophil actin dysfunction (NAD) syndrome shows poor polymerization dynamics of F-actin and is characterized by recurrent bacterial infections, marked neutrophilic leukocytosis, diminished neutrophil migration from blood to the inflammation site, and reduced phagocytosis (Keszei and Westerberg 2014; Pontow et al. 2004; Tangye et al. 2019). Another example of neutropenia arising due to actin cytoskeleton dysfunction is severe congenital neutropenia (SCN), which is an X-linked neutropenia (XLN) characterized by fast dividing mitoticgranulocytic progenitor cells and aberrant cell division. Accumulation of cytoplasmic F-actin was observed in SCN patients (Keszei and Westerberg 2014).

2.5 Role in Fertilization of Eggs

During fertilization, the sperm is required to stay in the human female reproductive tract for several hours, during which they go through a series of changes called as capacitation. The capacitated sperms are capable of undergoing the acrosome reaction upon binding to the egg zona pellucida (Karsenti et al. 2020; Poulter et al. 2010; Santella and Chun 2011). This process helps the sperm to penetrate the egg to fertilize it (Santella et al. 2018). The underlying mechanism for series of capacitation process is the polymerization dynamics of actin. Polymerization of globular (G)-actin to filamentous (F)-actin is dependent upon protein kinase A activation, protein tyrosine phosphorylation, and phospholipase D activation (Santella et al. 2018, 2020; Veselská 2001) and is important to translocate phospholipase C from the cytosol to the sperm plasma membrane, during capacitation (Andersen et al. 2019; Mcavey et al. 2002; Santella et al. 2014, 2018, 2020; Veselská 2001). F-actin undergoes depolymerisation before the acrosome reaction. This enables the outer acrosomal membrane and the overlying plasma membrane to come into close proximity, which leads to their fusion (Mcavey et al. 2002; Veselská 2001). The binding to zona pellucida induced rapid increase in sperm's intracellular calcium, which in turn activates actin severing proteins to break down the actin fibers. This, thus, allows the faster acrosome reaction (Santella et al. 2020; Veselská 2001). The understanding of the polymerization dynamics in fertilization of eggs during sperm capacitation can lead to generation or production of antibodies that can be used as potential contraceptive in males (Santella et al. 2018, 2020; Veselská 2001).

2.6 Role in Neuronal Plasticity

Actin filaments are involved in complex process of spinogenesis and experience-dependent plasticity, due to their dynamic property to assemble into various different structures. This helps them to orchestrate an elaborate response (Pelucchi et al. 2020). Mutations in genes encoding regulators of the synaptic cytoskeletons are involved in human neurodevelopmental and psychiatric disorders (Gordon-Weeks and Fournier 2014; Svennberg 2006). Disruptions in specific actin filament structures are known to result in the phenotypic manifestation of synaptic and behavioral changes. These phenotypic or physiological changes are dependent upon the remodeling of dendritic spines, rich in actin protrusions. The dendritic spines are known to act as major receiving sites for the excitatory synaptic transmission in the brain (Gordon-Weeks and Fournier 2014; Pelucchi et al. 2020; Spence and Soderling 2015; Szabó et al. 2016). Anomalies in dendritic spine morphology and density are linked to neurological disorders, such as schizophrenia (SZ) 2 and intellectual disability (ID) (Gordon-Weeks and Fournier 2014; Szabó et al. 2016).

The polymerization dynamics of actin occur as a function of external stimuli pushes for the formation and maintenance of spine, synaptic adhesion, receptor endo and exocytosis, and synaptic plasticity (Gordon-Weeks and Fournier 2014;

Pelucchi et al. 2020; Spence and Soderling 2015; Szabó et al. 2016). There are several regulators of actin that help in maintaining the equilibrium of the filament dynamics, which are indirectly modulated by host of spine-enriched activators (GEFs) and inhibitors (GAPs). Mutations in these two gene's lead to neurological disorders, such as schizophrenia (SZ)2 and intellectual disability (ID) (Karsenti et al. 2020; Muñoz-Lasso et al. 2020; Spence and Soderling 2015; Szabó et al. 2016). Actin treadmilling drives essential spine functions through localized control of Rho family GTPase activity. The polymerization of actin is controlled via three major regulators in spine, including Arp2/3, Formin and Profilin, while the disassembly is promoted via actin-depolymerizing factors (ADF)/cofilins (Gordon-Weeks and Fournier 2014; Karsenti et al. 2020; Muñoz-Lasso et al. 2020; Pelucchi et al. 2020; Spence and Soderling 2015; Svennberg 2006; Szabó et al. 2016). The accumulation and localization of all the above actin-binding proteins is observed at post-synaptic density (PSD), which isa protein-rich compartment in the spines that receives neurotransmitters from the pre-synapse (Gordon-Weeks and Fournier 2014).

Other regulators like capping protein (CP-α and CP-β) and Eps8 localizes to dendritic spines during chemical Long-term potentiation (LTP) (Spence and Soderling 2015; Szabó et al. 2016). Eps8 is downregulated in autism spectrum disorder (ASD) patients, due to reduced spine enlargement and structural plasticity. Structural plasticity is an actin-dependent process, which brings about the functional changes associated with long-term potentiation (LTP) and long-term depression (LTD) in spines. Actin filament might possibly play four major roles during LTP, (1) to split the active zone into multiple independent zones; (2) to undergo reversible gel–sol–gel transition during LTP induction; (3) To cluster extra-synaptic AMPARs in the synapse; (4) to be involved in the vesicle fusion process to deliver AMPAR to the synapse during LTP (Gordon-Weeks and Fournier 2014; Szabó et al. 2016).

Dendritic spines have the presence of three sub-populations of actin filaments based on their rate of turnover (Bourne and Harris 2008; Okada and Soderling 2009). This includes dynamic actin filaments present at the tip of the PSD, largely stable actin filaments present at the base of the spine and actin filaments whose dynamics are altered by synaptic activity are usually present at the central "spine-enlargement" pool (Bourne and Harris 2008; Okada and Soderling 2009). Activation of actin filaments at the central enlargement pool is enhanced during LTP and subsequently controlled by CaMKII activity (Fukazawa et al. 2003; Kim and Lisman 1999; Luo 2002; Okamoto et al. 2009). The activation of CaMKII leads to the downstream activation of Rho-GTPases such as Cdc42 and Rac via GEFs which in turn lead to activation of cytoskeleton reorganization bringing about a change in dendritic spine which are known to last longer. It is important to study actin reorganizational dynamics in dendritic spine as they play a crucial role in both presynaptic and postsynaptic processes, contributing to basal synaptic transmission (Fukazawa et al. 2003; Kim and Lisman 1999; Luo 2002; Okamoto et al. 2009). It was found that post synaptic actin is required for maintenance of AMPAR-mediated synaptic transmission but not NMDAR-mediated synaptic transmission (Bourne and Harris 2008; Fukazawa et al. 2003; Kim and Lisman 1999; Luo 2002; Okada and Soderling 2009; Okamoto et al. 2009).

2.7 Role in Neurodegenerative Diseases

Neurodegenerative diseases (NDs) causes the highest mortality in the elderly population (Bourne and Harris 2008). The increasing frequency of neurodegenerative diseases is due to the lack of proper understanding of these ailment and due to lack of targeted approach of therapeutic towards them. This is due to the fact that most of these therapeutic agents lack the ability to cross the BBB (Van Bulck et al. 2019; Hussain et al. 2018). It causes the distressing impact on suffering individuals, their families and society. Thus there is a need to have rapid and enough discoveries in order to overcome the incidences and development of these aberrations (Bourne and Harris 2008). NDs are characterized by annihilation of neuronal cells which are involved in diverse functional phenomenon (Cowan and Mudher 2013; Lambrechts et al. 2004; Khan 2013). The Major NDs that affects the population around the world is Alzheimer's diseases, Parkinson's disease, Pick's disease, Guam amylotrophic lateral sclerosis and parkinsonism-dementia complex (Van Bulck et al. 2019; Ramanan and Saykin 2013). These neurodegenerative diseases are major causes of dementia in ageing individuals which is characterized by irreparable weakening of intellectual functions like memory, language skills, and concentration (Ross and Poirier 2004). The evolution of neurodegeneration is reliant upon Amyloid precursor proteins (APP) and aggregation of tau which are microtubules binding proteins (Bórquez and González-Billault 2012; Chu et al. 2007; Kumar et al. 2015; Pelucchi et al. 2020; Sun et al. 2012). However, the understanding of Hirano bodies and its role in proper functioning of the brain is of prime importance as very less has been known. Hirano bodies are paracrystalline structures which are highly ordered and made up of filamentous actin (F-actin) (Lazaro-Dieguez et al. 2008; Spears et al. 2014). They are formed as a result of abnormal cellular response and regulation actin cytoskeleton by ABPs (Griffin et al. 2014). Hirano bodies are found in brains of people suffering from various ND sand are known to act as a protective shield towards tau-independent and tau-dependent apoptotic pathway in various NDs (Dong et al. 2013; Furgerson et al. 2012, 2014; Maselli et al. 2003; Myre 2012; Pathak et al. 2020). Hirano bodies formed in CA1 region of the hippo campus and frontal cortex of mice model elicits an inflammatory response resulting in diminished spatial working memory (Furgerson et al. 2014; Pathak et al. 2020). However, the role that Hirano bodies play in the development of a particular neurodegeneration (Goldman 1983; Griffin et al. 2014; Melidone et al. 2011; Santa-Mara et al. 2008). They also are found in aged individuals and alcoholic individuals (Gourlay and Ayscough 2005). Hirano bodies are also formed as a consequence of buildup of tau protein such as C31 and 352PHPtau and co-localize with APP binding adaptor protein such as Fe65 (Gourlay and Ayscough 2005). However, phenomenon of their formation, regulation and its subsequent role in the intonation of neuronal cell remains highly non-evaluated (Gourlay and Ayscough 2005).

As shown in Fig. 5, actin polymerization dynamics which when unregulated undergoes changes resulting in aggregate formation. This forms as one of the underlying phenomenon of disease progression during NDs (Dong et al. 2013, 2016; Furgerson

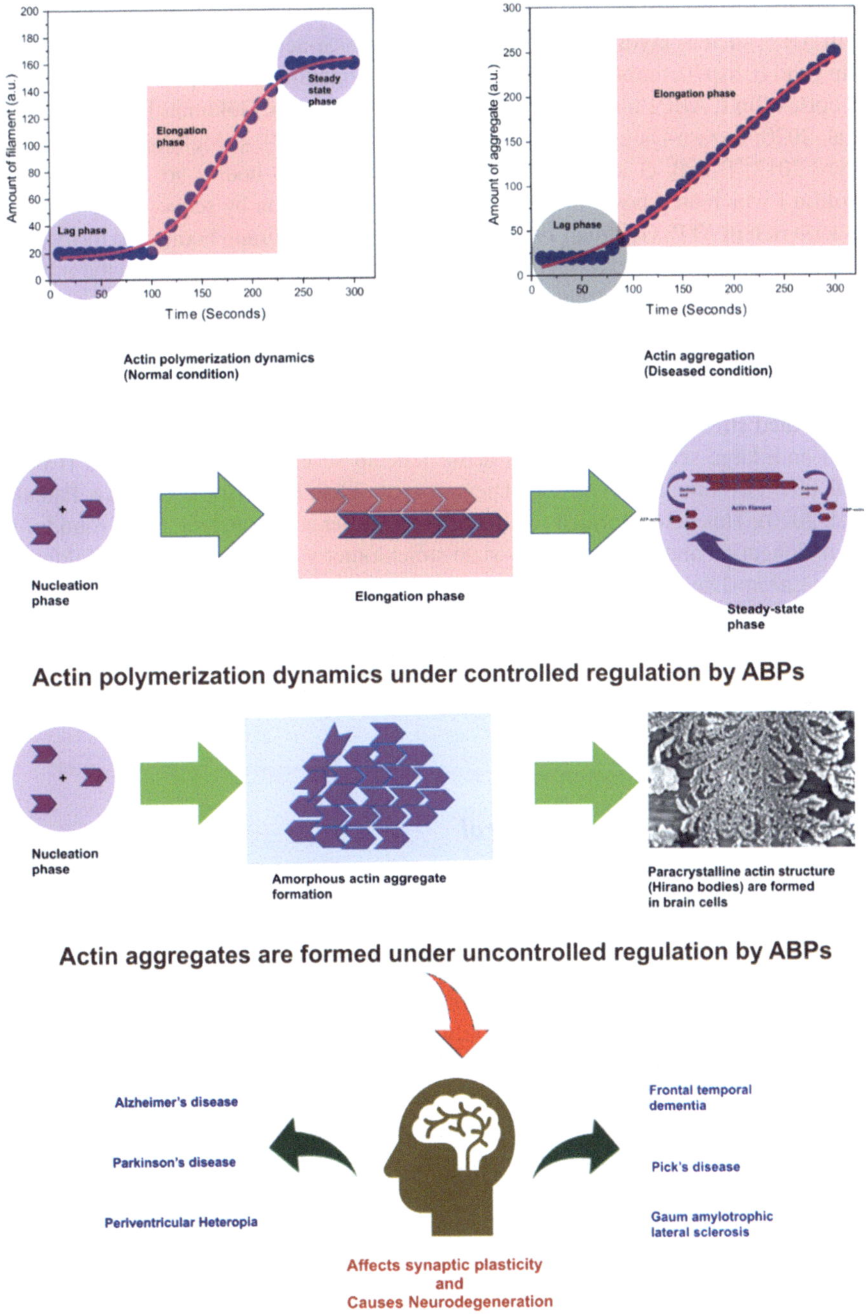

Fig. 5 Actin polymerization dynamics that occur under normal condition as well as during the aggregation leading to various neurodegenerative diseases. Actin aggregates formed in the brain lead to the production of inclusion bodies called as "Hirano body". They also affect the synaptic plasticity and structural integrity of the synapses thereby affecting learning and memory function of the brain

et al. 2014; Myre 2012). Hirano bodies, rich in actin inclusion bodies co-localizes with ABPs such as myosin II heavy and light chains,α-actinin, Arp2/3 complex, profilin I and II, hsp70, fimbrin, coactosin, EF1 alpha, vinculin-related protein, histophilin II, cortexillin I, ABP1 and filamin (Dong et al. 2013, 2016; Goldman 1983; Karsenti et al. 2020; Lazaro-Dieguez et al. 2008; Furgerson et al. 2014; Melidone et al. 2011; Myre 2012; Pathak et al. 2020). The size of the Hirano bodies are controlled by Profilin I which are known to control the F-actin regulation by sequestering pool of G-actin rich in ATP (Goldman 1983; Griffin et al. 2014). Hirano bodies are formed as a result of heightened cross-linking of the actin filament. This then caused the inhibition of the normal severing and regulation of the actin filament formation (Dong et al. 2013; Furgerson et al. 2014; Melidone et al. 2011; Myre 2012). Furthermore, F-actin filament reorganizes itself into a highly ordered arrays forming a patch of deposition in the neurons forming hirano bodies (Dong et al. 2013; Furgerson et al. 2014).During an elevated stress level response, Hirano bodies in Alzheimer patients get accumulated on a large scale as a result of actin filament severing by ADF/Cofilin. (Dong et al. 2013; Goldman 1983; Lazaro-Dieguez et al. 2008; Furgerson et al. 2014; Pathak et al. 2020). This newly formed smaller actin oligomer act as nuclei for new rounds of actin elongation and leads to steady-state stoichiometry (Dong et al. 2013; Goldman 1983; Lazaro-Dieguez et al. 2008; Furgerson et al. 2014; Pathak et al. 2020). Due to the constant accrual of ADP-actin thereby increasing the prospect of amorphous aggregate formation and are very tough to break(Dong et al. 2013, 2016; Goldman 1983; Lazaro-Dieguez et al. 2008; Furgerson et al. 2014; Melidone et al. 2011; Myre 2012; Pathak et al. 2020).

2.8 *Role in Epigenetic Control*

In higher eukaryotes, actin is encoded by six genes which are 95–99% identical to each other (Kashina 2020; Vedula and Kashina 2018). However, all the encoded isoforms differ in their functions *in vivo*. Actin is regulated during transcription at the gene-level, followed by mRNA regulation, and protein-level regulation (Dugina et al. 2019). Different actin isoforms are known to arbitrate every type of cellular movement and perform different process, such as embryogenesis, tissue integrity, cell division and contractility of all types of muscle and non-muscle cells. Nuclear actin dynamics regulate transcription, mediate chromosome positioning, DNA rearrangements, and DNA repair as shown in Fig. 6 (Goodson and Hawse 2002; Hatanaka et al. 1996; Hyrskyluoto and Vartiainen 2020; Karsenti et al. 2020; Ono et al. 1993). The actin was observed in ATP-dependent chromatin remodeling complex (Hurst et al. 2019; Hyrskyluoto and Vartiainen 2020; Xie et al. 2018; Zetka et al. 2020) and was capable of binding to various transcription sites by co-transcriptionally getting added to nascent ribonucleoproteins (RNPs). Actin binds to the subset of nuclear ribonucleoproteins (hnRNPs) to facilitate co-transcriptional assembly of precursor - messenger RNPs (premRNPs) (Goodson and Hawse 2002; Gordon et al. 2005; Hurst et al. 2019; Hyrskyluoto and Vartiainen 2020; Karsenti et al. 2020; Xie et al. 2018; Zetka et al. 2020).

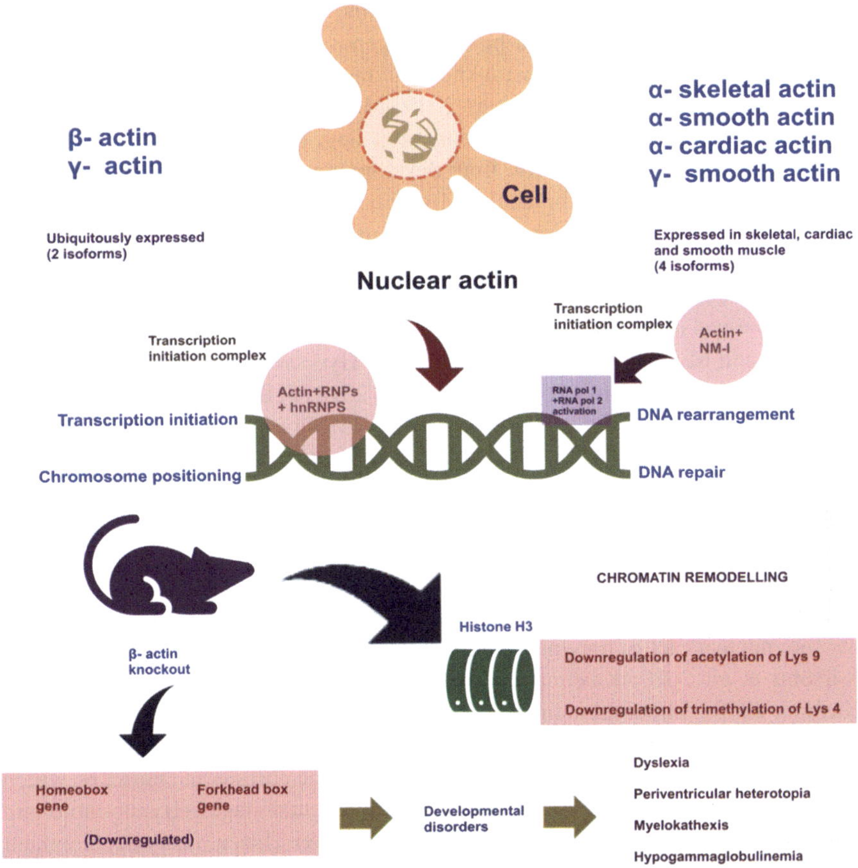

Fig. 6 Cartoon representation of role of nuclear actin in chromosome remodeling and transcription initiation. The absence of β-actin in knockout mouse model shows downregulation of various important genes leading to developmental disorders

Actin and nuclear myosin 1 (NM1) help in the initiation of transcription process by RNA polymerases I and II. Actin and its regulators like ARP2/3 complex and N-WASP, which synergies to promote actin filament branching, are known to be involved in regulating the nuclear structure thereby facilitating transcription (Hurst et al. 2019; Hyrskyluoto and Vartiainen 2020; Lambert 2019; Subramanian et al. 2013). Actin polymerization thus drives the mechanism of nuclear reprogramming, wherein multiple genes are regulatedsimultaneously (Chang et al. 2019; Estrem and Moore 2020; Lambert 2019; Moujaber and Stochaj 2020; Vedula and Kashina 2018). The trans-positioning of one cell type with the another forms the basis of cellular development and differentiation. The plausible mechanism for this trans positioning is that chromatin loops rich in various genes responsible for the developmental stages of an organism move towards the interior regions on the chromosomes to define gene activation or repression. Homeobox (Hox) and Forkhead box (Fox) genes were downregulated in Mouse embryonic fibroblast (MEF) β-actin knockout models (Hyrskyluoto and Vartiainen 2020; Karsenti et al. 2020; Lambert 2019; Moujaber and Stochaj 2020). These genes are important during the developmental stages of an embryo as they help in the formation of nervous system, limb development, development of the respiratory system with bronchial branching, formation of an initial genital system, thymus development and, development of an initial cardiovascular system with the atrial separation (Lambert 2019). Downregulation of epigenetic markers in active chromatin, such as acetylation on lysine 9 of histone H3 (H3K9ac) and trimethylation of lysine 4 on histone H3 (H3K4me3) was observed in the absence of β-actin (Fletcher and Mullins 2010; Hurst et al. 2019; Hyrskyluoto and Vartiainen 2020; Karsenti et al. 2020; Lambert 2019; Xie et al. 2018; Zetka et al. 2020). However, monomethylation of lysine 4 on histone H3 (H3K4me1) in the repressive genes were found to be upregulated. Also, absence of β-actin reduces the accessibility of the rDNA to regions like promoter-proximal enhancer (T0), Sal boxes (T1-T10) and a fraction of the intergenic region sequence (IGS) which are located immediately after the Sal boxes (Fletcher and Mullins 2010; Karsenti et al. 2020; Kashina 2020; Moujaber and Stochaj 2020; Vedula and Kashina 2018). Dynamic reorganization of the chromatin is an essential factor for cell differentiation, which relies on the nuclear actin polymerization dynamics (Fletcher and Mullins 2010; Karsenti et al. 2020; Kashina 2020; Moujaber and Stochaj 2020; Vedula and Kashina 2018). Impaired actin polymerization dynamics in yeast are associated with reduction in subtelomere dynamics and confined at the nuclear periphery. This hampers the global consolidation of the dynamic chromatin landscape which are essential for gene expression during developmental stages (Hyrskyluoto and Vartiainen 2020; Karsenti et al. 2020; Xie et al. 2018). Hence, many development disorders are linked to the impairment of actin dynamics, such as congenital immunodeficiencies, warts, hypogammaglobulinemia, infections, and myelokathexis (WHIM syndrome) (Keszei and Westerberg 2014), periventricular heterotopia (PH). These disorders represented by seizures, dyslexia and psychiatric disturbances (Lian and Sheen 2015). Thus, studying epigenetic control of nuclear actin should help us explore the essential targets that can be used for drug design against such developmental defects (Lian and Sheen 2015).

2.9 Role in Muscle Contraction

The muscle wasting is attributed to the sarcomere region, which largely comprises of actin filaments(Gautel and Djinović-Carugo 2016). Cardiac and skeletal muscles are intricately designed for muscle contraction to facilitate various muscular activity in humans (Skwarek-Maruszewska et al. 2009). These muscle cells are largely composed of sarcomeres, the basic unit of contraction (Gautel and Djinović-Carugo 2016). Various cytoskeletal proteins, such as actin, intermediate filaments, myosin, titin and nebulin provide proper contractile function to these muscles (Gautel and Djinović-Carugo 2016; Henderson et al. 2017). The muscle contraction is attributed to the sliding filament theory which involves the specific way myosin thick filament is able to pull the thinner actin filament in order to shorten the sarcomere (Clarke 2010; Crick and Watson 1953; Goody 2003; Graham 1969; Huxley and Hanson 1954; Hynes et al. 1987). The myosin S1 region readily binds to the actin filament, thereby releasing the actin monomer while contracting (Clarke 2010; Goody 2003; Graham 1969; Huxley and Hanson 1954; Hynes et al. 1987). This is a repetitive cycle called as myosin-actin cycling, wherein myosin again reaches to bind actin in the new cycle (Clarke 2010; Goody 2003; Graham 1969; Huxley and Hanson 1954; Hynes et al. 1987). This process helps in the formation of cross bridges, which are known to extend from the thick myosin filaments to the thin actin filaments (Clarke 2010; Goody 2003; Graham 1969; Huxley and Hanson 1954; Hynes et al. 1987). The contraction of myosin's S1 filament, also known as power stroke, requires hydrolysis of ATP molecule to provide free energy to the myosin to perform the cycling process (Clarke 2010; Crick and Watson 1953; Goody 2003; Graham 1969; Huxley and Hanson 1954; Hynes et al. 1987). This free energy is thus used to release actin, bring about a conformational changes to perform contraction and repeat the cycle (Clarke 2010; Crick and Watson 1953; Goody 2003; Graham 1969; Huxley and Hanson 1954). When ATP is not released, myosin will remain bound to actin indefinitely leading to stiffness and onset of rigor mortis (Clarke 2010; Crick and Watson 1953; Goody 2003; Graham 1969; Huxley and Hanson 1954; Hynes et al. 1987). Thus, any non-coordinated action of any of these protein may result into the disruption of the integrity of the sarcomere with subsequent muscle loss (Henderson et al. 2017). It is also reported that regulation of the dynamics of actin polymerization at the pointed end of the filament is one of the underlying mechanisms of maintaining constant length of sarcomeric actin filaments. While being a part of striated muscle cells, the half-life of actin was found to be in the order of days as compared to exchange rate within myofibrils with free G-actin which was estimated to be on the order of hours. Hence, the actin filaments within the muscle cells are more resistant to the external depolymerizing agents for hours. However, it has also been an observed factor that actin filament at end of sarcomere are capable of undergoing polymerization and depolymerization even in the presence of certain regulatory proteins of actin which in turn can disturbs the dynamics of actin at the pointed end. Two different types of actin binding proteins viz. Actin-dynamics enhancers (e.g., ADF/cofilin) and actin-filament stabilizers (e.g., tropomyosin) often regulate the actin turnover antagonistically thereby maintaining a balance for proper assembly and maintenance of sarcomeric actin organizations (Ono 2010; Littlefield and Fowler 2008; Skwarek-Maruszewska et al. 2009; Zak et al. 1977).

Further, chicken striated muscle consists of all the three isoforms of actin with abundant β-actin, followed by γ and lowest level of α-actin (Shimizu and Obinata

1986). G-actin concentration in embryonic muscle stage was much higher than the critical concentration required for polymerization of purified actin (Shimizu and Obinata 1986). The amount of actin filaments increased significantly during development, with a subsequent drop in the G-actin concentration. In the adult stage of the muscle development the presence of actin monomer was around the same as the critical concentration required for actin polymerization, thereby suggesting an important role of actin filament dynamics, during all stages of muscle development in eukaryotes (Shimizu and Obinata 1986). Thus, any changes in the actin cytoskeleton will lead to the disruption of the integrity of the sarcomere with subsequent muscle loss, and malnourishment (Clarke 2010; Goody 2003; Henderson et al. 2017; Huxley and Hanson 1954).

3 Conclusions

Actin polymerization dynamics is one of the crucial phenomena responsible for plethora of cellular functions in higher eukaryotes. About 100–150 regulatory proteins, known as actin binding proteins (ABPs),modulate these dynamics. Actin monomer is a highly unstable protein due to its intrinsic ATPase activity. The constant polymerization and depolymerisation of actin filament, known as treadmilling is involved in several cellular activities like locomotion, transport of the signalling protein, gliding motility, cancer metastasis etc. However, molecular details of the polymerization phenomenon remain highly unexplored.

As actin is highly ubiquitous protein, less-complex eukaryotes have lower number of ABPs. One of the majorly studied smaller system of actin regulation is in *Plasmodium* comprising of only 7 regulators such as 2 ADF isoforms, CAP, CP-α and β, Formin, Coronin, and Profilin Understanding these smaller systems will help us discern the more complex system and actin control in higher eukaryotes. Aberration in actin either at the protein or at the gene level causes various developmental diseases. We provide a comprehensive list of actin's functions in different cellular process, which when mis-regulated result in disorder. The polymerization of actin filaments and their association with actin regulatory proteins construct a variety of structures, such as branched or cross-linked meshwork in the lamellipodia, parallel bundles in filopodia and antiparallel structures in contractile fibres. Hence, the perpetual modification of actin filaments decides the fate of the cell within an organism. G-actin monomer binding molecules have been used to unravel the actin-ABP interactions in different cellular systems. Most of the study now has been focused upon arresting actin as a monomer in order to study its interaction with different actin binding protein at the atomistic level.

Acknowledgement The authors would like to thanks Tata Institute of Fundamental Research (TIFR), Mumbai for allowing to collect SEM images on actin molecule purified from pig thigh muscle.

References

Alberts, B., Johnson, A., Lewis, J., et al. (2002). The self-assembly and dynamic structure of cytoskeletal filaments. In *Molecular biology of the cell* (4th ed.). New York: Garland Science. https://www.ncbi.nlm.nih.gov/books/NBK26862/.

Andersen, A. H. Y., Wolfner, M. F., Wood, B. W., & Weil, T. T. (2019). A calcium—Mediated actin redistribution at egg activation in Drosophila. *Molecular Reproduction and Development, 2020*, 293–304. https://doi.org/10.1002/mrd.23311.

Arjonen, A., Kaukonen, R., & Ivaska, J. (2011). Filopodia and adhesion in cancer cell motility. *Cell Adhesion and Migration, 5*(5), 421–430. https://doi.org/10.4161/cam.5.5.17723.

Bachir, A. I., Horwitz, A. R., Nelson, W. J., & Bianchini, J. M. (2017). Actin-based adhesion modules mediate cell interactions with the extracellular matrix and neighboring cells. *Cold Spring Harbor Perspectives in Biology, 9*(7). https://doi.org/10.1101/cshperspect.a023234.

Balcer, H. I., Goodman, A. L., Rodal, A. A., Smith, E., Kugler, J., Heuser, J. E., & Goode, B. L. (2003). Coordinated regulation of actin filament turnover by a high-molecular-weight Srv2/CAP complex, cofilin, profilin, and Aip1. *Current Biology, 13*(24), 2159–2169. https://doi.org/10.1016/j.cub.2003.11.051.

Barron, G. A., Goua, M., Wahle, K. W. J., & Bermano, G. (2017). Circulating levels of angiogenesis-related growth factors in breast cancer: A study to profile proteins responsible for tubule formation. *Oncology Reports, 38*(3), 1886–1894. https://doi.org/10.3892/or.2017.5803.

Bayless, K. J., & Johnson, G. A. (2011). Role of the cytoskeleton in formation and maintenance of angiogenic sprouts. *Journal of Vascular Research, 48*(5), 369–385. https://doi.org/10.1159/000324751.

Bielenberg, D. R., & Zetter, B. R. (2015). The contribution of angiogenesis to the process of metastasis. *Cancer Journal (United States), 21*(4), 267–273. https://doi.org/10.1097/PPO.0000000000000138.

Blanchoin, L., Boujemaa-Paterski, R., Sykes, C., & Plastino, J. (2014). Actin dynamics, architecture, and mechanics in cell motility. *Physiological Reviews, 94*(1). https://doi.org/10.1152/physrev.00018.2013.

Bórquez, D. A., & González-Billault, C. (2012). The amyloid precursor protein intracellular domain-Fe65 multiprotein complexes: A challenge to the amyloid hypothesis for Alzheimer's disease? *International Journal of Alzheimer's Disease, 2012*, 1–10. https://doi.org/10.1155/2012/353145.

Bourne, J. N., & Harris, K. M. (2008). Balancing structure and function at hippocampal dendritic spines. *Annual Review of Neuroscience, 31*, 47–67. https://doi.org/10.1146/annurev.neuro.31.060407.125646.Balancing.

Bugyi, B., & Carlier, M.-F. (2010). Control of actin filament treadmilling in cell motility. *Annual Review of Biophysics, 39*, 449–470. https://doi.org/10.1146/annurev-biophys-051309-103849.

Burkhardt, J. K., Carrizosa, E., & Shaffer, M. H. (2008). The actin cytoskeleton in T Cell activation. *Annual Review of Immunology, 26*(1), 233–259. https://doi.org/10.1146/annurev.immunol.26.021607.090347.

Carlier, M. F., & Shekhar, S. (2017). Global treadmilling coordinates actin turnover and controls the size of actin networks. Nature Reviews Molecular Cell Biology 18(6). doi: https://doi.org/10.1038/nrm.2016.172.

Carlsson, L. (n.d.). Crystallization of a non-muscle actin. *Journal of Molecular Biology, 105*(3), 353–366. https://doi.org/10.1016/0022-2836(76)90098-X.

Carpenter, C. L. (2000). Actin cytoskeleton and cell signaling. *Critical Care Medicine, 28*(4 Suppl). https://doi.org/10.1097/00003246-200004001-00011.

Chang, W., Wang, Y., Gant Luxton, G. W., Östlund, C., Worman, H. J., & Gundersen, G. G. (2019). Imbalanced nucleocytoskeletal connections create common polarity defects in progeria and physiological aging. *Proceedings of the National Academy of Sciences of the United States of America, 116*(9), 3578–3583. https://doi.org/10.1073/pnas.1809683116.

Chen, A., Arora, P. D., McCulloch, C. A., & Wilde, A. (2017). Cytokinesis requires localized β-actin filament production by an actin isoform specific nucleator. *Nature Communications, 8*(1). https://doi.org/10.1038/s41467-017-01231-x.

Chen, Y., Gou, X., Ke, X., Cui, H., & Chen, Z. (2012). Human tumor cells induce angiogenesis through positive feedback between CD147 and insulin-like growth factor-I. *PLoS One, 7*(7), 1–11. https://doi.org/10.1371/journal.pone.0040965.

Chu, C. T., Plowey, E. D., Wang, Y., Patel, V., & Jordan-Sciutto, K. L. (2007). Location, location, location: Altered transcription factor trafficking in neurodegeneration. *Journal of Neuropathology and Experimental Neurology, 66*(10), 873–883. https://doi.org/10.1097/nen.0b013e318156a3d7.

Clarke, M. (2010). Muscle sliding filaments. *Nature Reviews Molecular Cell Biology, 9*(S1), s7. https://doi.org/10.1038/nrm2581.

Cooper, J. A., & Schafer, D. A. (2000). Control of actin assembly and disassembly at filament ends. *Current Opinion in Cell Biology, 12*(1), 97–103. https://doi.org/10.1016/S0955-0674(99)00062-9.

Cowan, C. M., & Mudher, A. (2013). Are tau aggregates toxic or protective in tauopathies? *Frontiers in Neurology, 4*, 1–13. https://doi.org/10.3389/fneur.2013.00114.

Crick, F., & Watson, J. (1953). Adenosine triphosphate creatine transphophorylase as relaxing factor of muscle.

Daly, M. E., Makris, A., Reed, M., & Lewis, C. E. (2003). Hemostatic regulators of tumor angiogenesis: A source of antiangiogenic agents for cancer treatment? *Journal of the National Cancer Institute, 95*(22), 1660–1673. https://doi.org/10.1093/jnci/djg101.

Disanza, A., & Scita, G. (2008). Cytoskeletal regulation: Coordinating actin and microtubule dynamics in membrane trafficking. *Current Biology, 18*(18), 873–875. https://doi.org/10.1016/j.cub.2008.07.059.

Dong, Y., Shahid-Salles, S., Sherling, D., Fechheimer, N., Iyer, N., Wells, L., Fechheimer, M., & Furukawa, R. (2016). De novo actin polymerization is required for model Hirano body formation in Dictyostelium. *Biology Open, 5*(6), 807–818. https://doi.org/10.1242/bio.014944.

Dong, Y., Shahid-Salles, S., Sherling, D., Fechheimer, N., Iyer, N., Wells, L., Fechheimer, M., Furukawa, R., Mitchison, T. J., Cramer, L. P., Murray, S. T., Shapiro, D. E., Lucas, E. P., Khanal, I., Gaspar, P., Fletcher, G. C., Polesello, C., Tapon, N., Thompson, B. J., et al. (2013). The Hippo pathway polarizes the actin cytoskeleton during collective migration of Drosophila border cells. *Traffic, 5*(6), 371–379. https://doi.org/10.1111/j.1600-0854.2004.00196.x.

Dugina, V. B., Shagieva, G. S., & Kopnin, P. B. (2019). Biological role of actin isoforms in mammalian cells. *Biochemistry (Moscow), 84*(6), 583–592. https://doi.org/10.1134/S0006297919060014.

Estrem, C., & Moore, J. K. (2020). Help or hindrance: How do microtubule-based forces contribute to genome damage and repair? *Current Genetics, 66*(2), 303–311. https://doi.org/10.1007/s00294-019-01033-2.

Fletcher, D. A., & Mullins, R. D. (2010). Cell mechanisms and cytoskeleton. *Nature, 463*(7280), 485–492. https://doi.org/10.1038/nature08908.Cell.

Frénal, K., Dubremetz, J.-F., Lebrun, M., & Soldati-Favre, D. (2017). Gliding motility powers invasion and egress in Apicomplexa. *Nature Reviews Microbiology, 15*(11), 645–660. https://doi.org/10.1038/nrmicro.2017.86.

Fukazawa, Y., Saitoh, Y., Ozawa, F., Ohta, Y., Mizuno, K., & Inokuchi, K. (2003). Hippocampal LTP is accompanied by enhanced F-actin content within the dendritic spine that is essential for late LTP maintenance in vivo. *Neuron, 38*, 447–460.

Furgerson, M., Clark, J. K., Crystal, J. D., Wagner, J. J., Fechheimer, M., & Furukawa, R. (2014). Hirano body expression impairs spatial working memory in a novel mouse model. *Acta Neuropathologica Communications, 2*, 131.

Furgerson, M., Fechheimer, M., & Furukawa, R. (2012). Model Hirano bodies protect against Tau-independent and Tau-dependent cell death initiated by the amyloid precursor protein intracellular domain. *PLoS One, 7*(9). https://doi.org/10.1371/journal.pone.0044996.

Gautel, M., & Djinović-Carugo, K. (2016). The sarcomeric cytoskeleton: From molecules to motion. *The Journal of Experimental Biology, 219*(2), 135–145. https://doi.org/10.1242/jeb.124941.

Goldman, J. E. (1983). The association of actin with Hirano bodies. *Journal of Neuropathology and Experimental Neurology, 42*(2), 146–152. https://doi.org/10.1097/00005072-198303000-00004.

Goodson, H. V., & Hawse, W. F. (2002). Molecular evolution of the actin family. *Journal of Cell Science, 115*(13), 2619–2622. http://jcs.biologists.org/content/115/13/2619.abstract.

Goody, R. S. (2003). The missing link in the muscle cross-bridge cycle. *Nature Structural Biology, 10*(43), 773–775. https://doi.org/10.1074/jbc.M304323200.

Gordon, J., Sibley, L. D., Nahlen, B., Korenromp, E., Miller, J., Shibuya, K., Billiouw, M., Vercruysse, J., Marcotty, T., Speybroeck, N., Chaka, G., Berkvens, D., Kaplan, J., Jones, J., Dykewicz, C., Sibley, L., Dubremetz, J., Schwartzman, J., Dubremetz, J., et al. (2005). Comparative genome analysis reveals a conserved family of actin-like proteins in apicomplexan parasites. *BMC Genomics, 6*(1), 179. https://doi.org/10.1186/1471-2164-6-179.

Gordon-Weeks, P. R., & Fournier, A. E. (2014). Neuronal cytoskeleton in synaptic plasticity and regeneration. *Journal of Neurochemistry, 129*(2), 206–212. https://doi.org/10.1111/jnc.12502.

Gourlay, C. W., & Ayscough, K. R. (2005). The actin cytoskeleton in ageing and apoptosis. *FEMS Yeast Research, 5*(12), 1193–1198. https://doi.org/10.1016/j.femsyr.2005.08.001.

Graham, B. (1969). Comparative aspects of muscle. *American Zoologist*.

Griffin, P., Furukawa, R., Piggott, C., Maselli, A., & Fechheimer, M. (2014). Requirements for Hirano body formation. *Eukaryotic Cell, 13*(5), 625–634. https://doi.org/10.1128/ec.00044-14.

Hatanaka, H., Ogura, K., Moriyama, K., Ichikawa, S., Yahara, I., & Inagaki, F. (1996). Tertiary structure of destrin and structural similarity between two actin-regulating protein families. *Cell, 85*(7), 1047–1055. https://doi.org/10.1016/S0092-8674(00)81305-7.

Henderson, C. A., Gomez, C. G., Novak, S. M., Mi-Mi, L., & Gregorio, C. C. (2017). Overview of the muscle cytoskeleton. *Comprehensive Physiology, 7*(3), 891–944. https://doi.org/10.1002/cphy.c160033.

Hurst, V., Shimada, K., & Gasser, S. M. (2019). Nuclear actin and actin-binding proteins in DNA repair. *Trends in Cell Biology, 29*(6), 462–476. https://doi.org/10.1016/j.tcb.2019.02.010.

Hussain, R., Zubair, H., Pursell, S., & Shahab, M. (2018). Neurodegenerative diseases: regenerative mechanisms and novel therapeutic approaches. *Brain Sciences, 8*(9), 177. https://doi.org/10.3390/brainsci8090177.

Huxley, H. E., & Hanson, J. (1954). Changes in the crosss-striations of muscle during contraction and stretch and their structural interpretation. *Nature, 173*, 973–976.

Hynes, T. R., Block, S. M., White, B. T., & Spudich, J. A. (1987). Movement of myosin fragments in vitro: Domains involved in force production. *Cell, 48*(6), 953–963. https://doi.org/10.1016/0092-8674(87)90704-5.

Hyrskyluoto, A., & Vartiainen, M. K. (2020). Regulation of nuclear actin dynamics in development and disease. *Current Opinion in Cell Biology, 64*, 18–24. https://doi.org/10.1016/j.ceb.2020.01.012.

Joo, Y. Y., Jang, J. W., Lee, S. W., Yoo, S. H., Kwon, J. H., Nam, S. W., Bae, S. H., Choi, J. Y., & Yoon, S. K. (2019). Circulating pro- and anti-angiogenic factors in multi-stage liver disease and hepatocellular carcinoma progression. *Scientific Reports, 9*(1), 1–8. https://doi.org/10.1038/s41598-019-45537-w.

Karsenti, E., & Vernos, I. (2001). Cell cycle: The mitotic spindle: A self-made machine. *Science, 294*(5542), 543–547. https://doi.org/10.1126/science.1063488.

Karsenti, E., Vernos, I., Zetka, M., Paouneskou, D., Jantsch, V., Mermelstein, C. S., Rebello, M. I. L., Amaral, L. M., Costa, M. L., Taratuto, A. L., Lariviere, R. C., Julien, J. P., Pelucchi, S., Stringhi, R., Marcello, E., Porter, R. M., Lane, E. B., Hyrskyluoto, A., Vartiainen, M. K., et al. (2020). Cytokinesis requires localized β-actin filament production by an actin isoform specific nucleator. *Current Opinion in Cell Biology, 21*(1), 1–12. https://doi.org/10.1038/jid.2014.371.

Kashina, A. S. (2020). Regulation of actin isoforms in cellular and developmental processes. *Seminars in Cell and Developmental Biology, 102*(July 2019), 113–121. https://doi.org/10.1016/j.semcdb.2019.12.003.

Keszei, M., & Westerberg, L. S. (2014). Congenital defects in neutrophil dynamics. *Journal of Immunology Research, 2014*. https://doi.org/10.1155/2014/303782.

Khan, N. A. (2013). Involvement of actin pathology in Alzheimer's disease. *Cell & Developmental Biology, 02*(01), 2–4. https://doi.org/10.4172/2168-9296.1000e121.

Kim, C., & Lisman, J. E. (1999). A role of actin filament in synaptic transmission and long-term potentiation. *The Journal of Neuroscience, 19*(11), 4314–4324.

Klingler-Hoffmann, M., Mittal, P., & Hoffmann, P. (2019). The emerging role of cytoskeletal proteins as reliable biomarkers. *Proteomics, 19*(21–22). https://doi.org/10.1002/pmic.201800483.

Kudryashov, D. S., Sawaya, M. R., Adisetiyo, H., Norcross, T., Hegyi, G., Reisler, E., & Yeates, T. O. (2005). The crystal structure of a cross-linked actin dimer suggests a detailed molecular interface in F-actin. *Proceedings of the National Academy of Sciences of the United States of America, 102*(37), 13105–13110. https://doi.org/10.1073/pnas.0506429102.

Kumar, A., Singh, A., & Ekavali. (2015). A review on Alzheimer's disease pathophysiology and its management: An update. *Pharmacological Reports, 67*(2), 195–203. https://doi.org/10.1016/j.pharep.2014.09.004.

Lamalice, L., Le Boeuf, F., & Huot, J. (2007). Endothelial cell migration during angiogenesis. *Circulation Research, 100*(6), 782–794. https://doi.org/10.1161/01.RES.0000259593.07661.1e.

Lambert, M. W. (2019). Cytoskeletal and nucleoskeletal interacting protein networks play critical roles in cellular function and dysfunction. *Experimental Biology and Medicine, 244*(15), 1233–1239. https://doi.org/10.1177/1535370219884875.

Lambrechts, A., Van Troys, M., & Ampe, C. (2004). The actin cytoskeleton in normal and pathological cell motility. *International Journal of Biochemistry and Cell Biology, 36*(10), 1890–1909. https://doi.org/10.1016/j.biocel.2004.01.024.

Lariviere, R. C., & Julien, J. P. (2004). Functions of intermediate filaments in neuronal development and disease. *Journal of Neurobiology, 58*(1), 131–148. https://doi.org/10.1002/neu.10270.

Lazaro-Dieguez, F., Aguado, C., Mato, E., Sanchez-Ruiz, Y., Esteban, I., Alberch, J., Knecht, E., Egea, G., Griffin, P., Furukawa, R., Piggott, C., Maselli, A., & Fechheimer, M. (2008). Requirements for Hirano body formation. *Eukaryotic Cell, 13*(5), 1415–1425. https://doi.org/10.1128/ec.00044-14.

Li, W. W., Li, V. W., Hutnik, M., & Chiou, A. S. (2012). Tumor angiogenesis as a target for dietary cancer prevention. *Journal of Oncology, 2012*. https://doi.org/10.1155/2012/879623.

Lian, G., & Sheen, V. L. (2015). Cytoskeletal proteins in cortical development and disease: Actin associated proteins in periventricular heterotopia. *Frontiers in Cellular Neuroscience, 9*, 1–13. https://doi.org/10.3389/fncel.2015.00099.

Liotta, L. A., & Stetler-Stevenson, W. G. (1991). Tumor invasion and metastasis: An imbalance of positive and negative regulation. *Cancer Research, 51*(18 Suppl), 327–336.

Littlefield, R. S., & Fowler, V. M. (2008). Thin filament length regulation in striated muscle sarcomeres: Pointed-end dynamics go beyond a nebulin ruler. *Seminars in Cell & Developmental Biology, 19*(6), 511–519. https://doi.org/10.1016/j.semcdb.2008.08.009.

Logan, C. M., & Menko, A. S. (2019). Microtubules: Evolving roles and critical cellular interactions. *Experimental Biology and Medicine, 244*(15), 1240–1254. https://doi.org/10.1177/1535370219867296.

Lugano, R., Ramachandran, M., & Dimberg, A. (2020). Tumor angiogenesis: Causes, consequences, challenges and opportunities. *Cellular and Molecular Life Sciences, 77*(9), 1745–1770. https://doi.org/10.1007/s00018-019-03351-7.

Luo, L. (2002). Actin cytoskeleton regulation in neuronal morphogenesis and structural plasticity. *Annual Review of Cell and Developmental Biology, 2002*(18), 601–635. https://doi.org/10.1146/annurev.cellbio.18.031802.150501.

Mahadevan, V., & Hart, I. R. (1990). Metastasis and angiogenesis. *Acta Oncologica, 29*(1), 97–103. https://doi.org/10.3109/02841869009089997.

Mangione, M. S. C., & Gould, K. L. (2019). Molecular form and function of the cytokinetic ring. *Journal of Cell Science, 132*(12). https://doi.org/10.1242/jcs.226928.

Maselli, A., Furukawa, R., Thomson, S. A. M., Davis, R. C., & Fechheimer, M. (2003). Formation of Hirano bodies induced by expression of an actin cross-linking protein with a gain-of-function mutation. *Eukaryotic Cell, 2*(4), 778–787. https://doi.org/10.1128/ec.2.4.778-787.2003.

Mcavey, B. A., Wortzman, G. B., Williams, C. J., & Evans, J. P. (2002). Involvement of calcium signaling and the actin cytoskeleton in the membrane block to polyspermy in mouse eggs. *Biology of Reproduction, 67*(4), 1342–1352. https://doi.org/10.1095/biolreprod.102.004630.

Melidone, R., Keating, J. H., Pfannl, R., & Alroy, J. (2011). Cerebral Hirano-like bodies in an Alpaca (*Vicugna pacos*). *Veterinary Pathology, 49*(4), 723–726. https://doi.org/10.1177/0300985811406886.

Mermelstein, C. S., Rebello, M. I. L., Amaral, L. M., & Costa, M. L. (2003). Changes in cell shape, cytoskeletal proteins and adhesion sites of cultured cells after extracellular Ca^{2+} chelation. *Brazilian Journal of Medical and Biological Research, 36*(8), 1111–1116. https://doi.org/10.1590/S0100-879X2003000800018.

Miller, A. L. (2011). The contractile ring. *Current Biology, 21*(24), R976–R978. https://doi.org/10.1016/j.cub.2011.10.044.

Mitchison, T. J. (1992). Compare and contrast actin filaments and microtubules. *Molecular Biology of the Cell, 3*(12), 1309–1315. https://doi.org/10.1091/mbc.3.12.1309.

Moujaber, O., & Stochaj, U. (2020). The cytoskeleton as regulator of cell signaling pathways. *Trends in Biochemical Sciences, 45*(2), 96–107. https://doi.org/10.1016/j.tibs.2019.11.003.

Moulding, D. A., Record, J., Malinova, D., & Thrasher, A. J. (2013). Actin cytoskeletal defects in immunodeficiency. *Immunological Reviews, 256*(1), 282–299. https://doi.org/10.1111/imr.12114.

Muñoz-Lasso, D. C., Romá-Mateo, C., Pallardó, F. V., & Gonzalez-Cabo, P. (2020). Much more than a scaffold: Cytoskeletal proteins in neurological disorders. *Cell, 9*(2), 358. https://doi.org/10.3390/cells9020358.

Murrell, M., Oakes, P. W., Lenz, M., & Gardel, M. L. (2015). Forcing cells into shape: The mechanics of actomyosin contractility. *Nature Reviews Molecular Cell Biology, 16*(8), 486–498. https://doi.org/10.1038/nrm4012.

Myre, M. A. (2012). Clues to γ-secretase, huntingtin and Hirano body normal function using the model organism *Dictyostelium discoideum*. *Journal of Biomedical Science, 19*(1), 1–11. https://doi.org/10.1186/1423-0127-19-41.

Nguyen, M. P., Lee, D., Lee, S. H., Lee, H. E., Lee, H. Y., & Lee, Y. M. (2015). Deguelin inhibits vasculogenic function of endothelial progenitor cells in tumor progression and metastasis via suppression of focal adhesion. *Oncotarget, 6*(18), 16588–16600. https://doi.org/10.18632/oncotarget.3752.

Nishida, N., Yano, H., Nishida, T., Kamura, T., & Kojiro, M. (2006). Angiogenesis in cancer. *Vascular Health and Risk Management, 2*(3), 213–219. https://doi.org/10.2147/vhrm.2006.2.3.213.

Okada, H., & Soderling, S. H. (2009). Signaling through actin to regulate spine formation and function. *The Open Neuroscience Journal, 3*, 97–107.

Okamoto, K., Bosch, M., Hayashi, Y., & Tag, S. (2009). Reviews the roles of CaMKII and F-actin in the structural plasticity of dendritic spines : A potential molecular identity of a synaptic tag. *Physiology, 24*, 357–366. https://doi.org/10.1152/physiol.00029.2009.

Ono, S. (2010). Dynamic regulation of sarcomeric actin filaments in striated muscle. *Cytoskeleton, 67*(11), 677–692. https://doi.org/10.1002/cm.20476.

Ono, S., Abe, H., Nagaoka, R., & Obinata, T. (1993). Colocalization of ADF and cofilin in intranuclear actin rods of cultured muscle cells. *Journal of Muscle Research and Cell Motility, 14*(2), 195–204. https://doi.org/10.1007/BF00115454.

Öztekin, L., Nicole, M., & Badre, D. (2008). Immune pathology associated with altered actin cytoskeleton regulation. *Autoimmunity, 23*(1), 1–7. https://doi.org/10.1038/jid.2014.371.

Papakonstanti, E. A., & Stournaras, C. (2008). Cell responses regulated by early reorganization of actin cytoskeleton. *FEBS Letters, 582*(14), 2120–2127. https://doi.org/10.1016/j.febslet.2008.02.064.

Paran, H., Paran, D., Saaristo, A., Karpanen, T., & Alitalo, K. (1996). Mechanisms of angiogenesis and their use in the inhibition of tumor growth and metastasis. *Oncogene, 19*(53), 921. https://doi.org/10.1038/sj.onc.1203969.

Pathak, S., Tripathi, S., Deori, N., Ahmad, B., Verma, H., Lokhande, R., Nagotu, S., & Kale, A. (2020). Effect of tetracycline family of antibiotics on actin aggregation, resulting in the formation of Hirano bodies responsible for neuropathological disorders. *Journal of Biomolecular Structure and Dynamics*, 1–18. https://doi.org/10.1080/07391102.2020.1717629.

Pelucchi, S., Stringhi, R., & Marcello, E. (2020). Dendritic spines in alzheimer's disease: How the actin cytoskeleton contributes to synaptic failure. *International Journal of Molecular Sciences, 21*(3), 1–23. https://doi.org/10.3390/ijms21030908.

Pessoa-Pureur, R., Heimfarth, L., & Rocha, J. B. (2014). Signaling mechanisms and disrupted cytoskeleton in the diphenyl ditelluride neurotoxicity. *Oxidative Medicine and Cellular Longevity, 2014*. https://doi.org/10.1155/2014/458601.

Pontow, S. E., Heyden, N. V., Wei, S., & Ratner, L. (2004). Actin cytoskeletal reorganizations and coreceptor-mediated activation of Rac during human immunodeficiency virus-induced cell fusion. *Journal of Virology, 78*(13), 7138–7147. https://doi.org/10.1128/jvi.78.13.7138-7147.2004.

Poulter, N. S., Staiger, C. J., Rappoport, J. Z., & Franklin-Tong, V. E. (2010). Actin-binding proteins implicated in the formation of the punctate actin foci stimulated by the self-incompatibility response in papaver. *Plant Physiology, 152*(3), 1274–1283. https://doi.org/10.1104/pp.109.152066.

Prager, G. W., Poettler, M., Unseld, M., & Zielinski, C. C. (2012). Angiogenesis in cancer: Anti-VEGF escape mechanisms. *Translational Lung Cancer Research, 1*(1), 14–25. https://doi.org/10.3978/j.issn.2218-6751.2011.11.02.

Pralhad, T., Madhusudan, S., & Rajendrakumar, K. (2003). Concept, mechanisms and therapeutics of angiogenesis in cancer and other diseases. *Journal of Pharmacy and Pharmacology, 55*(8), 1045–1053. https://doi.org/10.1211/0022357021819.

Rajabi, M., & Mousa, S. A. (2017). The role of angiogenesis in cancer treatment. *Biomedicine, 5*(2). https://doi.org/10.3390/biomedicines5020034.

Ramanan, V. K., & Saykin, A. J. (2013). Pathways to neurodegeneration: Mechanistic insights from GWAS in Alzheimer's disease. *Parkinson's Disease, and Related Disorders, 2*(3), 145–175.

Reisler, E., & Egelman, E. H. (2007). Actin structure and function: What we still do not understand. *Journal of Biological Chemistry, 282*(50), 36133–36137. https://doi.org/10.1074/jbc.R700030200.

Rodríguez-Caso, L., Reyes-Palomares, A., Sánchez-Jiménez, F., Quesada, A. R., & Medina, M. Á. (2012). What is known on angiogenesis-related rare diseases? A systematic review of literature. *Journal of Cellular and Molecular Medicine, 16*(12), 2872–2893. https://doi.org/10.1111/j.1582-4934.2012.01616.x.

Ross, C. A., & Poirier, M. A. (2004). Protein aggregation and neurodegenerative disease. *Nature Medicine, 10*(7), S10. https://doi.org/10.1038/nm1066.

Saaristo, A., Karpanen, T., & Alitalo, K. (2000). Mechanisms of angiogenesis and their use in the inhibition of tumor growth and metastasis. *Oncogene, 19*(53), 6122–6129. https://doi.org/10.1038/sj.onc.1203969.

Santa-Mara, I., Santpere, G., MacDonald, M. J., Gomez De Barreda, E., Hernandez, F., Moreno, F. J., Ferrer, I., & Avila, J. (2008). Coenzyme Q induces tau aggregation, tau filaments, and Hirano bodies. *Journal of Neuropathology and Experimental Neurology, 67*(5), 428–434. https://doi.org/10.1097/NEN.0b013e31816fc9b6.

Santella, L., & Chun, J. T. (2011). Actin, more than just a housekeeping protein at the scene of fertilization. *Science China Life Sciences, 54*(8), 733–743. https://doi.org/10.1007/s11427-011-4202-x.

Santella, L., Limatola, N., & Chun, J. T. (2014). Actin cytoskeleton and fertilization in starfish eggs. *Sexual Reproduction in Animals and Plants*, 141–155. https://doi.org/10.1007/978-4-431-54589-7.

Santella, L., Limatola, N., & Chun, J. T. (2020). Cellular and molecular aspects of oocyte maturation and fertilization: A perspective from the actin cytoskeleton. *Zoological Letters, 6*, 1–21.
Santella, L., Limatola, N., Vasilev, F., & Chun, J. T. (2018). Biochemical and biophysical research communications maturation and fertilization of echinoderm eggs: Role of actin cytoskeleton dynamics. *Biochemical and Biophysical Research Communications*, 1–11. https://doi.org/10.1016/j.bbrc.2018.09.084.
Sawaya, M. R., Kudryashov, D. S., Pashkov, I., Adisetiyo, H., Reisler, E., & Yeates, T. O. (2008). Multiple crystal structures of actin dimers and their implications for interactions in the actin filament. *Acta Crystallographica Section D: Biological Crystallography, 64*(4), 454–465. https://doi.org/10.1107/S0907444908003351.
Senju, Y., & Lappalainen, P. (2019). Regulation of actin dynamics by PI(4,5)P2 in cell migration and endocytosis. *Current Opinion in Cell Biology, 56*, 7–13. https://doi.org/10.1016/j.ceb.2018.08.003.
Seo, J., & Kim, J. (2018). Regulation of Hippo signaling by actin remodeling. *BMB Reports, 51*(3), 151–156. https://doi.org/10.5483/BMBRep.2018.51.3.012.
Sept, D., Xu, J., Pollard, T. D., & McCammon, J. A. (1999). Annealing accounts for the length of actin filaments formed by spontaneous polymerization. *Biophysical Journal, 77*(6), 2911–2919. https://doi.org/10.1016/S0006-3495(99)77124-9.
Shimizu, N., & Obinata, T. (1986). Actin concentration and monomer-polymer ratio in developing chicken skeletal muscle. *Journal of Biochemistry, 99*(3), 751–759. https://doi.org/10.1093/oxfordjournals.jbchem.a135534.
Skwarek-Maruszewska, A., Hotulainen, P., Mattila, P. K., & Lappalainen, P. (2009). Contractility-dependent actin dynamics in cardiomyocyte sarcomeres. *Journal of Cell Science, 122*(12), 2119–2126. https://doi.org/10.1242/jcs.046805.
Spears, W., Furgerson, M., Sweetnam, J. M., Evans, P., Gearing, M., Fechheimer, M., & Furukawa, R. (2014). Hirano bodies differentially modulate cell death induced by tau and the amyloid precursor protein intracellular domain. *BMC Neuroscience, 15*, 0–20. https://doi.org/10.1186/1471-2202-15-74.
Spence, E. F., & Soderling, S. H. (2015). Actin out: Regulation of the synaptic cytoskeleton. *Journal of Biological Chemistry, 290*(48), 28613–28622. https://doi.org/10.1074/jbc.R115.655118.
Stevenson, R. P., Veltman, D., & Machesky, L. M. (2012). Actin-bundling proteins in cancer progression at a glance. *Journal of Cell Science, 125*(5), 1073–1079. https://doi.org/10.1242/jcs.093799.
Subramanian, D., Huang, J., Sevugan, M., Robinson, R. C., Balasubramanian, M. K., & Tang, X. (2013). Insight into actin organization and function in cytokinesis from analysis of fission yeast mutants. *Genetics, 194*(2), 435–446. https://doi.org/10.1534/genetics.113.149716.
Sun, B., Fang, Y., Li, Z., Chen, Z., & Xiang, J. (2015). Role of cellular cytoskeleton in epithelial-mesenchymal transition process during cancer progression. *Biomedical Reports, 3*(5), 603–610. https://doi.org/10.3892/br.2015.494.
Sun, N. A., Funke, S., & Willbold, D. (2012). A survey of peptides with effective therapeutic potential in Alzheimer's disease rodent models or in human clinical studies. *Mini-Reviews in Medicinal Chemistry, 12*(5), 388–398. https://doi.org/10.2174/138955712800493942.
Sund, M., & Kalluri, R. (2008). Endogenous inhibitors of angiogenesis. *Tumor Angiogenesis: Basic Mechanisms and Cancer Therapy, 10*, 215–231. https://doi.org/10.1007/978-3-540-33177-3_12.
Svennberg, K. (2006). Moisture buffering in the Indoor environment. *Physics, 67*, 184–192. https://doi.org/10.1016/j.brainresrev.2011.01.003.Impaired.
Szabó, E. C., Manguinhas, R., & Fonseca, R. (2016). The interplay between neuronal activity and actin dynamics mimic the setting of an LTD synaptic tag. *Scientific Reports, 6*, 1–14. https://doi.org/10.1038/srep33685.
Takaine, M., Numata, O., & Nakano, K. (2015). An actin-myosin-II interaction is involved in maintaining the contractile ring in fission yeast. *Journal of Cell Science, 128*(15), 2903–2918. https://doi.org/10.1242/jcs.171264.
Tangye, S. G., Bucciol, G., Casas-Martin, J., Pillay, B., Ma, C. S., Moens, L., & Meyts, I. (2019). Human inborn errors of the actin cytoskeleton affecting immunity: Way beyond WAS and WIP. *Immunology and Cell Biology, 97*(4), 389–402. https://doi.org/10.1111/imcb.12243.

Taratuto, A. L. (2002). Congenital myopathies and related disorders. *Current Opinion in Neurology, 15*(5), 553–561. https://doi.org/10.1097/00019052-200210000-00006.

Van Bulck, M., Sierra-Magro, A., Alarcon-Gil, J., Perez-Castillo, A., & Morales-Garcia, J. A. (2019). Novel approaches for the treatment of Alzheimer's and parkinson's disease. *International Journal of Molecular Sciences, 20*(3). https://doi.org/10.3390/ijms20030719.

Varland, S., Vandekerckhove, J., & Drazic, A. (2019). Actin post-translational modifications: The cinderella of cytoskeletal control. *Trends in Biochemical Sciences, 44*(6), 502–516. https://doi.org/10.1016/j.tibs.2018.11.010.

Vavylonis, D., Yang, Q., & O'Shaughnessy, B. (2005). Actin polymerization kinetics, cap structure, and fluctuations. *Proceedings of the National Academy of Sciences of the United States of America, 102*(24), 8543–8548. https://doi.org/10.1073/pnas.0501435102.

Vedula, P., & Kashina, A. (2018). The makings of the "actin code": Regulation of actin's biological function at the amino acid and nucleotide level. *Journal of Cell Science, 131*(9). https://doi.org/10.1242/jcs.215509.

Veselská, R. J. R. (2001). Cortical actin cytoskeleton in human oocytes: A comparison with mouse oocytes. *Scripta Medica (Brno), 74*, 265–272.

Wong, W., Skau, C. T., Marapana, D. S., Hanssen, E., Taylor, N. L., Riglar, D. T., Zuccala, E. S., Angrisano, F., Lewis, H., Catimel, B., Clarke, O. B., Kershaw, N. J., Perugini, M. A., Kovar, D. R., Gulbis, J. M., & Baum, J. (2011). Minimal requirements for actin filament disassembly revealed by structural analysis of malaria parasite actin-depolymerizing factor 1. *Proceedings of the National Academy of Sciences of the United States of America, 108*(24), 9869–9874. https://doi.org/10.1073/pnas.1018927108.

Worman, H. J., & Courvalin, J. (2004). How do mutations in lamins A and C cause disease. *The Journal of Clinical Investigation, 113*(3), 349–351. https://doi.org/10.1172/JCI200420832.a.

Xie, X., Almuzzaini, B., Drou, N., Kremb, S., Yousif, A., Farrants, A. K. Ö., Gunsalus, K., & Percipalle, P. (2018). β-Actin-dependent global chromatin organization and gene expression programs control cellular identity. *FASEB Journal, 32*(3), 1296–1314. https://doi.org/10.1096/fj.201700753R.

Yamaguchi, H., & Condeelis, J. (2007). Regulation of the actin cytoskeleton in cancer cell migration and invasion. *Biochimica et Biophysica Acta: Molecular Cell Research*. https://doi.org/10.1016/j.bbamcr.2006.07.001.

Yamazaki, D., Kurisu, S., & Takenawa, T. (2005). Regulation of cancer cell motility through actin reorganization. *Cancer Science, 96*(7), 379–386. https://doi.org/10.1111/j.1349-7006.2005.00062.x.

Yoon, Y., Pitts, K., & McNiven, M. (2002). Studying cytoskeletal dynamics in living cells using green fluorescent protein. *Applied Biochemistry and Biotechnology – Part B: Molecular Biotechnology, 21*(3), 241–250. https://doi.org/10.1385/MB:21:3:241.

Yu, Y., Smoligovets, A. A., & Groves, J. T. (2013). Modulation of t cell signaling by the actin cytoskeleton. *Journal of Cell Science, 126*(5), 1049–1058. https://doi.org/10.1242/jcs.098210.

Zak, R., Martin, A. F., Prior, G., & Rabinowitz, M. (1977). Comparison of turnover of several myofibrillar proteins and critical evaluation of double isotope method. *Journal of Biological Chemistry, 252*(10), 3430–3435.

Zetka, M., Paouneskou, D., & Jantsch, V. (2020). The nuclear envelope, a meiotic jack-of-all-trades. *Current Opinion in Cell Biology, 64*, 34–42. https://doi.org/10.1016/j.ceb.2019.12.010.

Zhang, Y. J., Zhu, C., Ding, Y., Yan, Z. W., Li, G. H., Lan, Y., Wen, J. F., & Chen, B. (2018). Subcellular stoichiogenomics reveal cell evolution and electrostatic interaction mechanisms in cytoskeleton. *BMC Genomics, 19*(1), 1–12. https://doi.org/10.1186/s12864-018-4845-0.

Zuazo-Gaztelu, I., & Casanovas, O. (2018). Unraveling the role of angiogenesis in cancer ecosystems. *Frontiers in Oncology, 8*, 1–13. https://doi.org/10.3389/fonc.2018.00248.

Diabetes Mellitus and it Management with Plant-Based Therapy

Mithun Rudrapal, Nazim Hussain, and Chukwuebuka Egbuna

1 Introduction

Diabetes mellitus (DM) is a metabolic illness which is caused due to shortcomings in the insulin secretion, its action, or both. Deficiency of insulin causes increase in blood glucose levels leading to chronic hyperglycemia with changes in metabolism of molecules providing energy (carbohydrate, fat and protein) (Yadav et al. 2014). With the progression of the disease, damage in the tissue occurs causing serious complications like neuropathy, retinopathy, nephropathy (Huang et al. 2002), ulceration (Centers for Disease Control and Prevention 2011) and cardiovascular complications. Diabetes is the one of the most prevalent disease around the globe and it was estimated that by the year 2025, more than 300 million people will subsequently have the disease.

DM is presumably one of the most debilitating disorder known to human being. It was first recognized in Egypt around 3000 years ago (Ahmed 2002) while the division between two types of DM was clearly made only in 1936. Type II DM

M. Rudrapal (✉)
Rasiklal M. Dhariwal Institute of Pharmaceutical Education and Research,
Chinchwad, Pune, Maharashtra, India

N. Hussain
Kingston Imperial Institute of Technology and Sciences, Dunga,
Dehradun, Uttarakhand, India

C. Egbuna
Africa Centre of Excellence for Public Health and Toxicological Research (ACE-PUTOR),
University of Port Harcourt, Port Harcourt, Nigeria

Department of Biochemistry, Faculty of Natural Science, University of Port Harcourt,
Port Harcourt, Choba, Nigeria

Department of Biochemistry, Faculty of Natural Sciences, Chukwuemeka Odumegwu
Ojukwu University, Anambra, Nigeria

C. Egbuna, S. Hassan (eds.), *Dietary Phytochemicals*,
https://doi.org/10.1007/978-3-030-72999-8_7

(known as non-insulin dependent DM) was firstly described in 1988 as a type of metabolic syndrome (Patlak 2002). It is the most widely recognized type of DM portrayed by insulin resistance, hyperglycemia and insulin insufficiency. Type II diabetes is caused from the interaction between hereditary, natural and different behavioral risk variables (Chen et al. 2012).Whereas Type-I diabetes mellius is caused due to the destruction of β-cells in the islets of langerhan of the pancreas, causing an absolute insulin deficiency. It is responsible for ~10% of all the individuals suffering from DM in USA. It is further divided into two types: The one involving auto-immune indicators like antibodies against glutamic acid decarboxylase (GAD), insulin auto-antibodies (IAAs), and islet cell antibodies (ICAs). Approximately 85–90% of patients having a fasting hyperglycemic condition, are affirmative for one of these markers. The second type also known as idiopathic diabetes is of unknown cause and only a handful of the patients fall in this category. It mainly occurs in people living in the African and Asian continent. It is strongly genetic in nature, but it does not involve the autoimmune indicators.

Individuals living with type II diabetes are more prone to different types of complications, which may lead to their premature mortality. This propensity of elevated morbidity and mortality is observed in patients with type II diabetes due to its frequency, its subtle onset and late identification, particularly in resource-ill nations (Azevedo and Alla 2008).

It has been estimated that almost 366 million people had been suffering diabetes in 2011 and by the year 2030, the numbers would increase to 552 million. The number of individuals with type II diabetes is expanding in every nation with about 80% of individuals in the low and middle income nations suffering from DM. This disease has caused 4.6 million mortalities in the year 2011 (Ali et al. 2010). It is also anticipated that approx. 439 million individuals would have type II DM by 2030 (Chamnan et al. 2011). The occurrence of type II DM varies considerably from one geographical area to the other on account of the different lifestyle and environmental variables (Zimmet et al. 2001).

Literary analysis has demonstrated that there are little data present on the occurrence of type II diabetes in the African nation as a whole. Investigations analyzing the data within the African nation indicate a dramatic rise in frequency, both in the rural and urban location, and influencing both the genders uniformly (Mbanya 2009). Most of the DM load in Africa is due to type II DM, with under 10% of the disease cases of type I diabetes (Mbanya 2009). The Centre for Disease Control and Prevention (CDC) report in 2011 predicts that DM influences about 25.8 million individuals in 2010 in the US (7.8% of the population) with almost 90–95% of the individuals suffering from type II DM (Centers for Disease Control and Prevention 2011).

It is anticipated that the occurrence of DM in adults will rise in the next few decades and much of this elevation will occur in the developing nations where most of the patients are between the age group of 45–64 years (Rathmann and Giani 2004). It is also predicted that the latter will equivalent or even go beyond the former in the developing countries, consequently concluding in form of a two-fold burden because of the current inclination of change from communicable disease to a non-communicable one.

2 Classification of Diabetes

2.1 *Type I Diabetes Mellitus (T1DM)*

The American Diabetes Association (ADA) has clearly defined different types of DM, their identification, and management strategies. T1DM is caused due to the destruction of β-cells in the islets of langerhan of the pancreas, causing an absolute insulin deficiency. It is responsible for ~10% of all the individuals suffering from DM in USA. It is further divided into two types: The one involving auto-immune indicators like antibodies against glutamic acid decarboxylase (GAD), insulin auto-antibodies (IAAs), and islet cell antibodies (ICAs). Approximately 85–90% of patients having a fasting hyperglycemic condition, are affirmative for one of these markers. The second type also known as idiopathic diabetes is of unknown cause and only a handful of the patients fall in this category. It mainly occurs in people living in the African and Asian continent. It is strongly genetic in nature, but it does not involve the autoimmune indicators.

While it can appear at any stage of life, T1DM is more frequent in people <30 years age. The rate of destruction of pancreatic cells is inconsistent and is usually quicker in infants and children but slow in the adults. Patients have a tendency to be highly symptomatic at the onset of the disease with regular complains of delay in wound healing, dryness of mouth, increased hunger, thirst and urination, along with leg cramps, pruritus, recurring skin, genitalia or urinary tract infections and unexplained weight reduction. The basic feature of T1DM is complete dependence on external insulin to avoid ketoacidosis.

2.2 *Type II Diabetes Mellitus (T2DM)*

Type II is a well-known type of DM around the world, and its occurrence is expanding. Its basic shortcomings can change from insulin resistance associated with insulin deficit to insulin secretory deficiency along with insulin resistance. But there is a lot of heterogeneity and most of the patients with T2DM don't need insulin therapy initially. T2DM accounts for around 90% of all the cases of diabetes in USA and it occurs in adults more often than in children. The incidence rises with age, (particularly after 40 years age). But, the occurrence of T2DM in children is rising, mainly in the risk-prone ethnicities of African, Asian, Hispanic and Native Americans. Many of these kids of the age range 10–19 years have had symptoms sing long time, with rare or mild ketoacidosis, obesity, and a clear family genetics of the disease. A typical observation is increased skin pigmentation (acanthosis nigricans) and there is an increase in the frequency of insulin resistance (Fagot-Campagna et al. 2000). As the onset of the disease is often subtle, numerous individuals with T2DM are without any symptoms and the disease stays undiscovered for quite a long time. Obesity in the upper part of the body is a familiar issue since it leads to insulin resistance. The β-cells balance for the insulin resistance by enhancing the insulin secretion and sustaining the standard glucose tolerance. Ultimately, the hyperglycemia gets worse with development of glucose toxicity and reduction in insulin production and its actions. The loss of β-cells can also prompt to the development of insulin dependency. So, now the extended description of the insulin resistance disorder incorporates coronary heart disease, diffused atherosclerosis, dyslipidemia (increased triglycerides, and low-density lipoprotein with reduced high-density cholesterol), glucose intolerance, hypertension, and increased plasminogen activator inhibitor-1 levels. These results might be the reason for the distinct rise in coronary artery disorder found in T2DM.

2.3 Gestational DM (GDM)

By description, GDM is any amount of glucose intolerance with initial or primary analysis during pregnancy. Frequently analyzed during the second or third trimester of pregnancy, GDM takes place in around 4% of pregnancies or 135,000 cases every year. The occurrence rate of 1–14% relies upon the population analyzed. Glucose tolerance typically comes back to normal in around 6 weeks after the last part of pregnancy, after which the woman should be reclassified. Numerous GDM patients don't show DM at the later stage of their life; however some patients will develop IGT, IFG, T2DM or even T1DM. Since the case of higher fetal mortality rates have been related with GDM, earlier recognition and aggressive management are crucial. GDM still remains a subset within the new arrangement, but the sorting criteria have been modified. Now, not all the pregnant women require screening; the ones exempted must comply with the following criteria:

1. Women belonging to an ethnic category with low frequency of DM,
2. Women under 25 years of age,
3. Women with standard weight before pregnancy,
4. Women without any history of poor obstetric result,
5. Women without any history of unusual glucose tolerance,
6. Women without known first-degree relatives with DM.

Risk evaluation ought to be performed prior to the pregnancy and glucose test must be performed immediately for those women which are more prone to risk. Follow-up analysis using an OGTT test should be carried out between 24–28 weeks for women with high risk attributes whose early analysis shows no indication of GDM.

3 Risk Factors of Diabetes

T2DM is caused mostly due to lifestyle and heredity attributes. Numerous lifestyle aspects like alcohol consumption, physical inertness, sedentary lifestyle, and smoking are known to be responsible for the development and progression of T2DM (Hu et al. 2001). Obesity is known to be responsible for approx. 55% of the cases of T2DM. The higher rate of obesity in children from 1960 to 2000 is suggested to be responsible for the increase in T2DM in them. Different environmental contaminants may add to the current rise in the frequency of T2DM. A little positive association has been observed between the urinary concentrations of bisphenol A and the frequency of T2DM.

There is a powerful hereditary association in T2DM. People having direct relation with the patients suffering from T2DM considerably elevate the risks of disease. Monozygotic twins resemble each other close to 100%, and around 25% of them with the illness have a family record of DM (Rother 2007). Lately, the genes found to be considerably related with T2DM, include CDKAL1, FTO, HHEX, IGF2BP2, JAZF1, KCNJ11, NOTCH2, PPARG, SLC30A8, TCF7L2, and WFS1. KCNJ11 codes the ATP-sensitive islet potassium channel (Kir6.2), and TCF7L2 (transcription factor-7 like 2) controls expression of proglucagon gene and consequently the synthesis of glucagon-like peptide-1 (McCarthy 2010). Furthermore, obesity which is considered as an autonomous risk aspect for T2DM is strongly acquired (Walley et al. 2006).

Monogenic types of diabetes such as Maturity-onset diabetes of young (MODY), comprises around 5% of the total cases (Camastra et al. 1999). There are various medical provisions like elevated cholesterol (combined hyperlipidemia), hypertension, obesity, metabolic syndrome also called Syndrome X or Reaven's syndrome etc., which can possibly cause or worsen T2DM (Alberti et al. 2005). Some other diseases include acromegaly, cancer, chronic pancreatitis, Cushing's syndrome, pheochromocytoma, and thyrotoxicosis. Other factors responsible for increasing the risk of T2DM comprise of aging, sedentary lifestyle and high-fat diet (Jack Jr et al. 2004; Lovejoy 2002).

4 Pathophysiology

T2DM is demonstrated by insulin insensitivity on account of insulin resistance, decreasing insulin synthesis, and ultimately failure of the pancreatic β-cells (Kahn 1994; Robertson and Bogardus 1995). This causes a reduction in the glucose movement into the fat, liver, and muscle cells and there is a rise in the fat breakdown process. The association of damaged α-cell function has lately been documented in the pathophysiology of T2DM (Fujioka 2007).

Subsequently of this dysregulation, glucagon and the hepatic levels of glucose that increase amid fasting are not reduced with a diet. Hyperglycemia occurs due to insufficient insulin levels and higher insulin resistance. The incretins are essential gut mediators of insulin secretion, and glucagon control. In spite of the fact that gastric inhibitory peptide (GIP) movement is debilitated in those with T2DM, glucagons-like peptide-1 (GLP-1) insulinotropic actions are protected, and hence GLP-1 demonstrates a conceivably useful restorative option. However, as GIP; GLP-1 is quickly inactivated by dipeptidyl peptidase-IV (DPP-IV) enzyme in vivo.

Two therapeutic ways to deal with this issue have been produced: GLP-1 analogs with higher half-lives, and DPP-IV inhibitors, which check the breaking of endogenous GLP-1 and GIP (Fujioka 2007). Both classes of specialists have demonstrated promising results, with potential to restore fasting and postprandial glucose levels alongwith enhancing β-cell activity. Researches are continuing on the role of mitochondrial dysregulation in the progression of insulin resistance and etiology of T2DM (Fujioka 2007; Garcia-Roves 2011).

Many people experiencing T2DM are overweight. Thus, the adipose tissue assumes an essential part in the pathogenesis of T2DM. Despite the fact that the prime hypothesis used to clarify this link is the portal/visceral theory giving a cheif part in increased non-esterified levels of fat, two new developing speculations are the ectopic fat storage disorder (accumulation of triglycerides in liver, muscle and cells of pancreas). These assumptions make the system for the investigation of the transaction between insulin resistance and dysregulation of β-cells in T2DM and also between our obesogenic condition and risk of diabetes in the subsequent decade (Fujioka 2007).

5 Screening and Diagnosis

Tests for diagnosis of diabetes are easily accessible. The examination suggested for analyzing is similar to that of performing diagnosis, with the outcome that affirmative screen is proportionate to a determination of pre-diabetes or diabetes (Cox and David 2009). Albeit around 25% of patients with T2DM as of now have microvascular complexities during the diagnosis proposing that they have had the disorder for over 5 years (Harris et al. 1992). It is based on the guidelines of ADA, 1997 or criteria of National diabetic group, for a lone raised glucose value with indications (polydipsia, polyphagia, polyuria and weight reduction), generally raised values on two instances, of either FPG ≥7.0 mmol/L (126 mg/dL) or with an OGTT, 2 h prior to the oral dosing a plasma glucose ≥11.1 mmol/L (200 mg/dL) (Cox and David 2009).

The ADA proposals of 1997 for the diabetes analysis focus on the FPG, while WHO concentrates on the OGTT (Cox and David 2009). The fructosamine and HbA1c is still helpful for deciding blood glucose control after sometime. Notwithstanding, doctors regularly utilize different measures along with those suggested. The International Expert Committee (IEC) in 2009 advised the additional symptomatic criteria of an HbA1c result ≥6.5% for diabetes. This board proposed that the utilization of the term pre-diabetes might be eliminated yet recognized the levels of HbA1c levels ≥6.0% and <6.5% to recognize those at increased risk of precipitating diabetes (IEC 2009). Just like the glucose-based tests, there is no clear edge of HbA1c at which the normality concludes with onset of diabetes (Cox and David 2009; IEC 2009).

6 Management

6.1 *Through Lifestyle and Diet Modification*

Research has demonstrated that there was huge decrement in the occurrence of T2DM with conservation of body mass list of 25 kg/m^2, eating fiber rich diet and unsaturated fat and food low in saturated and trans-fatty acids and glycemic index, routine exercise, restraint from smoking and direct utilization of liquor (Boffetta et al. 2011; Chen et al. 2012; Hu et al. 2001; Willi et al. 2007; Yoon et al. 2006). Proposing that bulk of T2DM can be averted by lifestyle adjustment. Patients with T2DM ought to get a medicinal assessment of diet; lifestyle advice ought to be customed as indicated by physical capacity (Chiniwala and Jabbour 2011).

6.2 *Pharmacological Agents*

This section presents the various pharmacological agents.

1. **Biguanides**—Biguanides (metformin) is generally utilized in obese patients which represses hepatic synthesis of glucose, raises insulin sensitivity, increases glucose uptake by phosphorylation of glucose transporter enhancing factors, enhances oxidation of unsaturated fat, and reduces the assimilation of glucose in the gastrointestinal tract. Research in 2008 demonstrates auxiliry mechanism of metformin as actuation of AMP-stimulated protein kinase. But, metformin ought to be utilized cautiously in aged patients with renal failure as compared to sulphonylureas due to lactic acidosis development.
2. **Sulfonylureas**—These are much endured medications but since they activate endogenous secretion of insulin, they can cause hypoglycemia (Chiniwala and Jabbour 2011). Aged patients suffering from diabetes who are treated with sulfonylureas have a 36% higher hypoglycemic risk in contrast with the younger ones (Van Staa et al. 1997). Glyburide is related with higher hypoglycemic rates in comparison with glipizide (Shorr et al. 1996). Some of the risk variables for hypoglycemia are age associated impairment in renal functioning, concurrent utilization of insulin or insulin sensitizers, more than >60 years of age, recent discharge from hospital, excess alcohol consumption, multiple prescriptions or drugs that potentiate the actions of sulfonylurea (Scheen 2005). Using long acting sulfonylurea, like glyburide ought to be circumvented in aged patients while use of glipizide (short-acting) ought to be chosen (Chiniwala and Jabbour 2011.
3. **Meglitinides**—These non-sulfonylurea secretagogues (repaglinide and nateglinide) act on the ATP-depending potassium channels present in the β pancreatic cells thus activating the insulin secretion (Fuhlendorff et al. 1998). These drugs have a quick onset but a brief span (4–6 h) of activity and consequently reduce the hypoglycemic risk. These drugs are given prior to meal for postprandial levels of blood glucose. The pre-prandial intake permits flexibility in case someone

misses a meal without hypoglycemic risk (Blicklé 2006). Repaglinide is mostly metabolized in the liver with little amount excreted from kidneys and therefore dose modification is not required in patients with renal impatrment except in patients suffering from end-stage renal disorder (Fuhlendorff et al. 1998).

4. **Thiazolidinediones**—These are insulin sensitizers, specific ligands transcription factors (peroxisomes proliferator-activated receptor-γ). They are the principal drugs to address the essential issue of insulin resistance in T2DM patients and incorporates pioglitazone after the confined utilization of rosiglitazone by Food and Drug Administration (FDA) because of higher risk of cardiovascular effects with it (Yoon et al. 2006). Pioglitazone utilization is not related with hypoglycemia and can be employed in renal failure and can be given to all elderly patients. Then again, because of the concerns in regards to fluid retention, peripheral edema, and risk of break in women, it can be restricted to higher age groups with diabetes. It should not be used in case of congestive heart failure and is also contraindicated in case of class III & IV heart failure.
5. **Alpha-glucosidase inhibitors**—These drugs (Acarbose, miglitilol and voglibose) have not been generally utilized in treating T2DM yet are considered effective and safe. These drugs are best for postprandial increase in blood glucose levels and ought to be circumvented in patients with considerable renal failure. These drugs are normally restricted because of high adverse effects, for example, flatulence, diarrhea etc. (Chiniwala and Jabbour 2011). The newest addition in this category (Voglibose) has been demonstrated to essentially enhance glucose resistance, in the patients with normal glucose levels in a study (Kawamori et al. 2009).
6. **Incretin-based therapies**—Analogues of GLP-1 (liraglutide, exenatide) are the basis of incretin-based treatments which result in sustained changes in glycemic control and enhanced control in body weight (Nauck et al. 2009). They are used as monotherapy, along with diet and physical exercise or along with other oral hypoglycemic drugs in adult patients with T2DM (Chiniwala and Jabbour 2011). There is no hypoglycemic risk while using GLP-1 drugs unless given in combination with insulin secretagogues. Besides, increasing proof proposes incretin-based treatments may positively affect hepatic and cardiovascular health, central nervous system, inflammation, and sleep (Stonehouse et al. 2012).
7. **DPP-IV inhibitors**—Inhibitors of DPP-IV hinder DPP-IV enzyme which is an omnipresent enzyme that quickly inactivates both GIP and GLP-1, raise their levels and, thus enhancing the functioning of islet and blood glucose levels in T2DM (Pratley and Salsali 2007). These inhibitors are novel class of anti-diabetic medicaments that offers efficacy tantamount to current therapies. They are compelling as monotherapy in patients not controlled with diet control and physical activity and as add on drug along with thiazolidinediones, metformin and insulin. These drugs are well-tolerated with low hypoglycemic risk but are relatively expensive (Barnett 2006; Pratley and Salsali 2007).
8. **Insulin**—Insulin is utilized as monotherapy or along with oral hypoglycemic drugs. If a number of β- cells are active then insulin therapy can be used to provide augmentation therapy. The substitution of basal bolus insulin is essential if β- cells are depleted. Rescue treatment by substitution is fundamental in cases of

increased glucose levels which ought to resemble the standard insulin secretion by the pancreatic β- cells (Mayfield and White 2004). It is available as injectables with fast, short, and long acting half-life. The long acting ones have less chance of causing hypoglycemia as compared to the short acting forms.

9. **Insulin analogues**—Insulin treatment was restricted in its capacity to resemble normal physiological insulin secretion. The traditional long and intermediate acting insulin types (lente insulin, ultralente insulin) are restricted by unpredictable assimilation and peaks of activity that may cause hypoglycemia (Burge and Schade 1997; Cameron and Bennett 2009). The pharmacokinetic profiles of the new insulin analogs are unique from those of the normal insulins, and their onset and time of action may be quick or prolonged. As of now, two fast acting analogs of insulin, insulin aspart and insulin lispro, and insulin glargine, a long-acting insulin analog are available easily (Burge and Schade 1997; Cameron and Bennett 2009).

6.3 *Plant Based Therapy*

Plants have been utilized as a remarkable source of food and drug all through the world from the earliest starting point of life. They have been used by the living creatures on the bases of practice and experience bringing about different traditional systems of medications. Ayurveda, Chinese, Homeopathy, Siddha, Unani etc. are the notable customary systems of medicine basically utilizing plants as a treatment solution. Among the global aggregate of 422,000 plant species, almost 5000 plant species are being utilized for therapeutic purposes. In these indigenous sytems of treatment (Ayurvedic, Unani, and so forth) huge number of herbs and home grown formulations have been attempted in the treatment of diabetes (Chopra 1953). Till date, >1200 blossoming plants have been found to have antidiabetic potential and approximately 33% have been experimentally analysed and documented (Chang et al. 2013). However, there is requirement to carry out the experimentation for less investigated herbs and their constituents to discover more valuable alternatives to manage the ailment.

This study is an endeavor to investigate the therapeutic plants having such impacts. WHO has equally supported and prescribed that traditional treatment for diabetes requires further assessment (Lemhadri et al. 2006).

6.3.1 Mechanisms Underlying Herbal Anti-Diabetic Therapies

Distinctive kinds of restorative herbs can be ordered in light of their methods of action, for example:

- Insulin resistance (category I plants)
- β-cell work (category II plants)
- GLP-1 (category III plants)
- Glucose (re)absorption (category IV plants)

6.3.2 Classification of Plant-Based Anti-Diabetics

On the premise of conceivable mechanism of activity of different plant constituents taken from various herbs, antidiabetic plants might be grouped as follows:-(Tripathi et al. 2011)

(a) **Insulin like acting drugs**
The protein extracts from *Momordica charantia* or bitter gourd have been found resembling the human insulin, and it offered steady hypoglycemic action in subjects like gerbils, human, langurs and rats. The polypeptides from the roots of *Panax ginseng* exhibited a decrease in the blood glucose and liver glycogen levels in rats.

(b) **Drugs enhancing the insulin secretion from the pancreatic β-cells**
The alcoholic concentrate from the roots of *Panax ginseng* reduced the glucose level by activating the secretion of insulin, particularly the glucose-induced insulin discharge from the pancreatic islets. The hypoglycemic activity of the Acacia seeds was due to the enhanced release of insulin from β-cells. The extracts of *Eugenia jambolana* and *Momordica charantia* have demonstrated to stimulate the secretion of insulin.

(c) **Drugs acting by regenerating the pancreatic β-cells**
Epicatechin obtained from *Pterocarpus marsupium*, demonstrated recovery in the β-cell number in alloxan induced damaged subjects. *Tinospora cordifolia* demonstrated insulin-like activity and can significantly reduce the blood glucose levels with no effect on the lipid levels in alloxan-treated rabbits. The fundamental constituent of *Gymnema sylvestre* (gymnemic acid) acts through activation in insulin discharge from the pancreas, but it does not directly affect the metabolism of carbohydrate.

(d) **Drugs inhibiting glucagon release from pancreatic β-cells**
A Chinese antidiabets medication (Ke-Tang-Ling) diminished the blood glucose levels in non-insulin dependent DM patients, proposing an inhibitory impact on the release of glucagons from the pancreas.

(e) **Drug reducing absorption of glucose from gastrointestinal tract**
Gower or *Cyamopsis tetragonolobus* has been demonstrated to diminish postprandial hyperglycemia in diabetic subjects. Oral treatment with alcoholic extract of tulsi (*Ocimum sanctum*) leaves decreased glycemia in normal, glucose-fed hyperglycemic and streptozotocin-treated diabetic rats.

(f) **Drugs inhibiting the activity of aldose reductase enzyme**
Paeonia latiflora (roots), *Glycyrrhiza glabra* (roots and rhizomes), *Atractyloides lancea* (rhizomes) and *Cinnamomum cortex* (bark) hindered the rat lens activities of aldose reductase enzyme.

(g) **Drugs enhancing the utilization of glucose**
Zingiber officinale or ginger, *Cyamopsis tetragonolobus* or Gowar and *Grewia asiatica* or phalsa are demonstrated to generate hypoglycaemia by adjusting the usage of glucose.

(h) **Drugs that decreases the lactic dehydrogenase and γ-glutamyl transpeptidase enzymes**

The extracts of *Lythrum salicaria* were tested for their hypoglycemic action in animals and it was observed that the ether extracts of the flowers and stem showed considerable hypoglycemic action in rats subjected to glucose and epinephrine-treated hyperglycemia.

(i) **Drugs suppressing the enzymes of glycogen metabolism**

The effect of *Bryonia alba* on the various enzymes of glycogen metabolism has been analyzed, and it was observed that its phytoconstituents exerted an immediate effect on glycogen phosphorylase, hexokinase and phosphoprotein phosphatase in the muscles and liver tissues of rats previously treated with alloxan.

(j) **Drugs enhancing the activity of glyoxalase-1-enzyme in liver**

The activity of the enzyme glyoxalase-1-enzyme present in the liver of diabetic rats decreases which leads to increased blood glucose levels and worsening of other biochemical considerations in the diabetic rats. When the powdered seed of *Trigonella foenumgraecum* were administered to these rats, a reversal in the activity of glyoxalase-1-enzyme to normal control values and restoration of the other biochemical parameters was observed.

(k) **Drugs enhancing the glucose uptake in the lipocytes**

Bellidifolin, obtained from *Swertia japonica*, was observed to be a powerful hypoglycemic drug in streptozotocin-induced diabetic animals. It activated the glucose uptake in the Rat 1 fibroblasts expressing the receptors of human insulin. Both bellidifolin and insulin significantly activated the uptake activity of 2-deoxy-D-glucose in the fibroblast cells.

(l) **Drugs hindering the glucose-6-phosphate arrangement**

Aqueous extract of the leaves of *Bauhinia megalandra* causes a ~ 50% reduction in the glucose generation, from the fructose or lactate of the liver slices of rats which were incubated in the Krebs-ringer buffer enhanced with oleate saturated albumin.

(m) **Drugs enhancing the levels of creatine kinase in tissues**

When compared with the normal control animals, activities of creatine kinase enzyme were found to be significantly reduced in heart, liver and skeletal muscle of the experimental-induced diabetes in rats. Treatment with *Trigonella foenumgraecum* restored these levels back to normal.

(n) **Drugs possessing free radical scavenging properties**

Juice of *Momordica charantia* fruit was observed to be a strong scavenger of hydroxyl and superoxide free radicals. Since these free radicals are involved in diabetes, the antidiabetic activity of *M. charantia* has been arbitrated via this mechanism. The effectiveness of *E. jambolana, M. charantia, M. pruriens* and *T. cordifolia* was accounted in the impediment of diabetic cataract.

Table 1 is a list of some plants scientifically proven to possess antidiabetic properties.

Table 1 List of some anti-diabetic plants with their therapeutic actions

Sl. no.	Parts used, botanical and traditional name	Antidiabetic and other important activities	References
1.	Aerial roots and stem bark of *Ficus bengalensis*, also known as banyan[a], bur[b] (Family—Moraceae)	Antioxidant and hypoglycemic action along with concentration dependant suppression of human α-amylase	Augusti et al. (1994), Ponnusamy et al. (2010), Singh et al. (2009)
2.	Bark of *Swetenia mahagoni* also known as West Indian mahogany[a] (Family—Meliaceae)	Antioxidant and antidiabetic activity	Panda et al. (2010)
3.	Flowers and peel of fruit of *Punica granatum* also known as Anar[b] (Family—Lythraceae)	Anti-hyperglycemic, and antioxidant effect, suppression of TNF-α stimulated release of free fatty acids and decreased TNF-α inhibition of secretion of adinopectin, avert diabetes via binding with the PPAR-γ receptor, generation of nitric oxide and suppression of α-glucosidase activity	Babish et al. (2010), Huang et al. (2005), Jafri et al. (2000), Katz et al. (2007), Li et al. (2005), Parmar and Kar (2008)
4.	Fruits of *Capparis spinosa* also known as caper bush[a] (Family—Capparaceae)	Antihyperglycemic action	Devasagayam et al. (1996)
5.	Fruits of *Emblica officinalis* also known as dhatriphala[b] or amla[b] (Family—Euphorbiaceae)	Antibacterial, antioxidant, antiulcerogenic, gastroprotective, hepatoprotective, and hypoglycemic activity	Krishnaveni and Mirunalini (2010), Bhattacharya et al. (1999), Devasagayam et al. (1995)
6.	Fruits of *Juniperus chinensis* also known as Chinese juniper[a] (Family—Cupressaceae)	Hypolipidemic and antihyperglycemic action	Ju et al. (2008)
7.	Grains of *Paspalum srobiculatum* also known as kodo millet[a], koda millet[a] or kodra millet[a] (Family—Poaceae)	Antihyperglycemic and hypolipidemic activities along with elevation in serum insulin levels	Jain et al. (2010)
8.	Leaves and flowers of *Cassia auriculata* also known as avaram senna[a], ranawara[b] or avaram[b] (Family—Caesalpiniaceae)	Hypolipidemic, antihyperglycemic and α-glucosidase inhibitory activity	Abesundara et al. (2004), Gupta et al. (2009)

(continued)

Table 1 (continued)

Sl. no.	Parts used, botanical and traditional name	Antidiabetic and other important activities	References
9.	Leaves and gel of *Aloe vera* also known as true or medicinal aloe[#] (Family—Liliaceae)	Anti-inflammatory, antioxidant, hypoglycemic, and hypolipidemic activity; elevation in glucose tolerance; activation of synthesis and/or secretion of insulin from the β-cells; increased insulin transport	Ajabnoor (1990), Al-Awadi and Gumaa (1987), Chen et al. (2009), Davis and Maro (1989), Rajasekaran et al. (2005), Rajasekaran et al. (2006)
10.	Leaves of *Bauhinia forficate* also known as pata de vaca[b] (Family—Fabaceae)	Antioxidant and hypoglycemic effect; used in averting complications of diabetes	Khalil et al. (2008), de Sousa et al. (2004)
11.	Leaves of *Gynostemma pentaphyllum* also known as jiaogulan (Family—Cucurbitaceae)	Change in the activities of different glucose metabolic enzymes	Yeo et al. (2008)
12.	Leaves of *Lagerstroemia speciosa* also known as banabá plant of philippines[a], giant crape-myrtle[a], pride of India[a], queen's crape-myrtle[a], (Family—Lythraceae)	α-glucosidase inhibition	Hou et al. (2009)
13.	Leaves of *Morus alba* also known as Mulberry[a], white mulberry[a] (Family—Moraceae)	Suppresion of α-glucosidase, maltase, sucrose enzyme, and porcine pancreatic α-amylase action with decreased starch hydrolysis	Sudha et al. (2011)
14.	Leaves of *Persea americana* also known as alligator pear[a] or avocado[a] (Family—Lauraceae)	Hypocholesterolemic and hypoglycemic action	Brai et al. (2007)
15.	Rhizomes of *Hedychium spicatum*, also known as sandharlika[b], spiked ginger lily[a] (Family—Zingiberaceae)	Antihyperglycemic action, inhibitory effect on α-glucosidase activity	Reddy et al. (2009)
16.	Roots and sprouts of *Raphanus sativus* also known as Japanese radish[a] (Family—Brassicaceae)	Hypoglycemic activity with enhanced lipid metabolism	Shukla et al. (2011), Taniguchi et al. (2006), Taniguchi et al. (2007)

(continued)

Table 1 (continued)

Sl. no.	Parts used, botanical and traditional name	Antidiabetic and other important activities	References
17.	Roots of *Acoru scalamus* also known as calamus[a] or sweet flag[a] (Family—Araceae)	Hypoglycemia; insulin sensitizing activity; elevation of glucose utilization; reduction in triglycerides and free fatty acid levels	Wu et al. (2009)
18.	Roots of *Berberis aristata* also known as pepperidge bushes[a] or barberries[a] (Family—Berberidaceae)	Antioxidant and antihyperglycemic activity	Singh and Kakkar (2009)
19.	Roots of *Paeonia lactiflora* also known as Chinese peony[a] (Family—Paeoniaceae)	Suppression of PTP1B	Baumgartner et al. (2010)
20.	Roots of *Polygonatum odoratum* also known as scented solomaon's-seal[a] or angular soloman's-seal[a] (Family—Ruscaceae)	Insulin sensitizing activity	Zhang et al. (2010)
21.	Roots of *Withania somnifera* also known winter cherry[a] or Ashvagandha[a] (Family—Solanaceae)	Antioxidant, diuretic, hypoglycemic and hypocholesterolemic activities, reduces metformin-induced hypothyroidism	Andallu and Radhika (2000); Jatwa and Kar (2009)
22.	Roots, stems, leafs, and fruits of *Vaccinium angustifolium* also knwn as Canadian low bush blueberry[a] (Family—Ericaceae)	Increased glucose transport, along with increase in glucose stimulated insulin secretion, decreased apoptosis, increased β-cells proliferation	Martineau et al. (2006)
23.	Seeds of *Acacia arabica* also known as guar gum[a], gum arabic[a], gum arabic[a] (Family—Leguminosae)	Activation of β-cells of pancreas for insulin release; hypoglycemia	Wadood et al. (1989)
24.	Seeds of *Linum usitatissimum* also known as flax[a], common flax[a] or linseed[a] (Family—Linaceae)	Inhibitory effect on pancreatic amylase	Sudha et al. (2011)

(continued)

Table 1 (continued)

Sl. no.	Parts used, botanical and traditional name	Antidiabetic and other important activities	References
25.	Seeds of *Phaseolus vulgaris* also known as white kidney bean[a], hulga[b], common bean[a] (Family—Fabaceae)	Hypolipidemic, antioxidant, hypoglycemic action, with inhibition of α-amylase activity; increased adinopectin synthesis; reduced TNF-α levels and alternation in the insulin receptor levels and GLU-4 mRNA in the skeletal muscle	Babish et al. (2010); Knott et al. (1992); Pari and Venkateswaran (2004)
26.	Seeds of *Swietenia macrophylla* also known as big leaf mahogany[a] (Family—Meliaceae)	Hypolipidemic and hypoglycemic activity	Dewanjee et al. (2009)
27.	Seeds of *Syagrus romanzoffiana* also known as coco palm[a] or Queen palm[a] (Family—Arecaceae)	Suppression of α-glucosidase activity and reduction in post-prandial blood glucose levels	Lam et al. (2008)
28.	Seeds of *Syzygium alternifolium* also known as Shahajire[a] (Family—Myrtaceae)	Hypoglycemic and antihyperglycemic activity	Rao and Rao (2001)
29.	Shoots of *Sutherlandia frutescens* also known as balloon grass or cancer bush[a] (Family—Fabaceae)	Anti-inflammatory and hypoglycaemic activity, reinstates the insulin levels and decreases intestinal glucose uptake	Chadwick et al. (2007), Ojewole and Analgesic (2004)
30.	Stem bark of *Sclerocarya birrea* also known as Murula[b] (Family—Anacardiacease)	Hypoglycemia, enhanced plasma insulin level; decreased weight loss; hypocholesterolemia, cardioprotection renoprotection with anticonvulsant, antidiabetic, antidiarrhoeal, antihypertensive, anti-inflammatory, antimicrobial, antinociceptive, antioxidant and antiplasmodial activities	Dimo et al. (2007); Ojewole et al. (2010); Ojewole (2003)
31.	Stems of *Dendrobium chrysotoxum*also known as fried-egg orchid[a] or golden-bow dendrobium[a] (Family—Orchidaceae)	Antihyperglycemic and antioxidant action	Zhao et al. (2007)
32.	Whole plant of *Tridax procumbens* also known as Dhaman grass[a] (Family—Asteraceae)	Antihyperglycemic and hypoglycaemic activity; correction in body weight	Pareek et al. (2009)

[a]English common name
[b]Regional name

7 Conclusion

Many different plants have been used individually or in combinations in various traditional preparations for treatment of diabetes and its complications. One of the major problems with this herbal formulation is that the active ingredients are not well defined. It is important to know the active component and their molecular interaction, which will help to analyze therapeutic efficacy of the product and also to standardize the product.

References

Abesundara, K. J. M., Matsui, T., & Matsumoto, K. (2004). α-Glucosidase inhibitory activity of some Sri Lanka plant extracts, one of which, *Cassia auriculata*, exerts a strong antihyperglycemic effect in rats comparable to the therapeutic drug acarbose. *Journal of Agricultural and Food Chemistry, 52*(9), 2541–2545.

Ahmed, A. M. (2002). History of diabetes mellitus. *Saudi Medical Journal, 23*(4), 373–378.

Ajabnoor, M. A. (1990). Effect of aloes on blood glucose levels in normal and alloxan diabetic mice. *Journal of Ethnopharmacology, 28*(2), 215–220.

Al-Awadi, F. M., & Gumaa, K. A. (1987). Studies on the activity of individual plants of an antidiabetic plant mixture. *Acta Diabetologica, 24*(1), 37–41.

Alberti, K., George, M. M., Zimmet, P., & Shaw, J. (2005). The metabolic syndrome: A new worldwide definition. *The Lancet, 366*(9491), 1059.

Ali, M. K., Weber, M. B., & Narayan, K. M. (2010). The global burden of diabetes. In *Textbook of diabetes* (4th ed., pp. 69–84). Wiley-Blackwell.

Andallu, B., & Radhika, B. (2000). Hypoglycemic, diuretic and hypocholesterolemic effect of winter cherry (Withania somnifera, Dunal) root. *Indian Journal of Experimental Biology, 38*(6), 607–609.

Augusti, K. T., Daniel, R. S., Cherian, S., Sheela, C. G., & Nair, C. R. (1994). Effect of leucoperalgonin derivative from Ficus benagalnesis Linn. on diabetic dogs. *Indian Journal of Medical Research, 99*, 82–86.

Azevedo, M., & Alla, S. (2008). Diabetes in sub-saharan Africa: Kenya, Mali, Mozambique, Nigeria, South Africa and Zambia. *International Journal of Diabetes in Developing Countries, 28*(4), 101.

Babish, J. G., Pacioretty, L. M., Bland, J. S., Minich, D. M., Hu, J., & Tripp, M. L. (2010). Antidiabetic screening of commercial botanical products in 3T3-L1 adipocytes and db/db mice. *Journal of Medicinal Food, 13*(3), 535–547.

Barnett, A. (2006). DPP-4 inhibitors and their potential role in the management of type 2 diabetes. *International Journal of Clinical Practice, 60*(11), 1454–1470.

Baumgartner, R. R., Steinmann, D., Heiss, E. H., Atanasov, A. G., Ganzera, M., Stuppner, H., & Dirsch, V. M. (2010). Bioactivity-guided isolation of 1, 2, 3, 4, 6-penta-O-galloyl-D--glucopyranose from *Paeonia lactiflora* roots as a PTP1B inhibitor. *Journal of Natural Products, 73*(9), 1578–1581.

Bhattacharya, A., Chatterjee, A., Ghosal, S., & Bhattacharya, S. K. (1999). Antioxidant activity of active tannoid principles of *Emblica officinalis* (amla). *The Journal of Experimental Biology, 37*, 676–680.

Blicklé, J. F. (2006). Meglitinide analogues: A review of clinical data focused on recent trials. *Diabetes & Metabolism, 32*(2), 113–120.

Boffetta, P., McLerran, D., Yu, C., Inoue, M., Sinha, R., & He, J. (2011). Prakash Chandra Gupta et al. body mass index and diabetes in Asia: A cross-sectional pooled analysis of 900,000 individuals in the Asia cohort consortium. *PLoS One, 6*(6), e19930.

Brai, B. I. C., Odetola, A. A., & Agomo, P. U. (2007). Hypoglycemic and hypocholesterolemic potential of *Persea americana* leaf extracts. *Journal of Medicinal Food, 10*(2), 356–360.

Burge, M. R., & Schade, D. S. (1997). Insulins. *Endocrinology and Metabolism Clinics of North America, 26*(3), 575–598.

Camastra, S., Bonora, E., Del Prato, S., Rett, K., Weck, M., & Ferrannini, E. (1999). Effect of obesity and insulin resistance on resting and glucose-induced thermogenesis in man. *International Journal of Obesity & Related Metabolic Disorders, 23*, 12.

Cameron, C. G., & Bennett, H. A. (2009). Cost-effectiveness of insulin analogues for diabetes mellitus. *Canadian Medical Association Journal, 180*(4), 400–407.

Centers for Disease Control and Prevention. (2011). *National diabetes fact sheet: National estimates and general information on diabetes and prediabetes in the United States, 2011*. Atlanta, GA: US Department of Health and Human Services, Centers for Disease Control and Prevention., 201(1).

Chadwick, W. A., Roux, S., van de Venter, M., Louw, J., & Oelofsen, W. (2007). Antidiabetic effects of Sutherlandia frutescens in Wistar rats fed a diabetogenic diet. *Journal of Ethnopharmacology, 109*(1), 121–127.

Chamnan, P., Simmons, R. K., Forouhi, N. G., Luben, R. N., Khaw, K.-T., Wareham, N. J., & Griffin, S. J. (2011). Incidence of type 2 diabetes using proposed HbA1c diagnostic criteria in the European prospective investigation of cancer–Norfolk cohort. *Diabetes Care, 34*(4), 950–956.

Chang, C. L. T., Lin, Y., Bartolome, A. P., Chen, Y.-C., Chiu, S.-C., & Yang, W.-C. (2013). Herbal therapies for type 2 diabetes mellitus: Chemistry, biology, and potential application of selected plants and compounds. *Evidence-Based Complementary and Alternative Medicine*.

Chen, L., Magliano, D. J., & Zimmet, P. Z. (2012). The worldwide epidemiology of type 2 diabetes mellitus—Present and future perspectives. *Nature Reviews Endocrinology, 8*(4), 228–236.

Chen, W., Lu, Z., Viljoen, A., & Hamman, J. (2009). Intestinal drug transport enhancement by *Aloe vera*. *Planta Medica, 75*(6), 587–595.

Chiniwala, N., & Jabbour, S. (2011). Management of diabetes mellitus in the elderly. *Current Opinion in Endocrinology, Diabetes and Obesity, 18*(2), 148–152.

Chopra, R. N. (1953). *A review of work on Indian medicinal plants*. New Delhi.

Cox, M. E., & David, E. (2009). Tests for screening and diagnosis of type 2 diabetes. *Clinical Diabetes, 27*(4), 132–138.

Davis, R. H., & Maro, N. P. (1989). Aloe vera and gibberellin. Anti-inflammatory activity in diabetes. *Journal of the American Podiatric Medical Association, 79*(1), 24–26.

De Sousa, E., Zanatta, L., Seifriz, I., Creczynski-Pasa, T. B., Pizzolatti, M. G., Szpoganicz, B., & Silva, F. R. M. B. (2004). Hypoglycemic effect and antioxidant potential of kaempferol-3, 7-O-(α)-dirhamnoside from Bauhinia f orficata Leaves. *Journal of Natural Products, 67*(5), 829–832.

Devasagayam, T. P. A., Kamat, J. P., Mohan, H., & Kesavan, P. C. (1996). Caffeine as an antioxidant: Inhibition of lipid peroxidation induced by reactive oxygen species. *Biochimica et Biophysica Acta (BBA)-Biomembranes, 1282*(1), 63–70.

Devasagayam, T. P. A., Subramanian, M., Singh, B. B., Ramanathan, R., & Das, N. P. (1995). Protection of plasmid pBR322 DNA by flavonoids against single-strand breaks induced by singlet molecular oxygen. *Journal of Photochemistry and Photobiology B: Biology, 30*(2), 97–103.

Dewanjee, S., Maiti, A., Das, A. K., Mandal, S. C., & Dey, S. P. (2009). Swietenine: A potential oral hypoglycemic from *Swietenia macrophylla* seed. *Fitoterapia, 80*(4), 249–251.

Dimo, T., Rakotonirina, S. V., Tan, P. V., Azay, J., Dongo, E., Kamtchouing, P., & Cros, G. (2007). Effect of *Sclerocarya birrea* (Anacardiaceae) stem bark methylene chloride/methanol extract on streptozotocin-diabetic rats. *Journal of Ethnopharmacology, 110*(3), 434–438.

Fagot-Campagna, A., Pettitt, D. J., Engelgau, M. M., Burrows, N. R., Geiss, L. S., Valdez, R., Beckles, G. L. A., et al. (2000). Type 2 diabetes among North adolescents: An epidemiologic health perspective. *The Journal of Pediatrics, 136*(5), 664–672.

Fuhlendorff, J., Rorsman, P., Kofod, H., Brand, C. L., Rolin, B., MacKay, P., Shymko, R., & Carr, R. D. (1998). Stimulation of insulin release by repaglinide and glibenclamide involves both common and distinct processes. *Diabetes, 47*(3), 345–351.

Fujioka, K. (2007). Pathophysiology of Type 2 Diabetes and the role of incretin hormones and beta-cell dysfunction. *Journal of the American Academy of Physician Assistants, 20*(12), 3–8.

Garcia-Roves, P. M. (2011). Mitochondrial pathophysiology and type 2 diabetes mellitus (Retraction of vol 117, pg 177, 2011). *Archives of Physiology and Biochemistry, 117*(5), 283–283.

Gupta, S., Sharma, S. B., Bansal, S. K., & Prabhu, K. M. (2009). Antihyperglycemic and hypolipidemic activity of aqueous extract of *Cassia auriculata* L. leaves in experimental diabetes. *Journal of Ethnopharmacology, 123*(3), 499–503.

Harris, M. I., Klein, R., Welborn, T. A., & Knuiman, M. W. (1992). Onset of NIDDM occurs at least 4–7 yr before clinical diagnosis. *Diabetes Care, 15*(7), 815–819.

Hou, W., Li, Y., Zhang, Q., Wei, X., Peng, A., Chen, L., & Wei, Y. (2009). Triterpene acids isolated from Lagerstroemia speciosa leaves as α-glucosidase inhibitors. *Phytotherapy Research, 23*(5), 614–618.

Hu, F. B., Manson, J. A. E., Stampfer, M. J., Colditz, G., Liu, S., Solomon, C. G., & Willett, W. C. (2001). Diet, lifestyle, and the risk of type 2 diabetes mellitus in women. *New England Journal of Medicine, 345*(11), 790–797.

Huang, C., Kim, Y., Caramori, M. L. A., Fish, A. J., Rich, S. S., Miller, M. E., Russell, G. B., & Mauer, M. (2002). Cellular basis of diabetic nephropathy. *Diabetes, 51*(12), 3577–3581.

Huang, T. H. W., Peng, G., Kota, B. P., Li, G. Q., Yamahara, J., Roufogalis, B. D., & Li, Y. (2005). Anti-diabetic action of *Punica granatum* flower extract: Activation of PPAR-γ and identification of an active component. *Toxicology and Applied Pharmacology, 207*(2), 160–169.

International Expert Committee. (2009). International Expert Committee report on the role of the A1C assay in the diagnosis of diabetes. *Diabetes Care, 32*(7), 1327–1334.

Jack, L., Jr., Boseman, L., & Vinicor, F. (2004). Aging Americans and diabetes. A public health and clinical response. *Geriatrics, 59*(4), 14–17.

Jafri, M. A., Aslam, M., Javed, K., & Singh, S. (2000). Effect of Punica granatum Linn.(flowers) on blood glucose level in normal and alloxan-induced diabetic rats. *Journal of Ethnopharmacology, 70*(3), 309–314.

Jain, S., Bhatia, G., Barik, R., Kumar, P., Jain, A., & Dixit, V. K. (2010). Antidiabetic activity of *Paspalum scrobiculatum* Linn. in alloxan induced diabetic rats. *Journal of Ethnopharmacology, 127*(2), 325–328.

Jatwa, R., & Kar, A. (2009). Amelioration of metformin-induced hypothyroidism by Withania somnifera and *Bauhinia purpurea* extracts in Type 2 diabetic mice. *Phytotherapy Research, 23*(8), 1140–1145.

Ju, J. B., Kim, J. S., Choi, C. W., Lee, H. K., Oh, T.-K., & Kim, S. C. (2008). Comparison between ethanolic and aqueous extracts from Chinese juniper berries for hypoglycaemic and hypolipidemic effects in alloxan-induced diabetic rats. *Journal of Ethnopharmacology, 115*(1), 110–115.

Kahn, C. R. (1994). Insulin action, diabetogenes, and the cause of type II diabetes. *Diabetes, 43*(8), 1066–1085.

Katz, S. R., Newman, R. A., & Lansky, E. P. (2007). Punica granatum: Heuristic treatment for diabetes mellitus. *Journal of Medicinal Food, 10*(2), 213–217.

Kawamori, R., Tajima, N., Iwamoto, Y., Kashiwagi, A., Shimamoto, K., Kaku, K., & Voglibose Ph-3 Study Group. (2009). Voglibose for prevention of type 2 diabetes mellitus: A randomised, double-blind trial in Japanese individuals with impaired glucose tolerance. *The Lancet, 373*(9675), 1607–1614.

Khalil, N. M., Pepato, M. T., & Brunetti, I. L. (2008). Free radical scavenging profile and myeloperoxidase inhibition of extracts from antidiabetic plants: *Bauhinia forficata* and *Cissus sicyoides*. *Biological Research, 41*(2), 165–171.
Knott, R. M., Grant, G., Bardocz, S., Pusztai, A., de Carvalho, A. F. F. U., & Hesketh, J. E. (1992). Alterations in the level of insulin receptor and GLUT-4 mRNA in skeletal muscle from rats fed a kidney bean (*Phaseolus vulgaris*) diet. *International Journal of Biochemistry, 24*(6), 897–902.
Krishnaveni, M., & Mirunalini, S. (2010). Therapeutic potential of *Phyllanthusemblica* (amla): The ayurvedic wonder. *Journal of Basic and Clinical Physiology and Pharmacology, 21*(1), 93–105.
Lam, S. H., Chen, J. M., Kang, C. J., Chen, C. H., Lee, S. S. (2008). alpha-Glucosidase inhibitors from the seeds of Syagrus romanzoffiana. *Phytochemistry, 69*(5), 1173–1178.
Lemhadri, A., Hajji, L., Michel, J.-B., & Eddouks, M. (2006). Cholesterol and triglycerides lowering activities of caraway fruits in normal and streptozotocin diabetic rats. *Journal of Ethnopharmacology, 106*(3), 321–326.
Li, Y., Wen, S., Kota, B. P., Peng, G., Li, G. Q., Yamahara, J., & Roufogalis, B. D. (2005). Punica granatum flower extract, a potent α-glucosidase inhibitor, improves postprandial hyperglycemia in Zucker diabetic fatty rats. *Journal of Ethnopharmacology, 99*(2), 239–244.
Lovejoy, J. C. (2002). The influence of dietary fat on insulin resistance. *Current Diabetes Reports, 2*(5), 435–440.
Martineau, L. C., Couture, A., Spoor, D., Benhaddou-Andaloussi, A., Harris, C., Meddah, B., Leduc, C., et al. (2006). Anti-diabetic properties of the Canadian lowbush blueberry Vaccinium angustifolium Ait. *Phytomedicine, 13*(9), 612–623.
Mayfield, J. A., & White, R. D. (2004). Insulin therapy for type 2 diabetes: Rescue, augmentation, and replacement of beta-cell function. *American Family Physician, 70*, 489–516.
Mbanya, J. C. (2009). The burden of type 2 diabetes mellitus in the African diaspora. *Available atwww. medscape. com/view article/560718_2.*
McCarthy, M. I. (2010). Genomics, type 2 diabetes, and obesity. *New England Journal of Medicine, 363*(24), 2339–2350.
Nauck, M. A., Vilsbøll, T., Gallwitz, B., Garber, A., & Madsbad, S. (2009). Incretin-based therapies. *Diabetes Care, 32*(Suppl. 2), S223–S231.
Ojewole, J. A., & Analgesic, O. (2004). Antiinflammatory and hypoglycemic effects of *Sutherlandia frutescens* R. BR.(variety Incana E. MEY.) (Fabaceae) shoot aqueous extract. *Methods and Findings in Experimental and Clinical Pharmacology, 26*(6), 409–416.
Ojewole, J. A. O. (2003). Hypoglycemic effect of Sclerocarya birrea {(A. Rich.) Hochst.} [Anacardiaceae] stem-bark aqueous extract in rats. *Phytomedicine, 10*(8), 675–681.
Ojewole, J. A. O., Mawoza, T., Chiwororo, W. D. H., & Owira, P. M. O. (2010). Sclerocarya birrea (A. Rich) Hochst.['Marula'](Anacardiaceae): A review of its phytochemistry, pharmacology and toxicology and its ethnomedicinal uses. *Phytotherapy Research, 24*(5), 633–639.
Pareek, H., Sharma, S., Khajja, B. S., Jain, K., & Jain, G. C. (2009). Evaluation of hypoglycemic and anti-hyperglycemic potential of *Tridax procumbens* (Linn.). *BMC Complementary and Alternative Medicine, 9*(1), 48.
Pari, L., & Venkateswaran, S. (2004). Protective role of Phaseolus vulgaris on changes in the fatty acid composition in experimental diabetes. *Journal of Medicinal Food, 7*(2), 204–209.
Panda, S. P., Haldar, P. K., Bera, S., Adhikary, S., Kandar, C. C. (2010). Antidiabetic and antioxidant activity of Swietenia mahagoni in streptozotocin-induced diabetic rats. *Pharmaceutical Biology, 48*(9), 974–979.
Parmar, H. S., & Kar, A. (2008). Possible amelioration of atherogenic diet induced dyslipidemia, hypothyroidism and hyperglycemia by the peel extracts of *Mangifera indica*, Cucumis melo and Citrullus vulgaris fruits in rats. *Biofactors, 33*(1), 13–24.
Patlak, M. (2002). New weapons to combat an ancient disease: Treating diabetes. *The FASEB Journal, 16*(14), 1853e.
Zimmet, P., Alberti, K. G. M. M., & Shaw, J. (2001). Global and societal implications of the diabetes epidemic, Nature., *414*(6865), 782–787.

Ponnusamy, S., Ravindran, R., Zinjarde, S., Bhargava, S., & Kumar, A. R. (2010). Evaluation of traditional Indian antidiabetic medicinal plants for human pancreatic amylase inhibitory effect in vitro. *Evidence-Based Complementary and Alternative Medicine, 2011*.

Pratley, R. E., & Salsali, A. (2007). Inhibition of DPP-4: A new therapeutic approach for the treatment of type 2 diabetes. *Current Medical Research and Opinion, 23*(4), 919–931.

Rajasekaran, S., Sivagnanam, K., & Subramanian, S. (2005). Modulatory effects of *Aloe vera* leaf gel extract on oxidative stress in rats treated with streptozotocin. *Journal of Pharmacy and Pharmacology, 57*(2), 241–246.

Rajasekaran, S., Ravi, K., Sivagnanam, K., & Subramanian, S. (2006). Beneficial effects of Aloe vera leaf gel extract on lipid profile status in rats with streptozotocin diabetes. *Clinical and Experimental Pharmacology and Physiology, 33*(3), 232–237.

Rao, B. K., & Rao, C. H. A. (2001). Hypoglycemic and antihyperglycemic activity of *Syzygium alternifolium* (Wt.) Walp. seed extracts in normal and diabetic rats, Phytomedicine., *8*(2), 88–93.

Rathmann, W., & Giani, G. (2004). Global prevalence of diabetes: Estimates for the year 2000 and projections for 2030. *Diabetes Care, 27*(10), 2568–2569.

Reddy, P. P., Tiwari, A. K., Rao, R. R., Madhusudhana, K., Rao, V. R. S., Ali, A. Z., Babu, K. S., & Rao, J. M. (2009). New Labdane diterpenes as intestinal α-glucosidase inhibitor from antihyperglycemic extract of *Hedychium spicatum* (Ham. Ex Smith) rhizomes. *Bioorganic & Medicinal Chemistry Letters, 19*(9), 2562–2565.

Robertson, R. P., & Bogardus, C. (1995). Antagonist: Diabetes and insulin resistance: Philosophy, science, and the multiplier hypothesis. Discussion. *The Journal of Laboratory and Clinical Medicine, 125*(5), 560–565.

Rother, K. I. (2007). Diabetes treatment—Bridging the divide. *The New England Journal of Medicine, 356*(15), 1499.

Scheen, A. J. (2005). Drug interactions of clinical importance with antihyperglycaemic agents. *Drug Safety, 28*(7), 601–631.

Shorr, R. I., Ray, W. A., Daugherty, J. R., & Griffin, M. R. (1996). Individual sulfonylureas and serious hypoglycemia in older people. *Journal of the American Geriatrics Society, 44*(7), 751–755.

Shukla, S., Chatterji, S., Mehta, S., Rai, P. K., Singh, R. K., Yadav, D. K., & Watal, G. (2011). Antidiabetic effect of *Raphanus sativus* root juice. *Pharmaceutical Biology, 49*(1), 32–37.

Singh, J., & Kakkar, P. (2009). Antihyperglycemic and antioxidant effect of Berberis aristata root extract and its role in regulating carbohydrate metabolism in diabetic rats. *Journal of Ethnopharmacology, 123*(1), 22–26.

Singh, R. K., Mehta, S., Jaiswal, D., Rai, P. K., & Watal, G. (2009). Antidiabetic effect of *Ficus bengalensis* aerial roots in experimental animals. *Journal of Ethnopharmacology, 123*(1), 110–114.

Staa, V., Tjeerd, L. A., & Monette, J. (1997). Rates of hypoglycemia in users of sulfonylureas. *Journal of Clinical Epidemiology, 50*(6), 735–741.

Stonehouse, A. H., Darsow, T., & Maggs, D. G. (2012). Incretin-based therapies. *Journal of Diabetes, 4*(1), 55–67.

Sudha, P., Zinjarde, S. S., Bhargava, S. Y., & Kumar, A. R. (2011). Potent α-amylase inhibitory activity of Indian Ayurvedic medicinal plants. *BMC Complementary and Alternative Medicine, 11*(1), 5.

Taniguchi, H., Kobayashi-Hattori, K., Tenmyo, C., Kamei, T., Uda, Y., Sugita-Konishi, Y., Oishi, Y., & Takita, T. (2006). Effect of Japanese radish (Raphanus sativus) sprout (Kaiware-daikon) on carbohydrate and lipid metabolisms in normal and streptozotocin-induced diabetic rats. *Phytotherapy Research, 20*(4), 274–278.

Taniguchi, H., Muroi, R., Kobayashi-Hattori, K., Uda, Y., Oishi, Y., & Takita, T. (2007). Differing effects of water-soluble and fat-soluble extracts from Japanese radish (Raphanus sativus) sprouts on carbohydrate and lipid metabolism in normal and streptozotocin-induced diabetic rats. *Journal of Nutritional Science and Vitaminology, 53*(3), 261–266.

Tripathi, A. K., Bhoyar, P. K., Baheti, J. R., Biyani, D. M., Khalique, M., Kothmire, M. S., Amgaonkar, Y. M., & Bhanarkar, A. B. (2011). Herbal antidiabetics: A review. *International Journal of Research in Pharmacy and Science, 2*(1), 30–37.

Wadood, A., Wadood, A. N., & Shah, S. A. (1989). Effects of *Acacia arabica* and *Caralluma edulis* on blood glucose levels of normal and alloxan diabetic rabbits. *The Journal of the Pakistan Medical Association, 39*(8), 208–212.

Walley, A. J., Blakemore, A. I. F., & Froguel, P. (2006). Genetics of obesity and the prediction of risk for health. *Human Molecular Genetics, 15*(Suppl. 2), R124–R130.

Willi, C., Bodenmann, P., Ghali, W. A., Faris, P. D., & Cornuz, J. (2007). Active smoking and the risk of type 2 diabetes: A systematic review and meta-analysis. *Journal of the American Medical Association, 298*(22), 2654–2664.

Wu, H.-S., Zhu, D.-F., Zhou, C.-X., Feng, C.-R., Lou, Y.-J., Yang, B., & He, Q.-J. (2009). Insulin sensitizing activity of ethyl acetate fraction of L. in vitro and in vivo. *Journal of Ethnopharmacology, 123*(2), 288–292.

Yadav, M., Chatterji, S., Gupta, S. K., & Watal, G. (2014). Preliminary phytochemical screening of six medicinal plants used in traditional medicine. *International Journal of Pharmacy and Pharmaceutical Sciences, 6*(5), 539–542.

Yeo, J., Kang, Y.-J., Jeon, S.-M., Jung, U. J., Lee, M.-K., Song, H., & Choi, M.-S. (2008). Potential hypoglycemic effect of an ethanol extract of *Gynostemma pentaphyllum* in C57BL/KsJ-db/db mice. *Journal of Medicinal Food, 11*(4), 709–716.

Yoon, K.-H., Lee, J.-H., Kim, J.-W., Cho, J. H., Choi, Y.-H., Ko, S.-H., Zimmet, P., & Son, H.-Y. (2006). Epidemic obesity and type 2 diabetes in Asia. *The Lancet, 368*(9548), 1681–1688.

Zhang, H., Yang, F., Qi, J., Song, X.-C., Hu, Z.-F., Zhu, D.-N., & Yu, B.-Y. (2010). Homoisoflavonoids from the fibrous roots of *Polygonatum odoratum* with glucose uptake-stimulatory activity in 3T3-L1 adipocytes. *Journal of Natural Products, 73*(4), 548–552.

Zhao, Y., Son, Y. O., Kim, S. S., Jang, Y. S., Lee, J. C. (2007). Antioxidant and anti-hyperglycemic activity of polysaccharide isolated from Dendrobium chrysotoxum Lindl. BMB Reports, 40(5), 670-677.

Fruits and Vegetables as Sources of Functional Phytochemicals for the Prevention and Management of Obesity, Diabetes, and Cancer

Samah Ramadan and Amira Abd Allah Ibrahim

1 Introduction

Fresh or processed fruits and vegetables are important sources of phytochemicals for human consumption. So far, out of 200,000 known phytochemicals, 20,000 are present in fruits, vegetables, and cereals (Patra 2012; Altemimi et al. 2017). Phytochemicals are bioactive plant compounds that are not considered as energy supplies or essential nutrients, but they have health benefits beyond their use as macronutrients or micronutrients. Plants normally produce such low-molecular-weight ingredients to protect themselves from pests and diseases, to regulate their growth, or as pigments, essences, or odors (Perez-Vizcaino et al. 2006; Gupta et al. 2015). They are a defensive system for the plant against unfavorable environmental conditions. The more biotic and abiotic stresses in plants there are, the more secondary substances could be produced as shown in Fig. 1. Consequently, the content of phytochemicals may change with growing conditions (Reyes & Cisneros-Zevallos 2003; Isah 2019).

Tens of thousands of different plant metabolites have been identified as phytochemicals, but scientists predict that there is still a lot in food not discovered yet. Although the biggest groups of phytochemicals as isoflavones, flavonoids, or anthocyanidins are homogeneous groups, each group has different compounds, each individual compound has different chemical structures, which can be metabolized by different ways in the body, and also may have different effects on health

S. Ramadan (✉)
Botany Department, Faculty of Science, Mansoura University, Mansoura, Egypt
e-mail: samahmahmoud@mans.edu.eg

A. A. A. Ibrahim
Arid Lands Cultivation Research Institute, City of Scientific Research and Technological Applications, Universities and Research Center District,
New Borg El-Arab, Alexandria, Egypt

C. Egbuna, S. Hassan (eds.), *Dietary Phytochemicals*,
https://doi.org/10.1007/978-3-030-72999-8_8

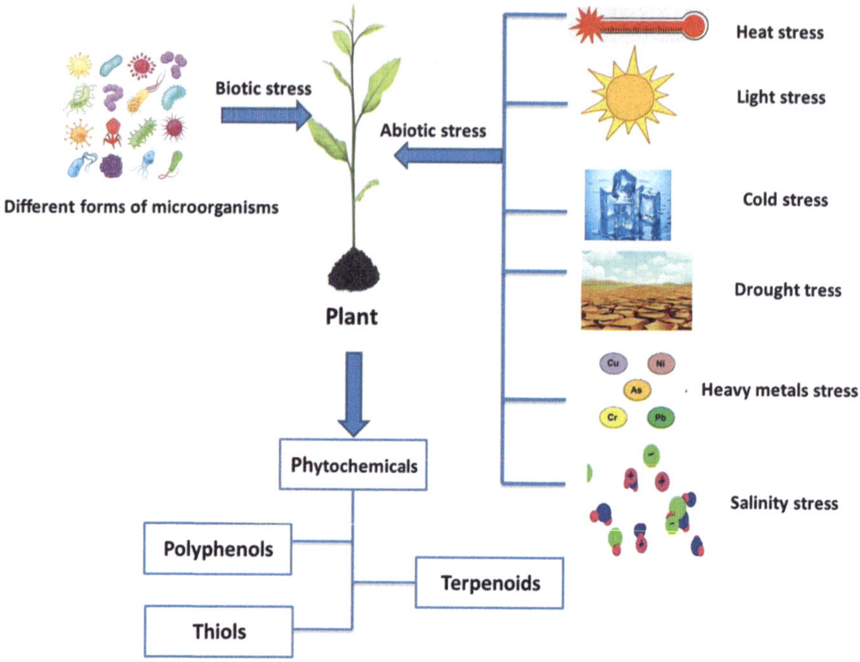

Fig. 1 A diagrammatic scheme shows the production of phytochemical compounds in the plant under biotic and abiotic stress

(Erdman Jr. et al. 2007; Kallassy 2017). Flavonoids are the largest group of phytochemicals. They are a highly diverse group, best investigated, and the best chemically identified. In fact, above 6000 flavonoids in plant human food have been defined (Arts and Hollman 2005; Tungmunnithum et al. 2018). Fruits and vegetables contain different compounds of phytochemicals such as carotenoids, phenolic compounds and ascorbic acid (Yahia and Ornelas-Paz 2010; Sancho et al. 2011; Liu 2013). Plant polyphenols exist in all diets and sources that are rich in wine, tea, vegetables and fruits; show significant antioxidant activity in vitro that can reflect significantly in clinical therapy and health systems (Wootton-Beard and Ryan 2011; Skotti et al. 2014).

Phytochemicals possess antioxidant, anti-inflammatory or antiproliferative activity, so they have many health effects (Wang et al. 2020). There is a strong correlation between the consumption of antioxidants with less risk for many diseases as cardiovascular, cancer, diabetes and hypertensive diseases and other human diseases (Kant 2004; Thaipong et al. 2006). The major types of antioxidant phytochemicals are polyphenols and carotenoids that provide most plants or food their antioxidant activities (Sun et al. 2002; Stagos 2020). The total antioxidant activity of phytochemical extracts of different plants or foods can be directly linked to its content of the total phenol. Higher antioxidant activity in fruits related to the high content of total phenols in it (Sun et al. 2002; Cai et al. 2004; Do et al. 2014). The phytochemical group that is responsible for food color is known as the carotenoids. Carotenoids give the food the yellow, orange, and red colors. The human body needs a certain main type of carotenoids represented in α-carotene, β-carotene, lycopene, lutein, and cryptoxanthin. Fruits and vegetables are the vital and essential natural sources of carotenoids in the human diet (Zhang et al. 2015).

Many well-identified phytochemicals are thought to work against cancer include soy genistein, curcumin from the roots of turmeric, resveratrol from grapes, and lycopene from tomatoes, yet there are various others found in vegetables, fruits, and herbs. In addition, great interest has been expressed around the world about the ability of phytochemicals to fight obesity (Rayalam et al. 2008; Paula Santos et al. 2010; Williams et al. 2013). Some studies recommended the use of polyphenols as anti-obesity medications since they can alter fat cells' life cycle (Rayalam et al. 2008; Nabavi et al. 2015).

Since diabetes proceeds from resistance to, or relative deficiency in, blood sugar regulating hormonal insulin (Alwan 2011). Many results conducted that the presence of different antioxidants may significantly contribute to increasing or decreasing sensitivity to insulin (Bryans et al. 2007). Considering the previously mentioned reviews, in this chapter, an overview of bioactive phytochemicals in fruits and vegetables as a treatment for cancer, obesity and diabetes are discussed in detail.

2 Classification of Phytochemicals

The classification of phytochemicals remains controversial. There are three main groups of phytochemicals: polyphenols, terpenoids and thiols (Fig. 2).

2.1 Polyphenols

The most common phytochemicals in human nutrition are polyphenols that consist of multi different chemical compounds with different chemical structures, varying from simple monomers to high molecular weight polymers (Del Rio et al. 2013). Polyphenols are classified into three groups; flavonoids, phenolic acids, and other non-flavonoid polyphenols compounds as lignans, and stilbenes (Manach et al. 2004; Kumar et al. 2019). Healthy benefits' effects of polyphenols are associated with regular daily intake in the human diet and bioavailability. Polyphenols give the food; color, flavor, bitter, and astringent, besides, they maintain stability protecting food from oxidation. The consumption of polyphenols has a profound effect on protection against a lot of chronic and persistent diseases such as cerebrovascular diseases, cardiovascular diseases, cancers, aging, neurodegenerative diseases, and diabetes, that had been explained in a lot of epidemiological studies and related meta-investigations (He and Sun 2016).

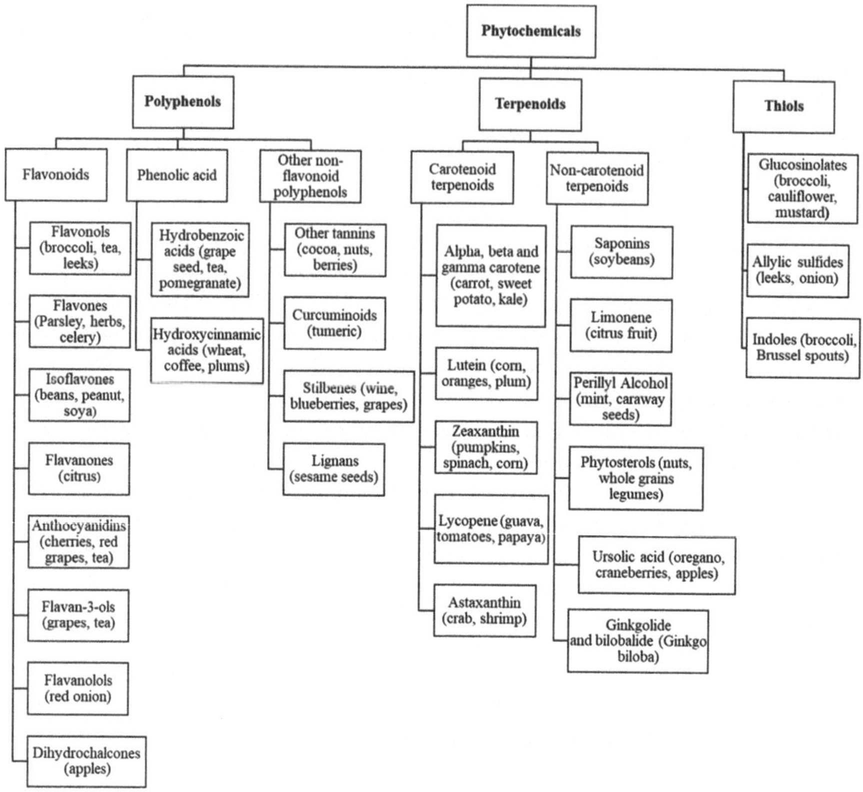

Fig. 2 Classification of phytochemical and their plants of origin (Thomas et al. 2015)

2.2 Terpenoids

Terpenoids; isoprenoids are a wide and diverse group of natural chemicals that originate from terpenes. Most of them are multi-cycled structures with functional groups of oxygen. They include non-carotenoid and carotenoid terpenoids. Carotenoids are the most popular and they are responsible for red, orange, and yellow colors in food. In addition to antioxidants, carotenoids protect people from certain types of cancer, coronary illness and ciliary, or age-related macular degeneration (AMD) (Yahia 2010; Yahia and Ornelas-Paz 2010; Meléndez-Martínez et al. 2019).

2.3 Thiols

Thiols are sulfhydryl containing compounds. They are one of the most beneficial antioxidants that protect cells from any kind of oxidative modification. Thiols include the glucosinolates, allyl sulfides and non-sulfur containing indoles (Sen and Packer 2000; Wodek 2002; Ulrich and Jakob 2019).

3 Health Benefits and Nutritional Value of Certain Fruits and Vegetables

3.1 Tomatoes

Lycopene pigment in tomato is the main phytochemical compound responsible for the dark red color in fruits and the important indicator of ripeness. It can be used for medicinal therapy, prevent cancer and heart disease due to its high antioxidant activity (Shi and Maguer 2000; Story et al. 2010). Despite the absence of provitamin A activity in lycopene, it shows a constant rate of body cooling, with oxygen only almost twice that of β-carotene. Several studies showed that tomato has antioxidant activity (Martínez-Valverde et al. 2002; Frusciante et al. 2007; Pirayesh Islamian and Mehrali 2015).

3.2 Grapes and Berries

Fresh fruits and juices or jam from the grape are the best natural sources of phenolic compounds with antioxidant activity (Frankel and Meyer 1998; Arnous et al. 2001; Kedage et al. 2007; Mulero et al. 2010; El-Kassas et al. 2014). High quality of grapes and wine depends on flavonoids, stilbenes and phenolic acid presence (Downey et al. 2006; Fernandes et al. 2017). It has been proved that different layers and huge amounts of phenolic compounds in Vitis vinifera play a significant role in people's health (Waterhouse 2002; Baydar et al. 2004; Baydar et al. 2006; Cosme et al. 2018). Natural products made from grapes have a high antioxidant activity (Alonso et al. 2002; Kammerer et al. 2004). Berries comprise a variety of phytochemical compounds such as phenolic acids, ascorbic acids, carotenoids and flavonoids. Mulberry polyphenols can exert potential cellular anti-cancer effects that are triggered by extracting redox active minerals or by removing reactive oxygen species (ROS) directly from cancer cells. Blackberry phenol (urine phenol) can interact with cellular receptors or enzymes, which changes the oxidation state in the cell and activates oxidation-related reactions (Forman et al. 2002; Scalbert et al. 2005).

3.3 Nuts

Nuts (hazelnuts, cashews, almonds, Brazilian nuts, macadamia, pistachios, nuts, walnuts and, pine nuts) and legumes (Baru and peanuts) are rich in unsaturated fatty acids, minerals, L-arginine, and phenolic compounds beside phytosterol, which have always been shown to have impacts on protecting the heart (Ros 2015; Souza et al. 2015). Some nuts are part of nutritional food with a high percentage of antioxidants. The highest levels of antioxidant compounds present in walnuts and chestnuts. Every 100 g of walnut contains more than 20 mmol of antioxidants, besides different varieties of walnuts showed the highest content of total phenolic compounds (1404 mg GAE/100 g) and antioxidant capacity 191μmol TE/g (Rosales-Martínez et al. 2014). Peanuts contribute substantially to the nutritional intake of antioxidants (Blomhoff et al. 2006; Ros 2010). Sang et al. (2002) found that eight phenolic compounds in almond exhibited strong antioxidant activity. In addition, almond shells have high concentrations of four different types of flavanol glycosides, which act as energetic antioxidants (Frison and Sporns 2002). Peanuts seeds showed phenolic compounds content as 457 (mg EAG/100 g dm) and oxidant capacity 8μmol TE/g (Rosales-Martínez et al. 2014).

3.4 Citrus

The bitter taste of all citrus fruits is due to the presence of limonoids like the nomilin and limonin which contain a large scale of biological activities that can be utilized as preservative and chemical protective agents (LAM et al. 1994; Gualdani et al. 2016). Hesperidin and naringin are the main types of flavonoids in oranges and grapefruits respectively, and hesperidin is also present in mandarin, lemon, and lime. It has been shown that polymethoxy and glycosylated flavones are the characteristic flavonoids for citrus fruit only, have biological activities with a broad spectrum, such as anti-inflammatory and analgesic effect, beside anti-cancer properties (Li et al. 2009; Chen et al. 2020). The highest antioxidant activity of orange juice is due to narirutin and hesperidin (Robards et al. 1999). Lemon, sour orange, sweet orange, tangerine, and lemon seeds have stronger antioxidant activity than peeled fruit (Ignat et al. 2011).

3.5 Brassica Vegetables

There are around 100 different types of glucosinolates in all plant families, where about ten types are found in Cruciferae. Different types of well-identified hydrolytic glucosinolates and isothiocyanates have the main toxic effect in goiter and general inhibition of thyroid uptake of iodine (Possenti et al. 2016). Several studies showed the carcinogenic effect of many hydrolytic products of phytochemicals: glucosinolates, glucoraphanin, gluconasturtiin, or glucobrassicin. Broccoli extracts, Brussels sprouts, white cabbage, and cauliflower have shown wide antioxidant capacity represented in anti-peroxidation of lipids (Podsędek 2007). However, it is believed that main significant of antioxidant activity of cruciferous or brassica vegetables is not as a result of the presence of high content of glucosinolates, but it may due to the content of phenol and hydroxylated polyphenols, as found in broccoli (Plumb et al. 1997; Cartea et al. 2011). Pro-oxidants effect of lipid peroxidation was found in microsomes of Brussels sprouts, cauliflower and cabbage, that containing a specific cytochrome P450 (Podsędek 2007). The antioxidant activity of broccoli, kale, and brussels sprouts is higher than the activity of cauliflower and certain other brassica vegetables (Podsędek 2007; Köksal and Gülçin 2008; Soengas Fernández et al. 2011).

3.6 Mushrooms

Some types of mushrooms are the best natural sources of fiber, minerals and vitamins C and B, besides its moderate content of high-quality protein (Cheung et al. 2003; Elmastas et al. 2007; Valverde et al. 2015). Wide pharmaceutical uses of different species of mushrooms, used orally or injected, have a significant value in medicine and disease therapy for cancer and viral diseases (flu and polio), high cholesterol, platelet aggregation, and high blood pressure (Breene 1990; Lull et al. 2005; Chang and Wasser 2012). Many researchers reported that adding some cultivated mushrooms, especially shiitake and enokitake, daily into the diet may offer great protection against certain manifestations of cancer (Mori et al. 1987; Wasser 2002; Shamtsyan et al. 2004).

3.7 Kiwi

Kiwi is rich in lutein, a very powerful antioxidant oxycarotene (Calvo 2005; Wolber et al. 2013). Kiwi has shown great effects on cardiovascular diseases (Duttaroy 2013) and helps in the reduction of angiotensin converting enzyme (ACE) activity (Duttaroy 2016). Daily consumption of kiwi fruit has the ability to regulate the genes responsible for DNA repair and control stress (Singletary 2012; Stonehouse et al. 2013). The human consumption of kiwifruit has prospective effects on hypotension in smokers (Karlsen et al. 2013) and in women and men with normal hypertension (Svendsen et al. 2015).

3.8 Cladodes

The vitamin C content of cladodes (called Nopal in Mexico) is around 100–150 mg/kg and the β-carotenes (provitamin A) around 300μg/kg. They are used in different clinical treatments due to their medical, therapeutic, bioactive and dermatological properties. Ethanolic extract of *Opuntia ficus-indica* has important medicinal roles because of its anti-inflammatory and analgesic properties (Park et al. 2001; Feugang et al. 2006). *Opuntia ficus-indica* has preventive and curative effects on mice infected with ethanol-induced ulcers (Galati et al. 2002; Akkol et al. 2020).

3.9 Carrots

Although cultivated *Daucus carota* plant has low concentrations of antioxidants compared to other vegetables, it is considered as one of the most vital sources of β-carotene and vitamin A (Cao et al. 1996; Ramarathnam et al. 1997; Kähkönen et al. 1999; Ahmad et al. 2019). However, the antioxidant activities of β carotene and linolenic acid in carrots significantly increase and enhance by the 30-min cooking process (Gazzani et al. 1998).

3.10 Potatoes

Solanum tuberosum contains considerable amounts of antioxidants such as ascorbic acid, alpha- tocopherol and polyphenolic compounds (Andre et al. 2007; Teow et al. 2007; Pihlanto et al. 2008). Potato peel has a high content of antioxidants (Singh and Rajini 2008; Mohdaly et al. 2010). Anthocyanin pigment, pelargonidin-3-rutinoside-5-glucoside, is responsible for antioxidant activity increasing in purple-colored potato compared to other white and yellow varieties (Brown 2005; Teow et al. 2007) (Brown 2005; Lee et al. 2016). The main carotenoids in potatoes are violaxanthin, followed by lutein, antheraxanthin, and others. Eight carotenoids (such as antheraxanthin, violaxanthin, lutein and zeaxanthin) were identified as the most common in most potato varieties (Breithaupt and Bamedi 2002; Mishra et al. 2020).

3.11 Onion and Garlic

Onions and garlic have become an integral component of the daily diet of different cultures. Many societies today believe that they have medicinal properties. They have a high content of phytochemical compounds such as allyl methyl trisulfide and diallyl sulfide (Yahia 2017). These compounds inhibit proliferation and growth of cells, improve the immune system, modify the activation of cancerous substances, stimulate detoxifying enzymes and reduce the binding of the carcinogen–DNA (Milner 2001). Oxygen radical absorbance capacity (ORAC) test showed that Allium sativum (garlic) has antioxidant activity four times more than onions (Morales-Soto et al. 2014).

4 The Effect of Consuming Fruits and Vegetables on Some Diseases

4.1 Cancer

Phytonutrients and derivatives contained in plants are promising options for improving the effectiveness of treatment in cancer patients and reducing side effects. Some phytochemicals are natural, biologically active substances with considerable anti-tumor activity. The development of active cancer therapy without side effects based on phytochemicals begins with tests of plant extracts to determine any anti-cancer activity, then followed by fractionation of active phytochemicals and examination for in- vitro in vivo impacts (Choudhari et al. 2020).

Chemoprevention is well-defined as the use up of natural or pharmacological compounds to inhibit the promotion, development, or growth of a tumor (Iqbal et al. 2019). Epidemiological studies show that food rich in vegetables and fruits and theoretically in phytochemicals can decrease the cancer risk. For example, a meta-analysis of observational studies found that eating more vegetables and fruits reduces the probability that eating more vegetables and fruits can reduce lung cancer risk, even among smokers (Wang et al. 2015; Vieira et al. 2016). Eating fruits especially is associated with less colorectal cancer (Ben et al. 2015). Vegetables that are not starchy can protect against mouth, throat, larynx, esophagus, and gastric tumors. Particularly, there is a few evidence and medicinal reports for using vegetables in protection against some cancer diseases as cancers of colon, lung, ovaries, rectum, nasopharynx, endometrium and liver (Yahia 2017). Generally, fruit consumption can protect against certain types of cancers like cancers of the mouth, pharynx, larynx, esophagus, lungs and stomach (WCRF/AICR 2007). A meta-analysis showed that eating fruits instead of vegetables was in converse correlation with the risk of stomach cancer (Wang et al. 2014a, b). Regular eating of garlic is correlated with decreased nitrite levels in the stomach and decreased gastric cancer mortality (Thomson and Ali 2003; Shukla and Kalra 2007). Garcia-Closas et al. (1999) conducted a study in Spain in 354 patients with stomach cancer and 354 controls, adapted to age, gender, place of residence and hospital. Food consumption was assessed using the nutritional history questionnaire. Conclusions of this study approved that the assumption that the established clinical effect of fruits and vegetables against stomach cancer may be partly owing to the presence of flavonoids (David et al. 2016). Consuming fruits and vegetables, as well as vitamin C, carotene, and fibers would reduce exposure to colon cancer (Koushik et al. 2007). Many sterols compounds are numerous in vegetables, a study by Baskar et al. (2012) demonstrated that β sitosterol offers a chemical protective ability in 1,2-dimethylhydrazine (DMH) stimulated trial colonic preneoplastic development in mice by its extreme scavenging ability.

Some studies indicate that eating certain kinds of fruit and vegetarian foods may reduce prostate cancer risk (Campbell et al. 2004; Kirsh et al. 2007; Hori et al. 2011). Soy isoflavones and their metabolites have a vital in the prevention or treatment of prostate cancer, the epidemiological study by Hwang et al. (2009) has shown that total serum levels of soy isoflavones are correlated with a lower risk of developing this cancer. Grapefruit and orange reduce the increase and metastases of MDA-MB-435 ER- human breast cancer cells introduced into nude rats (Guthrie et al. 2000). Breast cancer is inversely associated with high citrus consumption (Zhang et al. 2009; Song and Bae 2013). The useful effects of citrus on human nutrition are not only based on the presence of vitamin C, folic acid, fiber, and carotenoids but also on the antioxidant activity of flavonoids (Yang et al. 2001). The extracts of the whole apple have an efficient role in reducing the growth rate of breast cancer in rat models (Gerhauser 2008). A lot of studies conducted that presence of definite chemical constituents or different combination among phytochemical compounds in purple grapefruit has the ability to reduce the induction of carcinogenic cells and stop working of carcinogen that causes malformation and alteration in DNA in the mammary, besides retarding successive mammary tumorigenesis (Garcia-Villalba et al. 2015).

4.2 Obesity

Obesity is a complex disease that results from inadequate control of the body's energy balance due to overeating and/or a sedentary lifestyle. It contributes to the appearance of some chronic diseases, particularly certain types of cancers (i.e., colon, breast), diabetes and cardiovascular disease (Hill 2006; Wolin et al. 2010). On the cellular level, obesity is distinguished by the increase in the number and size of (fat storage cells (adipocytes), which are different from the pre adipocytes of adipose tissue (McGregor and Choi 2011). Polyphenols are the largest class of phytochemicals groups that can be used as an anti-obesity drug, as studies indicated that polyphenols can regulate the fat cell's life cycle (Meydani and Hasan 2010; Wang et al. 2014a, b). Some studies suggested that eating food containing phenolic compounds may be beneficial in inhibiting obesity caused by consuming high concentrations of fats rich diets. The observations of these studies can be linked to the bio mechanism prevent obesity in humans by using phenolic compounds (Han et al. 2003; Aoki et al. 2007). Useful evidence on the molecular mechanisms by which quercetin affects the control of adipocyte differentiation and apoptosis was provided by Ahn et al. (2008). Quercetin has been found by several studies to provide certain protective impacts against obesity associated inflammation (Al-Fayez et al. 2006; Chuang et al. 2010). Quercetin has been shown to mitigate macrophages, inflammation markers, and resistance of insulin in human adipocytes and decrease body weight (approximately 40%) and inhibited the expression of a lipogenic, adipogenic, and inflammation-related cytokines (Seo et al. 2015). Flavonoids have wide effects on the adipogenic syndrome. These effects represented in the regulation mechanism of several pathways such as stimulation of apoptosis, inhibition transcription factors for definite adipogenic cells, induction of the AMP-activated protein kinase (AMPK) and Wnt pathways, suppression of clonal expansion interruption of the cell cycle (Lee et al. 2015).

Adipocytes derived from human mesenchymal stem cells were manipulated to study adipogenesis (Gomez-Zorita et al. 2017; Casado-Díaz et al. 2019), the results showed that all three compounds; kaempferol, hesperidin, and apigenin reduce triacylglycerol accumulation by partially activating lipolysis and, in the case of hesperidin and apigenin, also by reducing the formation of fats. Yu et al. (2014) observed that cyanidin-3-glucoside can reduce body weight gain, obesity-related blood lipids, and insulin resistance in rats that are fed a high-fat diet via serum adiponectin regulating serum. Carotenoids, although not a polyphenol but a subclass of terpenoids are said to have obesity and anti-inflammatory properties (González-Castejón and Rodriguez-Casado 2011; Luisa Bonet et al. 2015). α-carotene is one of the most common carotenoids in food that can be transformed into the active form of vitamin A in the body. β-carotene inhibits the expression of inflammatory genes in macrophages stimulated by lipopolysaccharides. Possible roles against obesity were hypothesized for these two carotenes based upon the observation that the sera of obese children had considerably less concentrations of α-carotene and β-carotene than healthy body weight children (Burrows et al. 2009).

An additional group of organosulfur that is considered as an anti-obesity mediator is the hydrolysis products of glucosinolates (González-Castejón and Rodriguez-Casado 2011). Glucosinolates are natural herbal products containing sulfur found in cruciferous vegetables such as broccoli, cabbage and Brussels sprouts. Their presence in plants is correlated with the mechanism of the hydrolytic enzyme myrosinase. In case of damage and destruction of tissues (food preparation, chewing, or parasite infestation), myrosinases convert glucosinolates into a series of breakdown products, such as indoles, isothiocyanates, and nitriles (Martins et al. 2018). Sulforaphane isothiocyanate stimulates anti-obesity action by constraining adipogenesis via reducing expression of CCAAT/enhancer-binding protein α (C/EBPα) and peroxisome proliferator-activated receptor γ (PPARγ) and by inhibiting lipogenesis by stimulation of the AMPK pathway (Choi et al. 2014). Indeed, targeted multi-point treatment in the life cycle of adipocytes by nutritional phytochemicals has been suggested as an efficiently beneficial approach for obesity therapy (Badimon et al. 2010). However, precise evidence of the impacts of clinical tests is needed to verify the anti-obesogenic effects of foods rich in phytochemicals.

4.3 *Diabetes*

Diabetes mellitus defined as a chronic metabolic disorder characterized by elevated blood glucose and inadequate insulin production and function and is the seventh leading cause of death globally (Maritim et al. 2003; Thent and Abd Latiff 2018). Since diabetes is due to the relative deficiency of insulin or the body's cells do not respond adequately to the produced insulin (World Health Organization 2006), it has been suggested increasing or decreasing insulin sensitivity is based on the effective role of antioxidants (Bryans et al. 2007). A diversity of phytochemicals with anti-inflammatory activities has been displayed a substantial impact on adipose tissue, acts as antidiabetic mediators, or is effective in the therapy of chronic inflammatory illnesses (Leiherer et al. 2013). Approximately 80% of diabetic patients currently rely on herbal medicine for their consecutive cures (Sahoo et al. 2010). Therapeutic phytochemicals are applied alone or mixed with diabetes treatments (Ezuruike and Prieto 2014). Åsgård et al. (2007) study showed that consuming fruits and vegetables can reduce markers of oxidative stress and inflammation in 54 patients with type 2 diabetes.

A variety of polyphenols have been proved to reduce the activities of α-glucosidase and α-amylase in the laboratory (Hanhineva et al. 2010). Preclinical studies and some clinical studies show that eating catechin-containing green and white tea, fruits like blue grapes with anthocyanins, foods like red grape, wine with resveratrol, and spices such as curcumin-containing turmeric can offer many health benefits, including improvement of glucose and lipid levels in the blood, as well as increased insulin resistance (Meydani and Hasan 2010). Exposure to the second type of diabetes decreased for those who took 48–80 g of whole grains a day (3–5 servings) from those who never or rarely used them (Ye et al. 2012). A review of early health studies which included nearly 162,000 American women with no history of diabetes, observed that the possibility of acquiring type 2 diabetes was 37% lower among people who consumed the majority of cereals compared to those with less consumption (De Munter et al. 2007). The 2010 Dietary Guidelines Committee assumed that there is incomplete proof that the intake of whole grains is associated with a reduction of type 2 diabetes risk. On the other hand, the German Nutrition Society recommended that consuming grains can reduce type 2 diabetes risk (Hauner et al. 2012).

5 Conclusion

Different studies show that eating fruits and vegetables has positive health properties. Both fruits and vegetables contain various phytonutrients. Most evidence relating fruits and vegetables to health remains observable, and the data in some regions is still conflicting. Although it is obvious that consumption of fresh fruits and cultivated vegetables is important for health care, it is still necessary to carry out other controlled clinical studies to evaluate and confirm the importance of intake fruits and vegetables in daily human nutrition and its effects on various diseases.

References

Ahmad, T., Cawood, M., Iqbal, Q., Ariño, A., Batool, A., Tariq, R. M. S., et al. (2019). Phytochemicals in Daucus carota and their health benefits. *Food, 8*(9), 424.

Ahn, J., Lee, H., Kim, S., Park, J., & Ha, T. (2008). The anti-obesity effect of quercetin is mediated by the AMPK and MAPK signaling pathways. *Biochemical and Biophysical Research Communications, 373*(4), 545–549.

Akkol, E. K., Ilhan, M., Karpuz, B., Genç, Y., & Sobarzo-Sánchez, E. (2020). Sedative and anxiolytic activities of Opuntia ficus indica (L.) Mill.: An experimental assessment in mice. *Molecules, 25*(8), 1844.

Al-Fayez, M., Cai, H., Tunstall, R., Steward, W. P., & Gescher, A. J. (2006). Differential modulation of cyclooxygenase-mediated prostaglandin production by the putative cancer chemopreventive flavonoids tricin, apigenin and quercetin. *Cancer Chemotherapy and Pharmacology, 58*(6), 816–825.

Alonso, Á. M., Guillén, D. A., Barroso, C. G., Puertas, B., & García, A. (2002). Determination of antioxidant activity of wine byproducts and its correlation with polyphenolic content. *Journal of Agricultural and Food Chemistry, 50*(21), 5832–5836.

Altemimi, A., Lakhssassi, N., Baharlouei, A., Watson, D. G., & Lightfoot, D. A. (2017). Phytochemicals: Extraction, isolation, and identification of bioactive compounds from plant extracts. *Plants, 6*(4), 42.

Alwan, A. (2011). *Global status report on noncommunicable diseases 2010*. World Health Organization.

Andre, C. M., Ghislain, M., Bertin, P., Oufir, M., del Rosario Herrera, M., Hoffmann, L., et al. (2007). Andean potato cultivars (*Solanum tuberosum* L.) as a source of antioxidant and mineral micronutrients. *Journal of Agricultural and Food Chemistry, 55*(2), 366–378. https://doi.org/10.1021/jf062740i.

Aoki, F., Honda, S., Kishida, H., Kitano, M., Arai, N., Tanaka, H., et al. (2007). Suppression by licorice flavonoids of abdominal fat accumulation and body weight gain in high-fat diet-induced obese C57BL/6J mice. *Bioscience, Biotechnology, and Biochemistry, 71*(1), 206–214.

Arnous, A., Makris, D. P., & Kefalas, P. (2001). Effect of principal polyphenolic components in relation to antioxidant characteristics of aged red wines. *Journal of Agricultural and Food Chemistry, 49*(12), 5736–5742. https://doi.org/10.1021/jf010827s.

Arts, I. C., & Hollman, P. C. (2005). Polyphenols and disease risk in epidemiologic studies. *The American Journal of Clinical Nutrition, 81*(1), 317S–325S.

Åsgård, R., Rytter, E., Basu, S., Abramsson-Zetterberg, L., Möller, L., & Vessby, B. (2007). High intake of fruit and vegetables is related to low oxidative stress and inflammation in a group of patients with type 2 diabetes. *Scandinavian Journal of Food and Nutrition, 51*(4), 149–158.

Badimon, L., Vilahur, G., & Padro, T. (2010). Nutraceuticals and atherosclerosis: Human trials. *Cardiovascular Therapeutics, 28*(4), 202–215.

Baskar, A. A., Al Numair, K. S., Gabriel Paulraj, M., Alsaif, M. A., Muamar, M. A., & Ignacimuthu, S. (2012). β-sitosterol prevents lipid peroxidation and improves antioxidant status and histoarchitecture in rats with 1,2-dimethylhydrazine-induced colon cancer. *Journal of Medicinal Food, 15*(4), 335–343.

Baydar, N. G., Özkan, G., & Sağdiç, O. (2004). Total phenolic contents and antibacterial activities of grape (Vitis vinifera L.) extracts. *Food Control, 15*(5), 335–339. https://doi.org/10.1016/S0956-7135(03)00083-5.

Baydar, N. G., Sagdic, O., Ozkan, G., & Cetin, S. (2006). Determination of antibacterial effects and total phenolic contents of grape (*Vitis vinifera* L.) seed extracts. *International Journal of Food Science and Technology, 41*(7), 799–804.

Ben, Q., Zhong, J., Liu, J., Wang, L., Sun, Y., Yv, L., & Yuan, Y. (2015). Association between consumption of fruits and vegetables and risk of colorectal adenoma: A PRISMA-compliant meta-analysis of observational studies. *Medicine, 94*(42).

Blomhoff, R., Carlsen, M. H., Andersen, L. F., & Jacobs, D. R. (2006). Health benefits of nuts: Potential role of antioxidants. *British Journal of Nutrition, 96*(S2), S52–S60.

Breene, W. M. (1990). Nutritional and medicinal value of specialty mushrooms. *Journal of Food Protection, 53*(10), 883–894.

Breithaupt, D. E., & Bamedi, A. (2002). Carotenoids and carotenoid esters in potatoes (Solanum tuberosum L.): New insights into an ancient vegetable. *Journal of Agricultural and Food Chemistry, 50*(24), 7175–7181. https://doi.org/10.1021/jf0257953.

Brown, C. (2005). Antioxidants in potato. *American Journal of Potato Research, 82*(2), 163–172.

Bryans, J. A., Judd, P. A., & Ellis, P. R. (2007). The effect of consuming instant black tea on post-prandial plasma glucose and insulin concentrations in healthy humans. *Journal of the American College of Nutrition, 26*(5), 471–477.

Burrows, T. L., Warren, J. M., Colyvas, K., Garg, M. L., & Collins, C. E. (2009). Validation of overweight children's fruit and vegetable intake using plasma carotenoids. *Obesity, 17*(1), 162–168.

Cai, Y., Luo, Q., Sun, M., & Corke, H. (2004). Antioxidant activity and phenolic compounds of 112 traditional Chinese medicinal plants associated with anticancer. *Life Sciences, 74*(17), 2157–2184.

Calvo, M. M. (2005). Lutein: A valuable ingredient of fruit and vegetables. *Critical Reviews in Food Science and Nutrition, 45*(7–8), 671–696.

Campbell, J. K., Canene-Adams, K., Lindshield, B. L., Boileau, T. W.-M., Clinton, S. K., & Erdman, J. W., Jr. (2004). Tomato phytochemicals and prostate cancer risk. *The Journal of Nutrition, 134*(12), 3486S–3492S.

Cao, G., Sofic, E., & Prior, R. L. (1996). Antioxidant capacity of tea and common vegetables. *Journal of Agricultural and Food Chemistry, 44*(11), 3426–3431.

Cartea, M. E., Francisco, M., Soengas, P., & Velasco, P. (2011). Phenolic compounds in Brassica vegetables. *Molecules, 16*(1), 251–280.

Casado-Díaz, A., Dorado, G., & Quesada-Gómez, J. M. (2019). Influence of olive oil and its components on mesenchymal stem cell biology. *World Journal of Stem Cells, 11*(12), 1045.

Chang, S.-T., & Wasser, S. P. (2012). The role of culinary-medicinal mushrooms on human welfare with a pyramid model for human health. *International Journal of Medicinal Mushrooms, 14*(2).

Chen, Q., Wang, D., Tan, C., Hu, Y., Sundararajan, B., & Zhou, Z. (2020). Profiling of flavonoid and antioxidant activity of fruit tissues from 27 Chinese local citrus cultivars. *Plants, 9*(2), 196.

Cheung, L., Cheung, P. C., & Ooi, V. E. (2003). Antioxidant activity and total phenolics of edible mushroom extracts. *Food Chemistry, 81*(2), 249–255.

Choi, K.-M., Lee, Y.-S., Kim, W., Kim, S. J., Shin, K.-O., Yu, J.-Y., et al. (2014). Sulforaphane attenuates obesity by inhibiting adipogenesis and activating the AMPK pathway in obese mice. *The Journal of Nutritional Biochemistry, 25*(2), 201–207. https://doi.org/10.1016/j.jnutbio.2013.10.007.

Choudhari, A. S., Mandave, P. C., Deshpande, M., Ranjekar, P., & Prakash, O. (2020). Phytochemicals in cancer treatment: From preclinical studies to clinical practice. *Frontiers in Pharmacology, 10*, 1614.

Chuang, C.-C., Martinez, K., Xie, G., Kennedy, A., Bumrungpert, A., Overman, A., et al. (2010). Quercetin is equally or more effective than resveratrol in attenuating tumor necrosis factor-α-mediated inflammation and insulin resistance in primary human adipocytes. *The American Journal of Clinical Nutrition, 92*(6), 1511–1521.

Cosme, F., Pinto, T., & Vilela, A. (2018). Phenolic compounds and antioxidant activity in grape juices: A chemical and sensory view. *Beverages, 4*(1), 22.

David, A. V. A., Arulmoli, R., & Parasuraman, S. (2016). Overviews of biological importance of quercetin: A bioactive flavonoid. *Pharmacognosy Reviews, 10*(20), 84.

De Munter, J. S., Hu, F. B., Spiegelman, D., Franz, M., & van Dam, R. M. (2007). Whole grain, bran, and germ intake and risk of type 2 diabetes: A prospective cohort study and systematic review. *PLoS Medicine, 4*(8), e261.

Del Rio, D., Rodriguez-Mateos, A., Spencer, J. P., Tognolini, M., Borges, G., & Crozier, A. (2013). Dietary (poly) phenolics in human health: Structures, bioavailability, and evidence of protective effects against chronic diseases. *Antioxidants and Redox Signaling, 18*(14), 1818–1892.

Do, Q. D., Angkawijaya, A. E., Tran-Nguyen, P. L., Huynh, L. H., Soetaredjo, F. E., Ismadji, S., & Ju, Y.-H. (2014). Effect of extraction solvent on total phenol content, total flavonoid content, and antioxidant activity of *Limnophila aromatica*. *Journal of Food and Drug Analysis, 22*(3), 296–302.

Downey, M. O., Dokoozlian, N. K., & Krstic, M. P. (2006). Cultural practice and environmental impacts on the flavonoid composition of grapes and wine: A review of recent research. *American Journal of Enology and Viticulture, 57*(3), 257–268.

Duttaroy, A. K. (2013). Cardioprotective properties of kiwifruit. *Advances in Food and Nutrition Research, 68*, 273–282. Elsevier.

Duttaroy, A. K. (2016). Kiwifruit juice. In *Handbook of functional beverages and human health* (pp. 355–362). Boca Raton, FL: CRC Press.

El-Kassas, F., Ali, A., & Mostafa, S. (2014). Phenolic compounds as antioxidants of some products manufactured from two cultivated Egyptian varieties of seedless grapes. *Annals of Agricultural Sciences, 59*(2), 195–199.

Elmastas, M., Isildak, O., Turkekul, I., & Temur, N. (2007). Determination of antioxidant activity and antioxidant compounds in wild edible mushrooms. *Journal of Food Composition and Analysis, 20*(3–4), 337–345.

Erdman, J. W., Jr., Balentine, D., Arab, L., Beecher, G., Dwyer, J. T., Folts, J., et al. (2007). Flavonoids and heart health: Proceedings of the ILSI North America flavonoids workshop, May 31–June 1, 2005, Washington, DC. *The Journal of Nutrition, 137*(3), 718S–737S.

Ezuruike, U. F., & Prieto, J. M. (2014). The use of plants in the traditional management of diabetes in Nigeria: Pharmacological and toxicological considerations. *Journal of Ethnopharmacology, 155*(2), 857–924.

Fernandes, I., Pérez-Gregorio, R., Soares, S., Mateus, N., & De Freitas, V. (2017). Wine flavonoids in health and disease prevention. *Molecules, 22*(2), 292.

Feugang, J. M., Konarski, P., Zou, D., Stintzing, F. C., & Zou, C. (2006). Nutritional and medicinal use of Cactus pear (Opuntia spp.) cladodes and fruits. *Frontiers in Bioscience, 11*(1), 2574–2589.

Forman, H. J., Torres, M., & Fukuto, J. (2002). Redox signaling. *Molecular and Cellular Biochemistry, 234*(1), 49–62.

Frankel, E. N., & Meyer, A. M. B. S. (1998). Antioxidants in grapes and grape juices and their potential health effects. *Pharmaceutical Biology, 36*, 1–7.

Frison, S., & Sporns, P. (2002). Variation in the flavonol glycoside composition of almond seedcoats as determined by MALDI-TOF mass spectrometry. *Journal of Agricultural and Food Chemistry, 50*(23), 6818–6822.

Frusciante, L., Carli, P., Ercolano, M. R., Pernice, R., Di Matteo, A., Fogliano, V., & Pellegrini, N. (2007). Antioxidant nutritional quality of tomato. *Molecular Nutrition and Food Research, 51*(5), 609–617.

Galati, E., Pergolizzi, S., Miceli, N., Monforte, M., & Tripodo, M. (2002). Study on the increment of the production of gastric mucus in rats treated with Opuntia ficus indica (L.) Mill. cladodes. *Journal of Ethnopharmacology, 83*(3), 229–233.

Garcia-Closas, R., Gonzalez, C. A., Agudo, A., & Riboli, E. (1999). Intake of specific carotenoids and flavonoids and the risk of gastric cancer in Spain. *Cancer Causes and Control, 10*(1), 71–75.

Garcia-Villalba, R., Espín, J. C., Aaby, K., Alasalvar, C., Heinonen, M., Jacobs, G., et al. (2015). Validated method for the characterization and quantification of extractable and nonextractable ellagitannins after acid hydrolysis in pomegranate fruits, juices, and extracts. *Journal of Agricultural and Food Chemistry, 63*(29), 6555–6566.

Gazzani, G., Papetti, A., Massolini, G., & Daglia, M. (1998). Anti-and prooxidant activity of water soluble components of some common diet vegetables and the effect of thermal treatment. *Journal of Agricultural and Food Chemistry, 46*(10), 4118–4122.

Gerhauser, C. (2008). Cancer chemopreventive potential of apples, apple juice, and apple components. *Planta Medica, 74*(13), 1608–1624.

Gomez-Zorita, S., Lasa, A., Abendaño, N., Fernandez-Quintela, A., Mosqueda-Solís, A., Garcia-Sobreviela, M. P., et al. (2017). Phenolic compounds apigenin, hesperidin and kaempferol reduce in vitro lipid accumulation in human adipocytes. *Journal of Translational Medicine, 15*(1), 1–10.

González-Castejón, M., & Rodriguez-Casado, A. (2011). Dietary phytochemicals and their potential effects on obesity: A review. *Pharmacological Research, 64*(5), 438–455. https://doi.org/10.1016/j.phrs.2011.07.004.

Gualdani, R., Cavalluzzi, M. M., Lentini, G., & Habtemariam, S. (2016). The chemistry and pharmacology of citrus limonoids. *Molecules, 21*(11), 1530.

Gupta, C., Prakash, D., & Gupta, S. (2015). A biotechnological approach to microbial based perfumes and flavours. *Journal of Microbiology & Experimentation, 3*(1), 1–9.

Guthrie, N., Morley, K., Hasegawa, S., Manners, G. D., & Vandenberg, T. (2000). Inhibition of human breast cancer cells by citrus limonoids. In *Citrus limonoids* (Vol. 758, pp. 164–174). American Chemical Society.

Han, L. K., Sumiyoshi, M., Zhang, J., Liu, M. X., Zhang, X. F., Zheng, Y. N., et al. (2003). Anti-obesity action of *Salix matsudana* leaves (Part 1). Anti-obesity action by polyphenols of *Salix matsudana* in high fat-diet treated rodent animals. *Phytotherapy Research: An International Journal Devoted to Pharmacological and Toxicological Evaluation of Natural Product Derivatives, 17*(10), 1188–1194.

Hanhineva, K., Törrönen, R., Bondia-Pons, I., Pekkinen, J., Kolehmainen, M., Mykkänen, H., & Poutanen, K. (2010). Impact of dietary polyphenols on carbohydrate metabolism. *International Journal of Molecular Sciences, 11*(4), 1365–1402.

Hauner, H., Bechthold, A., Boeing, H., Brönstrup, A., Buyken, A., Leschik-Bonnet, E., et al. (2012). Evidence-based guideline of the German Nutrition Society: Carbohydrate intake and prevention of nutrition-related diseases. *Annals of Nutrition and Metabolism, 60*(Suppl. 1), 1–58.

He, X., & Sun, L.-M. (2016). Dietary intake of flavonoid subclasses and risk of colorectal cancer: Evidence from population studies. *Oncotarget, 7*(18), 26617.

Hill, J. O. (2006). Understanding and addressing the epidemic of obesity: An energy balance perspective. *Endocrine Reviews, 27*(7), 750–761.

Hori, S., Butler, E., & McLoughlin, J. (2011). Prostate cancer and diet: Food for thought? *BJU International, 107*(9), 1348–1359.

Hwang, Y. W., Kim, S. Y., Jee, S. H., Kim, Y. N., & Nam, C. M. (2009). Soy food consumption and risk of prostate cancer: A meta-analysis of observational studies. *Nutrition and Cancer, 61*(5), 598–606.

Ignat, I., Volf, I., & Popa, V. I. (2011). A critical review of methods for characterisation of polyphenolic compounds in fruits and vegetables. *Food Chemistry, 126*(4), 1821–1835. https://doi.org/10.1016/j.foodchem.2010.12.026.

Iqbal, J., Abbasi, B. A., Ahmad, R., Batool, R., Mahmood, T., Ali, B., et al. (2019). Potential phytochemicals in the fight against skin cancer: Current landscape and future perspectives. *Biomedicine and Pharmacotherapy, 109*, 1381–1393.

Isah, T. (2019). Stress and defense responses in plant secondary metabolites production. *Biological Research, 52*(1), 39.

Kähkönen, M. P., Hopia, A. I., Vuorela, H. J., Rauha, J.-P., Pihlaja, K., Kujala, T. S., & Heinonen, M. (1999). Antioxidant activity of plant extracts containing phenolic compounds. *Journal of Agricultural and Food Chemistry, 47*(10), 3954–3962. https://doi.org/10.1021/jf990146l.

Kallassy, H. (2017). Phytochemistry and biological activities of selected Lebanese plant species (*Crataegus azarolus* L. and Ephedra campylopoda).

Kammerer, D., Claus, A., Carle, R., & Schieber, A. (2004). Polyphenol screening of pomace from red and white grape varieties (*Vitis vinifera* L.) by HPLC-DAD-MS/MS. *Journal of Agricultural and Food Chemistry, 52*(14), 4360–4367.

Kant, A. K. (2004). Dietary patterns and health outcomes. *Journal of the American Dietetic Association, 104*(4), 615–635.

Karlsen, A., Svendsen, M., Seljeflot, I., Laake, P., Duttaroy, A., Drevon, C., et al. (2013). Kiwifruit decreases blood pressure and whole-blood platelet aggregation in male smokers. *Journal of Human Hypertension, 27*(2), 126–130.

Kedage, V. V., Tilak, J. C., Dixit, G. B., Devasagayam, T. P., & Mhatre, M. (2007). A study of antioxidant properties of some varieties of grapes (*Vitis vinifera* L.). *Critical Reviews in Food Science and Nutrition, 47*(2), 175–185.

Kirsh, V. A., Peters, U., Mayne, S. T., Subar, A. F., Chatterjee, N., Johnson, C. C., & Hayes, R. B. (2007). Prospective study of fruit and vegetable intake and risk of prostate cancer. *Journal of the National Cancer Institute, 99*(15), 1200–1209.

Köksal, E., & Gülçin, İ. (2008). Antioxidant activity of cauliflower (*Brassica oleracea* L.). *Turkish Journal of Agriculture and Forestry, 32*(1), 65–78.

Koushik, A., Hunter, D. J., Spiegelman, D., Beeson, W. L., Van Den Brandt, P. A., Buring, J. E., et al. (2007). Fruits, vegetables, and colon cancer risk in a pooled analysis of 14 cohort studies. *Journal of the National Cancer Institute, 99*(19), 1471–1483.

Kumar, V., Sharma, A., Kohli, S. K., Bali, S., Sharma, M., Kumar, R., et al. (2019). Differential distribution of polyphenols in plants using multivariate techniques. *Biotechnology Research and Innovation, 3*(1), 1–21.

Lam, L. T., Zhang, J., & Hasegawa, S. (1994). Citrus limonoid reduction of chemically induced tumorigenesis. *Food Technology (Chicago), 48*(11), 104–108.

Lee, D., Park, H. Y., Kim, S., Park, Y., Bang, M.-H., & Imm, J.-Y. (2015). Anti-adipogenic effect of oat hull extract containing tricin on 3T3-L1 adipocytes. *Process Biochemistry, 50*(12), 2314–2321.

Lee, S. H., Oh, S. H., Hwang, I. G., Kim, H. Y., Woo, K. S., Woo, S. H., et al. (2016). Antioxidant contents and antioxidant activities of white and colored potatoes (*Solanum tuberosum* L.). *Preventive Nutrition and Food Science, 21*(2), 110.

Leiherer, A., Mündlein, A., & Drexel, H. (2013). Phytochemicals and their impact on adipose tissue inflammation and diabetes. *Vascular Pharmacology, 58*(1), 3–20. https://doi.org/10.1016/j.vph.2012.09.002.

Li, S., Pan, M.-H., Lo, C.-Y., Tan, D., Wang, Y., Shahidi, F., & Ho, C.-T. (2009). Chemistry and health effects of polymethoxyflavones and hydroxylated polymethoxyflavones. *Journal of Functional Foods, 1*(1), 2–12.

Liu, R. H. (2013). Health-promoting components of fruits and vegetables in the diet. *Advances in Nutrition (Bethesda, MD), 4*(3), 384S–392S. https://doi.org/10.3945/an.112.003517.

Luisa Bonet, M., Canas, J. A., Ribot, J., & Palou, A. (2015). Carotenoids and their conversion products in the control of adipocyte function, adiposity and obesity. *Archives of Biochemistry and Biophysics, 572*, 112–125. https://doi.org/10.1016/j.abb.2015.02.022.

Lull, C., Wichers, H. J., & Savelkoul, H. F. (2005). Antiinflammatory and immunomodulating properties of fungal metabolites. *Mediators of Inflammation, 2005.*

Manach, C., Scalbert, A., Morand, C., Rémésy, C., & Jiménez, L. (2004). Polyphenols: Food sources and bioavailability. *The American Journal of Clinical Nutrition, 79*(5), 727–747.

Maritim, A., Dene, B., Sanders, R., & Watkins, J., III. (2003). Effects of pycnogenol treatment on oxidative stress in streptozotocin-induced diabetic rats. *Journal of Biochemical and Molecular Toxicology, 17*(3), 193–199.

Martínez-Valverde, I., Periago, M. J., Provan, G., & Chesson, A. (2002). Phenolic compounds, lycopene and antioxidant activity in commercial varieties of tomato (*Lycopersicum esculentum*). *Journal of the Science of Food and Agriculture, 82*(3), 323–330.

Martins, T., Colaço, B., Venâncio, C., Pires, M. J., Oliveira, P. A., Rosa, E., & Antunes, L. M. (2018). Potential effects of sulforaphane to fight obesity. *Journal of the Science of Food and Agriculture, 98*(8), 2837–2844.

McGregor, R. A., & Choi, M. S. (2011). microRNAs in the regulation of adipogenesis and obesity. *Current Molecular Medicine, 11*(4), 304–316.

Meléndez-Martínez, A. J., Mapelli-Brahm, P., Hornero-Méndez, D., & Vicario, I. M. (2019). *Structures, nomenclature and general chemistry of carotenoids and their esters.*

Meydani, M., & Hasan, S. T. (2010). Dietary polyphenols and obesity. *Nutrients, 2*(7), 737–751.

Milner, J. (2001). A historical perspective on garlic and cancer. *The Journal of Nutrition, 131*(3), 1027S–1031S.

Mishra, T., Raigond, P., Thakur, N., Dutt, S., & Singh, B. (2020). Recent updates on healthy phytoconstituents in potato: A nutritional depository. *Potato Research*, 1–21.

Mohdaly, A. A., Sarhan, M. A., Smetanska, I., & Mahmoud, A. (2010). Antioxidant properties of various solvent extracts of potato peel, sugar beet pulp and sesame cake. *Journal of the Science of Food and Agriculture, 90*(2), 218–226.

Morales-Soto, A., García-Salas, P., Rodríguez-Pérez, C., Jiménez-Sánchez, C., Cádiz-Gurrea, M. D. L. L., Segura-Carretero, A., & Fernández-Gutiérrez, A. (2014). Antioxidant capacity of 44 cultivars of fruits and vegetables grown in Andalusia (Spain). *Food Research International, 58*, 35–46. https://doi.org/10.1016/j.foodres.2014.01.050.

Mori, K., Toyomasu, T., Nanba, H., & Kuroda, H. (1987). Antitumor activities of edible mushrooms by oral administration. *Developments in Crop Science, 10*, 1–6. Elsevier.

Mulero, J., Pardo, F., & Zafrilla, P. (2010). Antioxidant activity and phenolic composition of organic and conventional grapes and wines. *Journal of Food Composition and Analysis, 23*(6), 569–574.

Nabavi, S. F., Russo, G. L., Daglia, M., & Nabavi, S. M. (2015). Role of quercetin as an alternative for obesity treatment: You are what you eat! *Food Chemistry, 179*, 305–310.

Park, E.-H., Kahng, J.-H., Lee, S. H., & Shin, K. H. (2001). An anti-inflammatory principle from cactus. *Fitoterapia, 72*(3), 288–290.

Patra, A. K. (2012). *Dietary phytochemicals and microbes*: Springer Science and Business Media.

Paula Santos, A., Rogero, M., Bastos, H. M., & D. (2010). Edible plants, their secondary metabolites and antiobesogenic potential. *Recent Patents on Food, Nutrition and Agriculture, 2*(3), 195–212.

Perez-Vizcaino, F., Duarte, J., & Andriantsitohaina, R. (2006). Endothelial function and cardiovascular disease: Effects of quercetin and wine polyphenols. *Free Radical Research, 40*(10), 1054–1065.

Pihlanto, A., Akkanen, S., & Korhonen, H. J. (2008). ACE-inhibitory and antioxidant properties of potato (*Solanum tuberosum*). *Food Chemistry, 109*(1), 104–112. https://doi.org/10.1016/j.foodchem.2007.12.023.

Pirayesh Islamian, J., & Mehrali, H. (2015). Lycopene as a carotenoid provides radioprotectant and antioxidant effects by quenching radiation-induced free radical singlet oxygen: An overview. *Cell Journal, 16*(4), 386–391. *Introduction Humans are exposed to ionizing radiation*.

Plumb, G. W., Price, K. R., Modes, M. J., & Williamson, G. (1997). Antioxidant properties of the major polyphenolic compounds in broccoli. *Free Radical Research, 27*(4), 429–435.

Podsędek, A. (2007). Natural antioxidants and antioxidant capacity of Brassica vegetables: A review. *LWT—Food Science and Technology, 40*(1), 1–11. https://doi.org/10.1016/j.lwt.2005.07.023.

Possenti, M., Baima, S., Raffo, A., Durazzo, A., Giusti, A. M., & Natella, F. (2016). Glucosinolates in food. In J.-M. Mérillon & K. G. Ramawat (Eds.), *Glucosinolates* (pp. 1–46). Cham: Springer International Publishing.

Ramarathnam, N., Ochi, H., & Takeuchi, M. (1997). Antioxidative defense system in vegetable extracts. *Natural Antioxidants. Chemistry, Health Effects, and Applications*, 76–87.

Rayalam, S., Della-Fera, M. A., & Baile, C. A. (2008). Phytochemicals and regulation of the adipocyte life cycle. *The Journal of Nutritional Biochemistry, 19*(11), 717–726.

Reyes, L. F., & Cisneros-Zevallos, L. (2003). Wounding stress increases the phenolic content and antioxidant capacity of purple-flesh potatoes (*Solanum tuberosum* L.). *Journal of Agricultural and Food Chemistry, 51*(18), 5296–5300.

Robards, K., Prenzler, P. D., Tucker, G., Swatsitang, P., & Glover, W. (1999). Phenolic compounds and their role in oxidative processes in fruits. *Food Chemistry, 66*(4), 401–436.

Ros, E. (2010). Health benefits of nut consumption. *Nutrients, 2*(7), 652–682.

Ros, E. (2015). Nuts and CVD. *British Journal of Nutrition, 113*(S2), S111–S120.

Rosales-Martínez, P., Arellano-Cárdenas, S., Dorantes-Álvarez, L., García-Ochoa, F., & López-Cortez, M. D. S. (2014). Comparison between antioxidant activities of phenolic extracts from Mexican peanuts, peanuts skins, nuts and pistachios. *Journal of the Mexican Chemical Society, 58*(2), 185–193.

Sahoo, N., Manchikanti, P., & Dey, S. (2010). Herbal drugs: Standards and regulation. *Fitoterapia, 81*(6), 462–471.
Sancho, L. E. G.-G., Yahia, E. M., & González-Aguilar, G. A. (2011). Identification and quantification of phenols, carotenoids, and vitamin C from papaya (*Carica papaya* L., cv. Maradol) fruit determined by HPLC-DAD-MS/MS-ESI. *Food Research International, 44*(5), 1284–1291.
Sang, S., Kikuzaki, H., Lapsley, K., Rosen, R. T., Nakatani, N., & Ho, C.-T. (2002). Sphingolipid and other constituents from almond nuts (*Prunus amygdalus* Batsch). *Journal of Agricultural and Food Chemistry, 50*(16), 4709–4712.
Scalbert, A., Johnson, I. T., & Saltmarsh, M. (2005). Polyphenols: Antioxidants and beyond. *The American Journal of Clinical Nutrition, 81*(1), 215S–217S.
Sen, C. K., & Packer, L. (2000). Thiol homeostasis and supplements in physical exercise. *The American Journal of Clinical Nutrition, 72*(2), 653S–669S.
Seo, M.-J., Lee, Y.-J., Hwang, J.-H., Kim, K.-J., & Lee, B.-Y. (2015). The inhibitory effects of quercetin on obesity and obesity-induced inflammation by regulation of MAPK signaling. *The Journal of Nutritional Biochemistry, 26*(11), 1308–1316. https://doi.org/10.1016/j.jnutbio.2015.06.005.
Shamtsyan, M., Konusova, V., Maksimova, Y., Goloshchev, A., Panchenko, A., Simbirtsev, A., & Denisova, N. (2004). Immunomodulating and anti-tumor action of extracts of several mushrooms. *Journal of Biotechnology, 113*(1–3), 77–83.
Shi, J., & Maguer, M. L. (2000). Lycopene in tomatoes: Chemical and physical properties affected by food processing. *Critical Reviews in Food Science and Nutrition, 40*(1), 1–42.
Shukla, Y., & Kalra, N. (2007). Cancer chemoprevention with garlic and its constituents. *Cancer Letters, 247*(2), 167–181. https://doi.org/10.1016/j.canlet.2006.05.009.
Singh, N., & Rajini, P. S. (2008). Antioxidant-mediated protective effect of potato peel extract in erythrocytes against oxidative damage. *Chemico-Biological Interactions, 173*(2), 97–104. https://doi.org/10.1016/j.cbi.2008.03.008.
Singletary, K. (2012). Kiwifruit: Overview of potential health benefits. *Nutrition Today, 47*(3), 133–147.
Skotti, E., Anastasaki, E., Kanellou, G., Polissiou, M., & Tarantilis, P. A. (2014). Total phenolic content, antioxidant activity and toxicity of aqueous extracts from selected Greek medicinal and aromatic plants. *Industrial Crops and Products, 53*, 46–54.
Soengas Fernández, M.D.P., Sotelo Pérez, T., Velasco Pazos, P., & Cartea González, M. E. (2011). *Antioxidant properties of Brassica vegetables.*
Song, J.-K., & Bae, J.-M. (2013). Citrus fruit intake and breast cancer risk: A quantitative systematic review. *Journal of Breast Cancer, 16*(1), 72–76.
Souza, R. G., Gomes, A. C., Naves, M. M., & Mota, J. F. (2015). Nuts and legume seeds for cardiovascular risk reduction: Scientific evidence and mechanisms of action. *Nutrition Reviews, 73*(6), 335–347.
Stagos, D. (2020). *Antioxidant activity of polyphenolic plant extracts*: Multidisciplinary Digital Publishing Institute.
Stonehouse, W., Gammon, C. S., Beck, K. L., Conlon, C. A., von Hurst, P. R., & Kruger, R. (2013). Kiwifruit: Our daily prescription for health. *Canadian Journal of Physiology and Pharmacology, 91*(6), 442–447.
Story, E. N., Kopec, R. E., Schwartz, S. J., & Harris, G. K. (2010). An update on the health effects of tomato lycopene. *Annual Review of Food Science and Technology, 1*, 189–210.
Sun, J., Chu, Y.-F., Wu, X., & Liu, R. H. (2002). Antioxidant and antiproliferative activities of common fruits. *Journal of Agricultural and Food Chemistry, 50*(25), 7449–7454.
Svendsen, M., Tonstad, S., Heggen, E., Pedersen, T. R., Seljeflot, I., Bøhn, S. K., et al. (2015). The effect of kiwifruit consumption on blood pressure in subjects with moderately elevated blood pressure: A randomized, controlled study. *Blood Pressure, 24*(1), 48–54.
Teow, C. C., Truong, V.-D., McFeeters, R. F., Thompson, R. L., Pecota, K. V., & Yencho, G. C. (2007). Antioxidant activities, phenolic and β-carotene contents of sweet potato geno-

types with varying flesh colours. *Food Chemistry, 103*(3), 829–838. https://doi.org/10.1016/j.foodchem.2006.09.033.

Thaipong, K., Boonprakob, U., Crosby, K., Cisneros-Zevallos, L., & Byrne, D. H. (2006). Comparison of ABTS, DPPH, FRAP, and ORAC assays for estimating antioxidant activity from guava fruit extracts. *Journal of Food Composition and Analysis, 19*(6–7), 669–675.

Thent, Z. C., & Abd Latiff, A. (2018). *Savior of diabetes: Antioxidants diabetes food plan.* IntechOpen.

Thomas, R., Butler, E., Macchi, F., & Williams, M. (2015). Phytochemicals in cancer prevention and management? *British Journal of Medical Practitioners, 8*(2), 1–8.

Thomson, M., & Ali, M. (2003). Garlic [*Allium sativum*]: A review of its potential use as an anticancer agent. *Current Cancer Drug Targets, 3*(1), 67–81.

Tungmunnithum, D., Thongboonyou, A., Pholboon, A., & Yangsabai, A. (2018). Flavonoids and other phenolic compounds from medicinal plants for pharmaceutical and medical aspects: An overview. *Medicine, 5*(3), 93.

Ulrich, K., & Jakob, U. (2019). The role of thiols in antioxidant systems. *Free Radical Biology and Medicine, 140*, 14–27.

Valverde, M. E., Hernández-Pérez, T., & Paredes-López, O. (2015). Edible mushrooms: Improving human health and promoting quality life. *International Journal of Microbiology, 2015.*

Vieira, A. R., Abar, L., Vingeliene, S., Chan, D., Aune, D., Navarro-Rosenblatt, D., et al. (2016). Fruits, vegetables and lung cancer risk: A systematic review and meta-analysis. *Annals of Oncology, 27*(1), 81–96.

Wang, M., Qin, S., Zhang, T., Song, X., & Zhang, S. (2015). The effect of fruit and vegetable intake on the development of lung cancer: A meta-analysis of 32 publications and 20,414 cases. *European Journal of Clinical Nutrition, 69*(11), 1184–1192.

Wang, Q., Chen, Y., Wang, X., Gong, G., Li, G., & Li, C. (2014a). Consumption of fruit, but not vegetables, may reduce risk of gastric cancer: Results from a meta-analysis of cohort studies. *European Journal of Cancer, 50*(8), 1498–1509.

Wang, S., Moustaid-Moussa, N., Chen, L., Mo, H., Shastri, A., Su, R., et al. (2014b). Novel insights of dietary polyphenols and obesity. *The Journal of Nutritional Biochemistry, 25*(1), 1–18.

Wang, X. J., Chen, J. Y., Fu, L. Q., & Yan, M. J. (2020). Recent advances in natural therapeutic approaches for the treatment of cancer. *Journal of Chemotherapy, 32*(2), 53–65. https://doi.org/10.1080/1120009x.2019.1707417.

Wasser, S. (2002). Medicinal mushrooms as a source of antitumor and immunomodulating polysaccharides. *Applied Microbiology and Biotechnology, 60*(3), 258–274.

Waterhouse, A. L. (2002). Determination of total phenolics. *Current Protocols in Food Analytical Chemistry, 6*(1), I1.1.1–I1.1.8.

WCRF/AICR. (2007). Breast cancer. In *Food, nutrition, physical activity, and the prevention of cancer: A global perspective* (pp. 289–295). Washington, DC: American Institute for Cancer Research.

Williams, D. J., Edwards, D., Hamernig, I., Jian, L., James, A. P., Johnson, S. K., & Tapsell, L. C. (2013). Vegetables containing phytochemicals with potential anti-obesity properties: A review. *Food Research International, 52*(1), 323–333.

Wodek, L. (2002). Beneficial and harmful effects of thiols. *Polish Journal of Pharmacology, 54*, 215–223.

Wolber, F. M., Beck, K. L., Conlon, C. A., & Kruger, M. C. (2013). Kiwifruit and mineral nutrition. *Advances in Food and Nutrition Research, 68*, 233–256. Elsevier.

Wolin, K. Y., Carson, K., & Colditz, G. A. (2010). Obesity and cancer. *The Oncologist, 15*(6), 556.

Wootton-Beard, P. C., & Ryan, L. (2011). Improving public health? The role of antioxidant-rich fruit and vegetable beverages. *Food Research International, 44*(10), 3135–3148.

World Health Organization. (2006). *Definition and diagnosis of diabetes mellitus and intermediate hyperglycaemia: Report of a WHO/IDF consultation.*

Yahia, E. M. (2010). The contribution of fruit and vegetable consumption to human health. *Fruit and Vegetable Phytochemicals*, 3–51.

Yahia, E. M. (2017). *Fruit and vegetable phytochemicals: Chemistry and human health* (Vol. 2). Wiley.

Yahia, E. M., & Ornelas-Paz, J. D. J. (2010). Chemistry, stability, and biological actions of carotenoids. In *Fruit and vegetable phytochemicals: Chemistry, nutritional value and stability* (pp. 177–222). Wiley Online Library.

Yang, C. S., Landau, J. M., Huang, M.-T., & Newmark, H. L. (2001). Inhibition of carcinogenesis by dietary polyphenolic compounds. *Annual Review of Nutrition, 21*(1), 381–406.

Ye, E. Q., Chacko, S. A., Chou, E. L., Kugizaki, M., & Liu, S. (2012). Greater whole-grain intake is associated with lower risk of Type 2 diabetes, cardiovascular disease, and weight gain. *The Journal of Nutrition, 142*(7), 1304–1313. https://doi.org/10.3945/jn.111.155325.

Yu, R.-Q., Wu, X.-Y., Zhou, X., Zhu, J., & Ma, L.-Y. (2014). Cyanidin-3-glucoside attenuates body weight gain, serum lipid concentrations and insulin resistance in high-fat diet-induced obese rats. *Zhongguo dang dai er ke za zhi = Chinese Journal of Contemporary Pediatrics, 16*(5), 534–538.

Zhang, C. X., Ho, S. C., Chen, Y. M., Fu, J. H., Cheng, S. Z., & Lin, F. Y. (2009). Greater vegetable and fruit intake is associated with a lower risk of breast cancer among Chinese women. *International Journal of Cancer, 125*(1), 181–188.

Zhang, Y.-J., Gan, R.-Y., Li, S., Zhou, Y., Li, A.-N., Xu, D.-P., & Li, H.-B. (2015). Antioxidant phytochemicals for the prevention and treatment of chronic diseases. *Molecules, 20*(12), 21138–21156.

Spices for Diabetes, Cancer and Obesity Treatment

Uswa Ahmad, Anum Nazir, Shiza Ahmad, and Nosheen Asghar

1 Introduction

Herbs and spices are derived from plants; it depends upon the part of the plant selected for the medicinal or culinary herb or for generating flavoring or adding color to food. The part of the plant can be a root, flower, petals, or leaves (Tapsell et al. 2006). Herbs and spices contain the special bioactive components called polyphenols such as hydroxybenzoic acid or flavonoids such as flavones or anthocyanidins (Kondratyuk and Pezzuto 2004; Scalbert and Williamson 2000).

Spices provide characteristic aroma, taste, and specific color. Furthermore, they are utilized (commonly dried seed, fruit, root, bark, or vegetative material) as preservatives, flavoring agents, colorants, and flavor replacement both in the old-style food preparations and in the food industry. These are also exploited for numerous medical purposes. Herbs can be found in aqueous solution, however; spices are widely used in dried form. Some examples of spices are cardamom, cinnamon, cloves, and ginger that have been used as a medicine from old ages. Many studies indicate that herbs and spices were used in the era of the Abbasid caliphate in the city of Baghdad. Apart from that, in the thirteenth century, it was observed that high-income families use spices to increase the shelf life of meat and meat products.

U. Ahmad (✉) · A. Nazir
Department of Nutrition and Dietetics, The University of Faisalabad, Faisalabad, Punjab, Pakistan
e-mail: uswa.ahmad@tuf.edu.pk; anum.nazir@tuf.edu.pk

S. Ahmad
Department of Home Economics, University of Agriculture, Faisalabad, Punjab, Pakistan

N. Asghar
Institute of Home and Food Sciences, Government College University, Faisalabad, Punjab, Pakistan

C. Egbuna, S. Hassan (eds.), *Dietary Phytochemicals*,
https://doi.org/10.1007/978-3-030-72999-8_9

Herbs and spices are safe and are utilized in many foods and also have the capability to manage and treat many diseases such as diabetes, obesity, cardiovascular diseases, and cancer due to the presence of natural compounds in them. Recent analyses identify that diabetes and cancer are the second leading cause of death; there are many types of cancer such as lung, breast, liver cancer, prostate cancer, stomach cancer. Mainly the people who are at risk are smokers, tobacco consumers, and a high trans-fat diet (WHO 2016). The major symptoms of cancer include a lump in the breast, blood in urine, pain during urination, redness of the skin, itchiness, loss of appetite, fatigue, fever, seizure, headache, and weight loss (National Cancer Institute 2019). It has been reported that 9.6 million deaths occur due to cancer in 2018 among them the high death occurs due to lung cancer, liver, breast, and colorectal cancer (WHO 2016).

The higher intake of calories and low energy expenditure results in obesity, clinical trials investigated certain factors that cause a high incidence of obesity; the main causes are inheritance, metabolic syndromes, and lifestyle (Power and Schulkin 2008). Moreover, obesity is known as high BMI affecting about 1.9 million people, according to research on 650 million overweight people (WHO 2016). Obesity is also involved in death-causing diseases like cardiovascular disease, cancer, and diabetes (Leitner et al. 2017).

Moreover, the irregular distribution of adipose tissue and high accumulation of fat in the peripheral regions lead to metabolic diseases such as obesity-related type 2 diabetes, high incidence of congestive heart failure, the occurrence of cerebrovascular accident, and ischemic stroke. Inflammation-induced by obesity results in a high incidence of osteoarthritis, unable to reproduce offspring, and other organ dysfunctions such as pulmonary dysfunctions (Kissebah and Krakower 1994). Many clinical studies revealed that BMI >30 kg m^{-2} is termed as obesity, however, the estimation for ideal body weight is 18.5–24.9 kg m^{-2} (WHO 2000).

The higher accumulation of adipose tissues results in the secretion of inflammatory mediators such as TNF-α, IL-6, and MCP-1, theses cytokinin triggers inflammation (Lafontan 2005). Many other studies indicate that fat cells synthesize different hormones like steroid hormones and angiotensin II. Furthermore, these hormones are responsible for chronic illness (Kershaw and Flier 2004). Herbs and spices mainly cause the apoptosis of malignant cells. Moreover, they prevent the proliferation of cell growth and also inhibit angiogenesis. Certain bioactive constituents of herbs and spices play an important role in suppressing tumor growth (Jie et al. 2016).

Moreover, herbs and spices burn fat by elevating basal metabolic rate, activating various protein and genes responsible for fat hydrolyzed, and inhibiting the absorption of fat (Arbo et al. 2008, 2009). Another study demonstrated that free radical scavengers are known as polyphenols such as curcumin, basils, and flavones inhibit inflammation by inactivating the inflammatory factors such as NF-KB and MAPK (Link et al. 2010).

2 Anti-diabetic Effect of Spices

Diabetes is a chief metabolic illness becoming increasingly common in the world. Diabetes continues to be a key health care problem worldwide and its occurrence is probable to rise from the current 382–471 million individuals by 2035 (Bi et al. 2017). Diabetes is characterized into three types. Type 1 diabetes that is an autoimmune disorder resulted due to damage to pancreatic beta-cells. The second type is type 2 diabetes that resulted in a reduction of glucose regulation due to defective pancreatic beta cells and insulin resistance. The third type is gestational diabetes. Various treatments like using insulin, pharmaceutical drugs, and treatment by using proper diet have been used for managing all types of diabetes (Kooti et al. 2016). Artificial anti-diabetic agents mostly used for the management of diabetes have many side effects such as they cannot be utilized during pregnancy due to toxic nature, have drug resistance, and are most expensive (Haque et al. 2011).

Due to the side effects and toxicity of synthetic drugs used for the treatment of diabetes, the treatment by using natural, non-toxic, and low-cost anti-diabetic agents like spices have become common nowadays. Spices have many medicinal benefits due to which they are utilized in the formation of many medicines. Spices can be the dried leaf, buds, bark, root/rhizome, berries, seeds, or even sigma of flower (Viuda-Martos et al. 2010). Most of the spices have antioxidants like carotenoids, flavonoids, terpenoids, alkaloids, and glycosides due to which they are utilized for the management of diabetes. The anti-diabetic effect of spices is due to their capacity to increase the efficiency of pancreatic tissue, by enhancing the secretions of insulin or decreasing the glucose intestinal absorption (Afrisham et al. 2015). Nowadays, the use of spices has become common to treat many diseases due to their biological activity. Previous researches proved that spices have anti-diabetic effects due to main active constituents. The most commonly used spices for the treatment of diabetes are ginger, turmeric, cinnamon, fenugreek, and garlic. Other spices such as cumin, coriander mustard, anise, onion, and black pepper are also effective in the treatment of diabetes according to previous studies (Şanlier and Gencer 2020).

2.1 Fenugreek

Fenugreek (*Trigonella foenum graecum L.*, family: Fabaceae) is one of the most advantageous prehistoric medicinal plants grown generally in India, Egypt, and Middle Eastern countries (Acharya et al. 2008). Its leaves and seeds have been generally utilized as a flavoring agent, vegetables, and natural drug in many medicinal and functional food industries (Zandi et al. 2015). Scholars have proved many health benefits of fenugreek seeds against many illnesses like diabetes, cancer, hyperlipidemia, and inflammation (Nagulapalli Venkata et al. 2017).

Fenugreek seeds are used for the treatment of diabetes mellitus (DM) all over the world. These seeds are high in soluble fibers, saponins, trigonelline, diosgenin, and

4-hydroxy isoleucine. Soluble fibers like galactomannan present in fenugreek seeds help in decreasing blood sugar by slowing down digestion and absorption of carbohydrates. The leaves and seeds of fenugreek are used either as extracts or powder form for medicinal use. In humans, fenugreek seeds have anti-hyperglycemic properties by exciting insulin secretion and also delaying the activity of sucrase and alpha-amylase (Sundaram et al. 2020).

Three (3) grams/kg body weight of fenugreek seeds powder for 21 days was given to diabetic rats that resulted in a decrease in blood glucose levels in diabetic rats. According to another study, the intake of an extract of fenugreek seeds is effective for lowering the glycemic level of blood (Abou El-Soud et al. 2007).

According to another research anti-diabetic consequence of fenugreek extract on blood serum glucose level in diabetic rats causing a decrease in blood glucose level in diabetic rats (Renuka et al. 2009).

Furthermore, Genet et al. (2009) indicated that an increase in blood glucose levels of diabetic patients was observed by the utilization of fenugreek seeds. Additionally, Vanitha et al. (2012) verified that the anti-hyperglycemic effect of fenugreek water extracts in diabetic rats from 314.25 to 159.00 mg/dL of blood. Likewise, Berroukche et al. (2018) also stated that fenugreek seeds are considered as an excellent therapeutic drug as an anti-diabetic agent for managing diabetes and its related health issues.

2.2 *Cinnamon*

Cinnamon (*Cinnamomum zeylanicum*) is commonly known as "Dalchini" in Hindi. Cinnamon is one of the most commonly utilized spice and is considered safe with many active ingredients. The phenolic extract of cinnamon has insulin potentiating action. Thus its administration is important for the management of glucose and insulin sensitivity of insulin in humans. Furthermore, it also indicates an increase in the breakdown of fat and antioxidant status. It contains antioxidants like alkaloids, tannins, glycosides, and saponins. An aqueous extract of cinnamon improved the ability to control glucose and prohibited fat defects in diabetic rats (Upasani et al. 2013).

Further researches have proved that cinnamon is rich in B-type procyanidin, cinnamaldehyde, and A-type 178 procyanidin due to which cinnamon has strong anti-oxidant action (Jayaprakasha et al. 2006). The use of cinnamon for the management of diabetes was first described in 1990 when Khan and his team recommended that cinnamon had insulin potentiating factor (IPF) and also confirmed that IPF may be involved in the control of blood glucose levels in diabetic patients. According to another study, different Cinnamomum species like *Cinnamomum cassia, Cinnamomum burmannii, Cinnamomum zeylanicum, and Cinnamomum verum* were tested for the improvement in glycemic response. From all the spices, *Cinnamomum cassia* showed the best result in the treatment of type 2 diabetes (Howard and White 2013).

The anti-diabetic properties of cinnamon were proved to be present due to cinnamaldehyde (Subash Babu et al. 2007). Moreover, it was described that cinnamon phenols present in cinnamon could increase the effectiveness of insulin (Cao et al. 2010). In another study, Lu et al. (2012) verified the decrease of fasting blood glucose and HbA1c level. Another study showed that treatment with 2 g of cinnamon resulted in a lowering of HbA1c of diabetic patients (Akilen et al. 2010). According to another study the effects of cinnamon on plasma glucose, HbA1c, and serum lipids in diabetic patients were studied (Mang et al. 2006). The results showed that cinnamon had a modest consequence in lowering fasting blood glucose (FBG) levels in diabetic patients. Taher et al. (2006) found that cinnamon contained cinnamtannin that is a proanthocyanidin that also improved insulin resistance. According to another research, cinnamon diminished HbA1c in diabetic patients. The consequences of the study found that cinnamon is safe can decrease HbA1c and fasting blood glucose levels. Therefore, cinnamon is considered as a natural drug to treat diabetes (Crawford 2009).

In another study, cinnamon utilization at a specific dosage reduced the serum insulin concentrations and glucagon-like peptide 1(GLP-1). Moreover, improved glucose transport across cell membranes reduced insulin resistance (Hlebowicz et al. 2009). According to another research, *Cinnamomum zeylanicum* extract enhances GLUT4 to the cell membrane of brown adipose tissue and muscle in a dose-dependent way (Shen et al. 2010). Similarly, Anand et al. (2010) found greater membrane translocation of GLUT4 in cinnamon treated rats than in the rats that were not given cinnamon.

Plaisier et al. (2011) confirmed that cinnamaldehyde raises glucose transporter-1 (GLUT-1) refereed glucose absorption. Meanwhile, in the presence of glucose shortage in the medium, cinnamaldehyde reduced the GLUT 1 refereed glucose absorption. Shen et al. (2014) investigated that cinnamon enhances the amount of glucose transporter 4 receptors as well as insulin receptors. Furthermore, it increases the quantity of insulin receptor substrates due to which glucose easily enters into the cells. Cinnamon and its active constituents like cinnamaldehyde, cinnamic acid, eugenol, and cinnamate have many health benefits. Cinnamon extracts also can treat different health disorders like hyperlipidemia, hypertension, heart diseases, obesity, and diabetes due to its strong antioxidant status. Cinnamon at a dose level of 2 g in diabetes patients is very effective in the reduction of blood pressure and blood glucose level (Goel and Mishra 2020).

2.3 Garlic and Onion

Onion (*Allium cepa* L.) is the most widely cultivated species of the genus Allium. Onion is a good source of several phytonutrients like flavonoids, fructooligosaccharides (FOS), and thiosulfinates, and other sulfur compounds due to which it is used for different diseases and their complications (Bindu and Podikunju 2015). The chemical composition and antioxidant properties of the *Allium cepa* bulb have shown that they contain secondary metabolites such as flavonoids, alkaloids, glycoside, and phenolic compounds due to which it has strong antioxidant potential (Liguori et al. 2017). According to another recent study, the decrease of fasting blood glucose levels showed the anti-diabetic potential of *A. cepa* bulb (Airaodion et al. 2019a, b, c). The decrease in the blood glucose level was due to slowing down the absorption of glucose through the inhibition of the carbohydrate hydrolyzing enzymes, α-amylase, and α-glucosidase, in the digestive tract (Airaodion et al. 2019a, b, c).

It was investigated that animals treated with 3 mL/100 g bodyweights of *A. cepa* juice reduced blood glucose as much as glibenclamide, a standard diabetic drug (Airaodion et al. 2019a, b, c).

Garlic (*Allium sativum* L.; Family: Amaryllidaceae) is another important herb that has been used from the earliest times as a source of the natural drug. It is considered as second most famous Allium species with onion (*Allium cepa* L.), which is utilized as a medicine to treat and manage diabetes, heart diseases and boost the immune system due to the presence of many active ingredients in them (Badal et al. 2019). Garlic had a strong effect on diabetic retinopathy in diabetic rats. The weight, blood glucose, and morphological changes of retinal tissue improved after 7 weeks of its administration in animal models (Al-Brakati 2016). Furthermore, another study accomplished on diabetic patients revealed that garlic addition considerably decreased fructosamine and glycosylated hemoglobin. This research proved that garlic was active in controlling type 2 diabetes mellitus (Wang et al. 2017). Thus, garlic and its bioactive components might be effective agents to help treat diabetes and diabetic complications.

According to an Experimental study, the anti-diabetic outcome of garlic pills given a dose of 900 mg per day in diabetic and hyperlipidemic patients showed that garlic pills reduced the cholesterol, serum lipids, and fasting blood sugar (Faroughi et al. 2018). Furthermore, Zhai et al. (2018) described that the action of garlic in decreasing diabetes in rats was related to the results of glibenclamide and insulin.

2.4 Turmeric

Turmeric is a spice that is beloved to India and South Asia with family as Zingiberaceae. The phenols are the abundant antioxidant present in them due to which it is utilized in food items and nutraceuticals (Jurenka 2009). Diabetes and obesity are the most prevalent. These diseases are resulted due to the intake of more calories, processed foods, and lack of physical activity (Han et al. 2010). Previous researches proved that the valuable properties of turmeric in diabetes mellitus and its related problems are due to its capability to control glucose level (Prasad et al. 2014; Kato et al. 2017). Also, chronic inflammation is linked with obesity and resulted in the start of diabetes, turmeric can manage diabetes (Hotamisligil 2017; Hajavi et al. 2017).

2.5 Cumin Seeds

Cumin (*Cuminum cyminum L.*) is a special spice and the seeds of this plant are used as flavor enhancers in the preparation of different dishes. The seed oil of cumin is also used in the food, perfumery, beverage, and drug industries. The specific aroma of cumin is due to cumin aldehyde and cumin alcohol. Furthermore, cumin seeds have a lot of antioxidants as shown in Fig. 1 due to which it is utilized in different industrial products as well as medicinal products (Chaudhry et al. 2020). Cumin is a spice that flavors many dishes from different areas of the world like India, North Africa, and the Middle East. Moreover, cumin seeds have been used for the treatment of different health issues like diarrhea, jaundice, and dyspepsia due to the active ingredient, cumin aldehyde. Current researches proved the anti-obese and anti-diabetic effects of cumin seeds (Iyer et al. 2009). Moreover, in another study done on rabbits, it was verified that cumin seeds were effective for controlling diabetes (Iyer et al. 2009).

Recent studies found that different levels of cumin seeds have a blood glucose-lowering effect in diabetic people (Bamosa et al. 2010). Furthermore, cumin addition was proved to be more effective than glibenclamide (an anti-diabetic drug) in the treatment of diabetes (Ismail et al. 2010a, b). Black cumins are also verified as more operative as anti-obesity, and hypoglycemic agent (Yaheya and Ismail 2010). According to another research, cumin seeds were found to have anti-obese and hypoglycemic activity in rats. In controlling diabetes, cumin seeds were proved more effective than a synthetic anti-diabetic drug such as glibenclamide (Yaheya

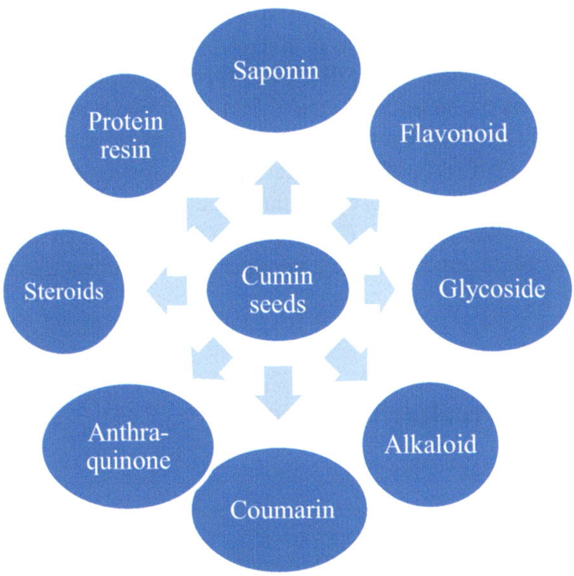

Fig. 1 Active components present in Cumin seed (Al-Snafi 2016)

and Ismail 2010). Another study confirmed that the methanolic extract of *C. cyminum* seed was capable of modulating blood glucose in diabetic rats (Jagtap and Patil 2010). In a study, El-Bahr et al. (2014) reported significant effects of turmeric and black cumin seed mixture on the body weight of diabetic rats, when compared to untreated diabetic rats.

2.6 Ginger

Ginger (*Zingiber officinale Roscoe*) is a very famous and useful spice with the family Zingiberaceae. It is utilized throughout the world due to its specific taste and smell. Furthermore, ginger is one of the most widespread natural therapeutic herbs used in different regions of the world for the treatment of different degenerative diseases like diabetes, heart diseases, etc. (Attokaran 2017). The previous studies verified that ginger considerably reduced the fasting blood glucose level of diabetic rats. Moreover, glucose intolerance is significantly enhanced with ginger (Madkor et al. 2011). The control of glucose is due to inhibition of α-glucosidase and α-amylase which are key enzymes in the digestion and absorption of complex carbohydrates (Priya Rani et al. 2011). Jafri et al. (2011) showed that the administration of ginger extract in the rats with diabetes resulted in a lowering of blood glucose levels.

Abdulrazaq et al. (2012) stated that the everyday addition of water extract of ginger in the diet for 30 days in diabetic rats resulted in lowering of blood glucose levels. Arabloo found that the daily addition of ginger in diabetic patients resulted in a lowering of fasting blood glucose and HbA1C level. Moreover, Lee et al. (2015) proved that 6-Gingerol a component of ginger exhibited strong action in exciting glucose metabolism. Shidfar et al. (2015) also reported a momentous improvement in the insulin level by the addition of ginger. Similarly, Wei et al. (2017) showed that 6-Paradol and 6-shogaol (pungent constituents of ginger) also have anti-diabetic potential. Zhu et al. (2018) found that ginger promotes insulin sensitivity. Moreover, it lowered insulin resistance in rats by regulating cell energy metabolism. According to another study, Vasconcelos et al. (2019) proved that ginger can increase the activity of hepatic glycolytic enzymes like glucokinase, phosphofructokinase, and pyruvate kinase.

3 Spices in the Treatment of Cancer

3.1 Basil

Basil commonly known as *Ocimum basilicum* by its scientific name is widely found in Iran, Asia, and the region of India. Moreover, many clinical studies investigated the anti-oxidative, antimicrobial, and anticancer properties of basil (Müller et al. 1994; Chiang et al. 2005; Makri and Kintzios 2007). Many clinical trials revealed that basil not only prevents bacterial infections but also lower oxidative stress. According to previous research done on mice for 15 days to analyze the oxidative lowering property of basil. The results revealed that basil not only increases the ratio of glutathione reductase but also helpful in increasing the endogenous antioxidant called superoxide dismutase (Dasgupta et al. 2004).

Other studies indicated that the extract of basil decreases skin tumors in mice induced by DMBA commonly known as 9, 10-dimethyl-1, 2-benzathracene. A diet containing basil extract ranging from 150 to 300 mg/kg reduces tumor burden (Dasgupta et al. 2004). Furthermore, another research conducted to investigate the anti-cancer property of basil found that it helps increase the ratio of MGMT, known as a healing protein; usually act against the DNA alkylation damage (Niture et al. 2006).

3.2 Caraway

The herb meridian fennel, usually renowned by its scientific name *Carum carvi.* It is widely found in areas of Asia, regions of Southern Africa, and Europe. The main constituents of oil extracted from caraway are limonene, *p*-mentha-1, 8-die-2-one, they are known as the derivative of anethofuran and carvone (Zheng et al. 1992). Moreover, caraway is renowned for its antioxidative trait. It has high reduced activity and free radical scavenging activity due to which caraway removes diphenyl picrylhydrazyl radicals and reduces the oxidative stress by donating its hydrogen atom and neutralizing the free radical (Kapoor et al. 2010). Moreover, it was found that less than 0.13 μM of caraway extract inhibits the action of CYP1A1 (Naderi-Kalali et al. 2005). Meanwhile, a 23-week study on mice examined the anti-oxidative effect of caraway oil. According to results, it was investigated that the incidence of skin tumors induced by DMBA reduces by intake of caraway oil that has carcinogenic gene blocking property (Shwaireb 1993).

3.3 Cardamom

Cardamom belongs to the family of the ginger family called Zingiberaceae. This special herb is mostly used in cooking in India. The clinical trial was performed on Wistar rats, this study investigated that there was a rapid decline in highly reactive substances found in heart tissue called 2-thiobarbituric acid by introducing a high-fat diet along with 10% of cardamom seed powder. However, it has also been reported that cardamom helps to increases the endogenous antioxidants' heart and liver like catalase and super dismutase (Dhuley 1999). On the other side, other studies also reported that cardamom can cause a significant decline in liver reactive constituents like CYP. Moreover, the anti-carcinogenic property of cardamom has also been investigated, it has been reported that there is a significant increase in antioxidant enzymes called GST (Banerjee et al. 1994). The studies also demonstrated that the anti-inflammatory and tumor-inhibiting properties of cardamom. Cardamom helps to decrease the incidence of colon cancer induced by azoxymethane. Other than this, it has been reported that cardamom helps to decrease lipid peroxidation and increases the toxin removing enzymes (Bhattacharjee et al. 2007).

3.4 *Rosemary*

The medicinal herb *Rosmarinus officinalis L.* belongs to the family of Lamiaceae, it has been reported that the rosemary extract contains the specialized polyphenolic compounds such as rosmarinic acid and diterpenes carosonic acid (Cuvelier et al. 1994). It has been reported that prostate tissue examination of mice fed with rosemary extract shows a significant decline in androgen receptors and demonstrates the decline in the level of prostate-specific antigens (Petiwala et al. 2014). Moreover, other studies reported that rosemary extract in a dosage of 5–20-Mis beneficial in the treatment of colon cancer as it decreases the COX2 initiator activity and reduces the level of COX2 protein in HT29 cancer treatment (Scheckel et al. 2008). Apart from that the active compound carnosic acid present in rosemary extract elevates the level of GSH and phase II enzyme termed as NADP (H)-quinone reductase, these specialized antioxidant shields the cell from various cariogenic toxins (Shabtay et al. 2008). However, the studies explored through the serum level of rats that rosemary extract reduces inflammation by alleviating the level of leptin, IGF-1, and insulin level (Kim et al. 2014a, b).

3.5 *Cumin*

Cumin is a native herb of India and the Mediterranean belongs to the family of Apiaceae. The major active component present in cumin seed is Thymoquinone, it is investigated that this active compound exhibits the free radical scavenging property. Moreover, it also reduces inflammation and prevents the action of chemo toxins (Allahghadri et al. 2010; Nader et al. 2010). Another research reported that thymoquinone decreases the chances of cancer development by altering the cell division of cancer cells by causing a decline in the synthesis of VEGF, cyclin D1 and BclxL (Aggarwal et al. 2008). Apart from this, the consumption of 6 mg/kg/day of thymoquinone has been reported to prevent prostate cancer by inhibiting tumor angiogenesis (Yi et al. 2008). Other epidemiological studies indicated that *Nigella sativa* exhibits anticancer properties and helps in the prevention of breast cancer by inactivating the breast cancer gene termed as MCF-7. Moreover, the mechanism behind this evaluated that inoculation of an aqueous solution of cumin seeds with the combination of melatonin and retinoic acid to rats inhibited the effect of DMBA present in mammary cancer cells (El-Aziz et al. 2005).

3.6 Turmeric

Turmeric is mainly used as a flavoring and coloring spice, mainly contain the active polyphenol termed as curcumin. Apart from this it has been reported the other part of turmeric mainly rhizomes exhibit the anti-neoplastic property (Kewitz et al. 2013; Devassy et al. 2015; Shanmugam et al. 2015). Many studies evaluated the protective effect of curcumin against nasopharyngeal cancer, as it is reported that curcumin alleviates the miR-125a-5p synthesis and elevates the synthesis of tumor protein 53 genes (Gao et al. 2014). Other clinical trials investigated that curcumin decreases the level of certain interleukins like interleukin-2 and interleukin-6. Apart from this curcumin increases the synthesis of enzymes like glutathione peroxidase, glutathione reductases, and superoxide dismutase. Therefore, turmeric has a beneficial effect on liver cancer development (Kadasa et al. 2015; Dai et al. 2013).

3.7 Garlic

Allium sativum is the widest spice used all over the world mainly characterized by its anticancer property, as it contains sulfur-containing compounds such as allicin, diallyl sulfide, and diallyl trisulfide (Czepukojc et al. 2014). Many studies revealed that garlic has a protective effect against colorectal cancer as it contains the natural selenium which is termed as selenomethionine and se-methyl-l-selenocysteine (MseC), it has been revealed that this organic selenium tends to exhibit anticancer property than inorganic selenium. Moreover, it has been analyzed that MseC has been involved in apoptosis of colo 205 cells and also regulates the apoptosis induced by endoplasmic reticulum stress (Tung et al. 2015). Other studies revealed that *Allium sativum* also has an inhibitory effect on breast cancer formation. Furthermore, diallyl sulfide in it prevents damage to DNA in mammary epithelial cells caused due to diethylstilbestrol, apart from this it also alleviates the lipid peroxidation (McCaskill et al. 2014). Similarly, it has been observed in another research that diallyl trisulfide inhibits the action of estrogen receptor -α activity (Hahm et al. 2014).

3.8 Black Pepper

Black pepper also is known as *Piper nigrum* contains the active constituents known as piperine. Many studies evaluated that piperine prevents the incidence of breast cancer by inhibiting breast cancer formation due to hormone alteration and abnormal growth of breast cancer cells. It promotes apoptosis of neoplastic cells through mitochondrial pathways (Greenshields et al. 2015).

Besides, it was also observed that the proliferation of prostate cancer cells like LNCaP and DU-145 was suppressed by piperine. Apart from this, it was also investigated that piperine inhibits both the tumor growth caused by androgen and tumor growth without androgen (Samykutty et al. 2013). Many studies indicated that piperine inhibits the undifferentiated growth of HT-29 colon neoplastic cells by inducing apoptosis and cell cycle arrest at G1phase (Yaffe et al. 2015).

3.9 Red Chili

Red chili (*Capsicum annum*) mainly contains an active compound called capsaicin. This compound is renowned for its anticancer property. It has been observed that capsaicin inhibits angiogenesis in lung cancer cells. Moreover, it also exerts pro-apoptotic activity (Chakraborty et al. 2014; Lau et al. 2014). Besides, the inhibitory effect of red chili water extracts it was examined that this extract has a protective effect on the growth of breast cancer cells. Moreover, this extract tends to decrease the level of MMP-2 and MMP-9. However; the level of E-cadherin was elevated by consuming red chili water extract (Kim et al. 2014a, b). Other studies revealed that capsaicin inhibits the growth of gastric cancer cells by promoting apoptosis through elevating the level of cleaved caspase-3 and alleviating the level of MAPK and phosphorylated ERK ½ (Park et al. 2014).

3.10 Ginger

Ginger also is known as *Zingiber officinale* contain the aromatic compound like gingerols, shogaols, and paradols. It has been observed that these non-volatile compounds exhibit anti-microbial, anti-inflammatory, and anticancer properties (Karna et al. 2012). It has been observed by many clinical trials that the main constitute of ginger known as 6-shogaol prevent breast cancer by preventing the production of proinflammatory chemokines (Hsu et al. 2015). Apart from this, it was also noticed that ginger elevates the level of adiponectin and GPX and mainly alleviates the level of MDA in women with high BMI and suffering from breast cancer (Karimi and Roshan 2013). Moreover, it was observed that the extract of ginger helps prevent colorectal cancer cells as it induces apoptosis and inhibits cell growth (Radhakrishnan et al. 2014). According to another study, it was also observed that 6-shogaol decreases the expression of genes such as interleukin −6, TNK- α, hence protecting against prostate cancer (Saha et al. 2014).

3.11 Saffron

Saffron the most expensive spice in the world also known as *Crocus sativus*, the main active compound present in saffron is crocin and crocetin. It was noticed that the main function of these bioactive compounds is to promote apoptosis and inhibit cell growth (Zhang et al. 2013; Gutheil et al. 2012). Another research observed that water extract of saffron activates the caspase-dependent pathway present in A549 cells which ultimately promote apoptosis of lung cancer cell and inhibit the undifferentiated cell growth (Samarghandian et al. 2013). Similarly, it was observed that the extract of saffron which mostly contains crocin prevents the proliferation of cells and promotes the Bax gene expression but decreases the Bcl-2 expression (D'Alessandro et al. 2013). Likewise, another research elaborated that saffron can inhibit ovarian cancer occurrence by inhibiting the growth of HO-8910 cells, induce apoptosis, and inhibit cell proliferation by regulating the apoptosis pathway of caspase −3.

4 Spices in the Treatment of Obesity

4.1 Ginger

Ginger prevents the hydrolysis fat and inhibits the absorption of fats, hence increasing the excretion of fats (Masuda et al. 2003). Several studies explored the mechanism of thermogenesis induced by garlic and it activates the specific receptor located on skeletal muscles and in hepatic cells termed as peroxisomes proliferator receptors pathway. They are responsible for fatty acid oxidation (Niu et al. 2011). Moreover, the extract of ginger inhibits synthesis of cyclooxygenase and lipo-oxygenase. Moreover, elevates the production of leukotrienes (Murase et al. 2009). The systematic review evaluates that ginger act as an appetite suppresser; it inhibits the absorptions of fats and increases the excretion of fats. Moreover, ginger increases lipolysis, thus being effective in weight loss (Ebrahimzadeh Attari et al. 2018). Similarly, certain results of studies indicated that 6-gingerol and 6-shogaol modulate the PPAR-γ pathway in adipocytes. Additionally, a recent analysis concluded that ginger regulates the expression of PPARδ gene, thus decrease the incidence of obesity (Misawa et al. 2015).

4.2 Turmeric

Curcumin (a bioactive compound in turmeric) prevents the phosphorylation of mitogen-activated protein kinase located in adipose tissue, Hence, suppressing the differentiation of adipose tissue (Di Pierro et al. 2015). Turmeric inhibits fatty acid synthase that is responsible for inducing lipogenesis in adipose cells (Ahn et al. 2010). Another previous research evaluated that curcumin inactivates the transcription factor gene termed as Egr-1, which regulates the expression of plasminogen activator inhibitor; it has been known that the Egr-1 gene causes obesity and insulin resistance (Pendurthi and Rao 2000). It has been known that this spice alleviates phosphorylation of leptin receptors, therefore, inhibiting the signaling pathway of leptin (Tang et al. 2009). It has also been studied that curcumin interferes with the pathway of Wny/β-catenin, which has a direct relation with obesity and metabolic disorders (Jaiswal et al. 2002). It was examined that turmeric decreases insulin resistance by alleviating the expression of transcription proteins such as TNF-α and IL-6, theses protein are reduced by inactivating the NF-KB. The study stated that TNF-α stimulates adipose tissue growth.

4.3 Garlic

Garlic inhibits fat cell differentiation by inactivating the transcription protein like PPAR-γ. Moreover, it also activates a gene called UCP-2 (Kim et al. 2007). Other studies indicate that the extract of garlic metabolizes lips and decreases cholesterol levels by alleviating the level of enzyme binding protein known as 1C(SREBP-1C) (Ha et al. 2015). The clinical findings evaluate that garlic is effective in the treatment of obesity as it reduces fat mass, total body cholesterol, and decreases the level of low-density lipoprotein (Zhang et al. 2019). Moreover, the studies revealed that the active component of garlic known as Ajoene promotes apoptosis, and inhibits the accumulation of lipids in adipocytes (Yang et al. 2006). The recent analysis demonstrates that Ajoene reduces weight making no change in appetite or food intake amount (Han et al. 2011). However, it has been reported that 1,2-vinyldithiin inhibits adiposes tissue differentiation (Keophiphath et al. 2009). Other than this, garlic increases UCP-1 expression which results cause a thermogenic effect and thus reducing weight (Pulinilkunnil et al. 2011).

4.4 Red Pepper

It has been revealed that capsaicin acts as an anti-obesity as by increasing the level of feeling of fullness, Hence, decreasing the consumption of food intake. Moreover, elevate the level of catecholamines, activate the receptors located in adipose tissues like UCP2 and UCP3. Other than this it has been reported it also activates AMMPK and inhibits high ghrelin production. Other studies revealed that red pepper increases the level of oxygen in the body. Hence, increasing the burning of calories (Masuda et al. 2003). A study in Denmark on 27 individuals demonstrated that by consumption of capsaicin helps to reduce appetite during high energy consumption. Moreover, it has also been reported that it alleviates hunger and promotes satiety, thus effective in treating obesity (Janssens et al. 2014). 10-week research conducted in Korea on mice concluded that the intake of capsaicin along with a high-fat diet reduced insulin insensitivity and suppresses the accumulation of hepatocytes in the liver of mice. Moreover, red pepper also decreases inflammatory mediators such as interleukin-6, MCP-1, and tumor-promoting factors (Kang et al. 2010). Other studies indicate that red pepper enhances the expression of PPAR and decreases leptin production. Moreover, it decreases triglycerides and improves lipid profile (Tan et al. 2014). Other than this, researchers concluded that capsaicin enhances thermogenesis (Joo et al. 2010).

5 Conclusion

Spices that are mostly utilized as coloring agents, flavoring agents, and also as preservatives in different dishes all over the world. Spices have various important constituents in them due to which they have many health benefits. The effect of Different spices on the prevention and treatment of obesity, diabetes, and cancer have been studied. Different studies on the anti-obesity, anti-diabetic and anti-cancer effects of different spices showed that they are more effective, safe, non-toxic, and also cheap as compared to artificial drugs available in the market that are expensive and when used for a long period then result in toxicity.

References

Abdulrazaq, N. B., Cho, M. M., Win, N. N., Zaman, R., & Rahman, M. T. (2012). Beneficial effects of ginger (Zingiber officinale) on carbohydrate metabolism in streptozotocin-induced diabetic rats. *British Journal of Nutrition, 108*(7), 1194–1201.

Abou El-Soud, N. H., Khalil, M. Y., Hussein, J. S., Oraby, F. S. H., & Farrag, A. H. (2007). Antidiabetic effects of fenugreek alkaliod extract in streptozotocin induced hyperglycemic rats. *Journal of Applied Sciences Research, 3*(10), 1073–1083.

Acharya, S. N., Thomas, J. E., & Basu, S. K. (2008). Fenugreek, an alternative crop for semiarid regions of North America. *Crop Science, 48*(3), 841–853.

Afrisham, R., Aberomand, M., Ghaffari, M. A., Siahpoosh, A., & Jamalan, M. (2015). Inhibitory effect of *Heracleum persicum* and *Ziziphus jujuba* on activity of alpha-amylase. *Journal of Botany*, Volume 2015, Article ID 824683, 1–8 Pages.

Aggarwal, B. B., Kunnumakkara, A. B., Harikumar, K. B., Tharakan, S. T., Sung, B., & Anand, P. (2008). Potential of spice-derived phytochemicals for cancer prevention. *Planta Medica, 74*(13), 1560–1569.

Ahn, J., Lee, H., Kim, S., & Ha, T. (2010). Curcumin-induced suppression of adipogenic differentiation is accompanied by activation of Wnt/β-catenin signaling. *American Journal of Physiology-Cell Physiology, 298*(6), C1510–C1516.

Airaodion, A. I., Adeniji, A. R., Ogbuagu, E. O., Ogbuagu, U., & Agunbiade, A. P. (2019a). Hypoglycemic and hypolipidaemic activities of methanolic extract of *Talinum triangulare* leaves in Wistar rats. *International Journal of Bio-Science and Bio-Technology, 11*(5), 1–13.

Airaodion, A. I., Airaodion, E. O., Ogbuagu, E. O., Ogbuagu, U., & Osemwowa, E. U. (2019b). Effect of oral intake of African locust bean on fasting blood sugar and lipid profile of albino rats. *Asian Journal of Research in Biochemistry*, *4*(4), 1–9.

Airaodion, A. I., Akinmolayan, J. D., Ogbuagu, E. O., Airaodion, E. O., Ogbuagu, U., & Awosanya, O. O. (2019c). Effect of methanolic extract of *Corchorus olitorius* Leaves on hypoglycemic and hypolipidaemic activities in albino rats. *Asian Plant Research Journal*, 2(4), 1–13.

Akilen, R., Tsiami, A., Devendra, D., & Robinson, N. (2010). Glycated haemoglobin and blood pressure-lowering effect of cinnamon in multi-ethnic type 2 diabetic patients in the UK: A randomized, placebo-controlled, double-blind clinical trial. *Diabetic Medicine, 27*, 1159–1167.

Al-Brakati, A. Y. (2016). Protective effect of garlic against diabetic retinopathy in adult albino rats. *Research Journal of Pharmaceutical, Biological and Chemical Sciences, 7*(5), 2748–2759.

Allahghadri, T., Rasooli, I., Owlia, P., Nadooshan, M. J., Ghazanfari, T., Taghizadeh, M., & Astaneh, S. D. A. (2010). Antimicrobial property, antioxidant capacity, and cytotoxicity of essential oil from cumin produced in Iran. *Journal of Food Science, 75*(2), H54–H61.

Al-Snafi, A. E. (2016). The pharmacological activities of *Cuminum cyminum*-A review. *IOSR Journal of Pharmacy, 6*(6), 46–65.

Anand, P., Murali, K. Y., Tandon, V., Murthy, P. S., & Chandra, R. (2010). Insulinotropic effect of cinnamaldehyde on transcriptional regulation of pyruvate kinase, phosphoenolpyruvate carboxykinase, and GLUT4 translocation in experimental diabetic rats. *Chemico-Biological Interactions, 186*(1), 72–81.

Arbo, M. D., Larentis, E. R., Linck, V. M., Aboy, A. L., Pimentel, A. L., Henriques, A. T., et al. (2008). Concentrations of p-synephrine in fruits and leaves of Citrus species (Rutaceae) and the acute toxicity testing of *Citrus aurantium* extract and p-synephrine. *Food and Chemical Toxicology, 46*(8), 2770–2775.

Arbo, M. D., Schmitt, G. C., Limberger, M. F., Charão, M. F., Moro, Â. M., Ribeiro, G. L., et al. (2009). Subchronic toxicity of *Citrus aurantium* L.(Rutaceae) extract and p-synephrine in mice. *Regulatory Toxicology and Pharmacology, 54*(2), 114–117.

Attokaran, M. (2017). *Natural food flavors and colorants*. Wiley.

Badal, D. S., Dwivedi, A. K., Kumar, V., Singh, S., Prakash, A., Verma, S., & Kumar, J. (2019). Effect of organic manures and inorganic fertilizers on growth, yield and its attributing traits in garlic (*Allium sativum* L.). *Journal of Pharmacognosy and Phytochemistry, 8*, 587–590.

Bamosa, A. O., Kaatabi, H., Lebdaa, F. M., Elq, A. M., & Al-Sultanb, A. (2010). Effect of *Nigella sativa* seeds on the glycemic control of patients with type 2 diabetes mellitus. *Indian Journal of Physiology and Pharmacology, 54*(4), 344–354.

Banerjee, S., Sharma, R., Kale, R. K., & Rao, A. R. (1994). Influence of certain essential oils on carcinogen-metabolizing enzymes and acid-soluble sulfhydryls in mouse liver. *Nutrition and Cancer, 21*(3), 263–269.

Berroukche, A., Terras, M., Kharoubi, M., Boudia, H., Kharoubi, O., Dellaoui, H., Lansari, W., & Zerarki, I. (2018). Assessing of funegreek (*Trigonella foenumgraecum*) seeds pharmacological effects against glucotoxicity and lipotoxicity in streptozotocine-induced diabetic rats. *Diabetes Updates, 5*, 1–5.

Bhattacharjee, S., Rana, T., & Sengupta, A. (2007). Inhibition of lipid peroxidation and enhancement of GST activity by cardamom and cinnamon during chemically induced colon carcinogenesis in Swiss albino mice. *Asian Pacific Journal of Cancer Prevention, 8*(4), 578–582.

Bi, X., Lim, J., & Henry, C. J. (2017). Spices in the management of diabetes mellitus. *Food Chemistry, 217*, 281–293.

Bindu, B., & Podikunju, B. (2015). Performance evaluation of onion (*Allium cepa* L. Var. Cepa) varieties for their suitability in kollam district. *International Journal of Research in Agricultural Science, 1*(1), 18–20.

Cao, H., Graves, D. J., & Anderson, R. A. (2010). Cinnamon extracts regulates glucose transporter and insulin-signaling gene expression in mouse adipocytes. *Phytomedicine, 17*, 1027–1032.

Chakraborty, S., Adhikary, A., Mazumdar, M., Mukherjee, S., Bhattacharjee, P., Guha, D., et al. (2014). Capsaicin-induced activation of p53-SMAR1 auto-regulatory loop down-regulates VEGF in non-small cell lung cancer to restrain angiogenesis. *PLoS One, 9*(6), e99743.

Chaudhry, Z., Khera, R. A., Hanif, M. A., Ayub, M. A., & Sumrra, S. H. (2020). Cumin. In *Medicinal plants of South Asia* (pp. 165–178). Elsevier.

Chiang, L. C., Ng, L. T., Cheng, P. W., Chiang, W., & Lin, C. C. (2005). Antiviral activities of extracts and selected pure constituents of *Ocimum basilicum*. *Clinical and Experimental Pharmacology and Physiology, 32*(10), 811–816.

Crawford, P. (2009). Effectiveness of cinnamon for lowering hemoglobin A1C in patients with type 2 diabetes: A randomized, controlled trial. *The Journal of the American Board of Family Medicine, 22*, 507–512.

Cuvelier, M. E., Berset, C., & Richard, H. (1994). Antioxidant constituents in sage (*Salvia officinalis*). *Journal of Agricultural and Food Chemistry, 42*(3), 665–669.

Czepukojc, B., Baltes, A. K., Cerella, C., Kelkel, M., Viswanathan, U. M., Salm, F., et al. (2014). Synthetic polysulfane derivatives induce cell cycle arrest and apoptotic cell death in human hematopoietic cancer cells. *Food and Chemical Toxicology, 64*, 249–257.

D'Alessandro, A. M., Mancini, A., Lizzi, A. R., De Simone, A., Marroccella, C. E., Gravina, G. L., et al. (2013). *Crocus sativus* stigma extract and its major constituent crocin possess significant antiproliferative properties against human prostate cancer. *Nutrition and Cancer, 65*(6), 930–942.

Dai, X. Z., Yin, H. T., Sun, L. F., Hu, X., Zhou, C., Zhou, Y., Zhang, W., Huang, X. E., & Li, X. C. (2013). Potential therapeutic efficacy of curcumin in liver cancer. *Asian Pacific Journal of Cancer Prevention,14*(6), 3855–9.

Dasgupta, T., Rao, A. R., & Yadava, P. K. (2004). Chemomodulatory efficacy of basil leaf (*Ocimum basilicum*) on drug metabolizing and antioxidant enzymes, and on carcinogen-induced skin and forestomach papillomagenesis. *Phytomedicine, 11*(2–3), 139–151.

Devassy, J. G., Nwachukwu, I. D., & Jones, P. J. (2015). Curcumin and cancer: Barriers to obtaining a health claim. *Nutrition Reviews, 73*(3), 155–165.

Dhuley, J. N. (1999). Anti-oxidant effects of cinnamon (*Cinnamomum verum*) bark and greater cardamom (*Amomum subulatum*) seeds in rats fed high fat diet. *Indian Journal of Experimental Biology, 37*(3), 238–242.

Di Pierro, F., Bressan, A., Ranaldi, D., Rapacioli, G., Giacomelli, L., & Bertuccioli, A. (2015). Potential role of bioavailable curcumin in weight loss and omental adipose tissue decrease: Preliminary data of a randomized, controlled trial in overweight people with metabolic syndrome. Preliminary study. *European Review for Medical and Pharmacological Sciences, 19*(21), 4195–4202.

Ebrahimzadeh Attari, V., Malek Mahdavi, A., Javadivala, Z., Mahluji, S., Zununi Vahed, S., & Ostadrahimi, A. (2018). A systematic review of the anti-obesity and weight lowering effect of ginger (*Zingiber officinale* Roscoe) and its mechanisms of action. *Phytotherapy Research, 32*(4), 577–585.

El-Aziz, M. A. A., Hassan, H. A., Mohamed, M. H., Meki, A. R. M., Abdel-Ghaffar, S. K., & Hussein, M. R. (2005). The biochemical and morphological alterations following administration of melatonin, retinoic acid and *Nigella sativa* in mammary carcinoma: An animal model. *International Journal of Experimental Pathology, 86*(6), 383–396.

El-Bahr, S. M., Taha, N. M., Korshom, M. A., Mandour, A. E., & Lebda, M. A. (2014). Influence of combined administration of turmeric and black cumin seed on selected biochemical parameters of diabetic rats. *Alexandria Journal of Veterinary Sciences, 41*(1), 19–27.

Faroughi, F., Charandabi, S. M. A., Javadzadeh, Y., & Mirghafourvand, M. (2018). Effects of garlic pill on blood glucose level in borderline gestational diabetes mellitus: A triple blind, randomized clinical trial. *Iranian Red Crescent Medical Journal, 20*(7), e60675.

Gao, W., Chan, J. Y. W., & Wong, T. S. (2014). Curcumin exerts inhibitory effects on undifferentiated nasopharyngeal carcinoma by inhibiting the expression of miR-125a-5p. *Clinical Science, 127*(9), 571–579.

Goel, B., & Mishra, S. (2020). Medicinal and nutritional perspective of cinnamon: A mini-review. *European Journal of Medicinal Plants, 31*(3), 10–16.

Greenshields, A. L., Doucette, C. D., Sutton, K. M., Madera, L., Annan, H., Yaffe, P. B., et al. (2015). Piperine inhibits the growth and motility of triple-negative breast cancer cells. *Cancer Letters, 357*(1), 129–140.

Gutheil, G. W., Reed, G., Ray, A., Anant, S., & Dhar, A. (2012). Crocetin: An agent derived from saffron for prevention and therapy for cancer. *Current Pharmaceutical Biotechnology, 13*(1), 173–179.

Ha, A. W., Ying, T., & Kim, W. K. (2015). The effects of black garlic (*Allium satvium*) extracts on lipid metabolism in rats fed a high fat diet. *Nutrition Research and Practice, 9*(1), 30–36.

Hajavi, J., Momtazi, A. A., Johnston, T. P., Banach, M., Majeed, M., & Sahebkar, A. (2017). Curcumin: A naturally occurring modulator of adipokines in diabetes. *Journal of Cellular Biochemistry, 118*(12), 4170–4182.

Han, C. Y., Ki, S. H., Kim, Y. W., Noh, K., Lee, D. Y., Kang, B., et al. (2011). Ajoene, a stable garlic by-product, inhibits high fat diet-induced hepatic steatosis and oxidative injury through LKB1-dependent AMPK activation. *Antioxidants & Redox Signaling, 14*(2), 187–202.

Han, J. C., Lawlor, D. A., & Kimm, S. Y. (2010). Childhood obesity. *The Lancet, 375*(9727), 1737–1748.

Haque, N., Salma, U., Nurunnabi, T. R., Uddin, M. J., Jahangir, M. F., Islam, S. M., & Kamruzzaman, M. (2011). Management of type 2 diabetes mellitus by lifestyle, diet and medicinal plants. *Pakistan Journal of Biological Sciences: PJBS, 14*(1), 13–24.

Hlebowicz, J., Hlebowicz, A., Lindstedt, S., Björgell, O., Höglund, P., Holst, J. J., et al. (2009). Effects of 1 and 3 g cinnamon on gastric emptying, satiety, and postprandial blood glucose, insulin, glucose-dependent insulinotropic polypeptide, glucagon-like peptide 1, and ghrelin concentrations in healthy subjects. *The American Journal of Clinical Nutrition, 89*(3), 815–821.

Hotamisligil, G. S. (2017). Inflammation, metaflammation and immunometabolic disorders. *Nature, 542*(7640), 177–185.

Howard, M. E., & White, N. D. (2013). Potential Benefits of Cinnamon in Type 2 Diabetes. *American Journal of Lifestyle Medicine, 7*(1), 23–26.

Hsu, Y. L., Hung, J. Y., Tsai, Y. M., Tsai, E. M., Huang, M. S., Hou, M. F., & Kuo, P. L. (2015). 6-shogaol, an active constituent of dietary ginger, impairs cancer development and lung metastasis by inhibiting the secretion of CC-chemokine ligand 2 (CCL2) in tumor-associated dendritic cells. *Journal of Agricultural and Food Chemistry, 63*(6), 1730–1738.

Ismail, M. Y. M., Assem, N. M., & Mohammad, Z. (2010a). Role of spices in diabetes mellitus. *Research Journal of Pharmaceutical, Biological and Chemical Sciences, 1*(3), 30–34.

Ismail, M. Y. M., Assem, N. M., & Zakriya, M. (2010b). Botanicals promoting oral and dental hygiene: A review. *Research Journal of Pharmaceutical, Biological and Chemical Sciences, 1*(2), 202–206.

Iyer, A., Panchal, S., Poudyal, H., & Brown, L. (2009). Potential health benefits of Indian spices in the symptoms of the metabolic syndrome: A review. *Journal of Biochemistry and Biophysics, 46*(6), 467–481.

Jafri, S. A., Abass, S., & Qasim, M. (2011). Hypoglycemic effect of ginger (*Zingiber officinale*) in alloxan induced diabetic rats (*Rattus norvagicus*). *Pakistan Veterinary Journal, 31*(2), 160–162.

Jagtap, A. G., & Patil, P. B. (2010). Antihyperglycemic activity and inhibition of advanced glycation end product formation by *Cuminum cyminum* in streptozotocin induced diabetic rats. *Food and Chemical Toxicology, 48*(8–9), 2030–2036.

Jaiswal, A. S., Marlow, B. P., Gupta, N., & Narayan, S. (2002). β-catenin-mediated transactivation and cell–cell adhesion pathways are important in curcumin (diferuylmethane)-induced growth arrest and apoptosis in colon cancer cells. *Oncogene, 21*(55), 8414–8427.

Janssens, P. L., Hursel, R., & Westerterp-Plantenga, M. S. (2014). Capsaicin increases sensation of fullness in energy balance, and decreases desire to eat after dinner in negative energy balance. *Appetite, 77*, 46–51.

Jayaprakasha, G. K., Ohnishi-Kameyama, M., Ono, H., Yoshida, M., & Jaganmohan, R. L. (2006). Phenolic constituents in the fruits of Cinnamomum zeylanicum and their antioxidant activity. *Journal of Agricultural and Food Chemistry, 54*, 1672–1679.

Jie, Z., Zhou, Y., Li, Y., Xu, D. P., Li, S., & Li, H. B. (2016). Spices for prevention and treatment of cancers. *Nutrients, 8*, 495.

Joo, J. I., Kim, D. H., Choi, J. W., & Yun, J. W. (2010). Proteomic analysis for antiobesity potential of capsaicin on white adipose tissue in rats fed with a high fat diet. *Journal of Proteome Research, 9*(6), 2977–2987.

Jurenka, J. S. (2009). Anti-inflammatory properties of curcumin, a major constituent of *Curcuma longa*: A review of preclinical and clinical research. *Alternative Medicine Review, 14*(2), 141–153.

Kadasa, N. M., Abdallah, H., Afifi, M., & Gowayed, S. (2015). Hepatoprotective effects of curcumin against diethyl nitrosamine induced hepatotoxicity in albino rats. *Asian Pacific Journal of Cancer Prevention, 16*(1), 103–108.

Kang, J. H., Tsuyoshi, G., Han, I. S., Kawada, T., Kim, Y. M., & Yu, R. (2010). Dietary capsaicin reduces obesity-induced insulin resistance and hepatic steatosis in obese mice fed a high-fat diet. *Obesity, 18*(4), 780–787.

Kapoor, I. P. S., Singh, B., Singh, G., De Heluani, C. S., De Lampasona, M. P., & Catalan C. A. (2010). Chemistry and antioxidant activity of essential oil and oleoresins of black caraway (*Carum bulbocastanum*) fruits. Part 69. *Journal of the Science of Food and Agriculture, 90*(3), 385–390.

Karimi, N., & Roshan, V. D. (2013). Change in adiponectin and oxidative stress after modifiable lifestyle interventions in breast cancer cases. *Asian Pacific Journal of Cancer Prevention, 14*(5), 2845–2850.

Karna, P., Chagani, S., Gundala, S. R., Rida, P. C., Asif, G., Sharma, V., et al. (2012). Benefits of whole ginger extract in prostate cancer. *British Journal of Nutrition, 107*(4), 473–484.

Kato, M., Nishikawa, S., Ikehata, A., Dochi, K., Tani, T., Takahashi, T., et al. (2017). Curcumin improves glucose tolerance via stimulation of glucagon-like peptide-1 secretion. *Molecular Nutrition & Food Research, 61*(3), 1600471.

Keophiphath, M., Priem, F., Jacquemond-Collet, I., Clément, K., & Lacasa, D. (2009). 1,2-vinyldithiin from garlic inhibits differentiation and inflammation of human preadipocytes. *The Journal of Nutrition, 139*(11), 2055–2060.

Kershaw, E. E., & Flier, J. S. (2004). Adipose tissue as an endocrine organ. *The Journal of Clinical Endocrinology & Metabolism, 89*(6), 2548–2556.

Kewitz, S., Volkmer, I., & Staege, M. S. (2013). Curcuma contra cancer? Curcumin and Hodgkin's lymphoma. *Cancer Growth and Metastasis, 6*, S11113.

Kim, H. A., Kim, M. S., Kim, S. H., & Kim, Y. K. (2014a). Pepper seed extract suppresses invasion and migration of human breast cancer cells. *Nutrition and Cancer, 66*(1), 159–165.

Kim, Y., Lee, M. S., Kim, J. S., & Lee, H. S. (2007). Garlic decreases body weight via decrease of serum lipid and increase of uncoupling proteins mRNA expression. *The FASEB Journal, 21*(6), LB59.

Kim, Y. J., Kim, J. S., Seo, Y. R., Park, J. H. Y., Choi, M. S., & Sung, M. K. (2014b). Carnosic acid suppresses colon tumor formation in association with antiadipogenic activity. *Molecular Nutrition & Food Research, 58*(12), 2274–2285.

Kissebah, A. H., & Krakower, G. R. (1994). Regional adiposity and morbidity. *Physiological Reviews, 74*, 761–811.

Kondratyuk, T. P., & Pezzuto, J. M. (2004). Natural product polyphenols of relevance to human health. *Pharmaceutical Biology, 42*(Suppl. 1), 46–63.

Kooti, W., Farokhipour, M., Asadzadeh, Z., Ashtary-Larky, D., & Asadi-Samani, M. (2016). The role of medicinal plants in the treatment of diabetes: A systematic review. *Electronic Physician, 8*(1), 1832.

Lafontan, M. (2005). Fat cells: Afferent and efferent messages define new approaches to treat obesity. *Annual Review of Pharmacology and Toxicology, 45*, 119–146.

Lau, J. K., Brown, K. C., Dom, A. M., Witte, T. R., Thornhill, B. A., Crabtree, C. M., et al. (2014). Capsaicin induces apoptosis in human small cell lung cancer via the TRPV6 receptor and the calpain pathway. *Apoptosis, 19*(8), 1190–1201.

Lee, J. O., Kim, N., Lee, H. J., Moon, J. W., Lee, S. K., Kim, S. J., et al. (2015). [6]-Gingerol affects glucose metabolism by dual regulation via the AMPKα2-mediated AS160–Rab5 pathway and AMPK-mediated insulin sensitizing effects. *Journal of Cellular Biochemistry, 116*(7), 1401–1410.

Leitner, D. R., Frühbeck, G., Yumuk, V., Schindler, K., Micic, D., Woodward, E., & Toplak, H. (2017). Obesity and type 2 diabetes: Two diseases with a need for combined treatment strategies-EASO can lead the way. *Obesity Facts, 10*(5), 483–492.

Liguori, L., Califano, R., Albanese, D., Raimo, F., Crescitelli, A., & Di Matteo, M. (2017). Chemical composition and antioxidant properties of five white onion (*Allium cepa* L.) landraces. *Journal of Food Quality, 2017*, 6873651.

Link, A., Balaguer, F., & Goel, A. (2010). Cancer chemoprevention by dietary polyphenols: Promising role for epigenetics. *Biochemical Pharmacology, 80*(12), 1771–1792.

Lu, T., Sheng, H., Wu, J., Cheng, Y., Zhu, J., & Chen, Y. (2012). Cinnamon extract improves fasting blood glucose and glycosylated hemoglobin level in Chinese patients with type 2 diabetes. *Nutrition Research, 32*, 408–412.

Madkor, H. R., Mansour, S. W., & Ramadan, G. (2011). Modulatory effects of garlic, ginger, turmeric and their mixture on hyperglycaemia, dyslipidaemia and oxidative stress in streptozotocin–nicotinamide diabetic rats. *British Journal of Nutrition, 105*(8), 1210–1217.

Makri, O., & Kintzios, S. (2007). Ocimum Sp. (Basil): Botany, Cultivation, Pharmaceutical Properties, and Biotechnology. *Journal of Herbs Spices and Medicinal Plants, 13*, 123–150.

Mang, B., Wolters, M., Schmitt, B., Kelb, K., Lichtinghagen, R., Stichtenoth, D. O., & Hahn, A. (2006). Effects of a cinnamon extract on plasma glucose, HbA, and serum lipids in diabetes mellitus type 2. *European Journal of Clinical Investigation, 36*, 340–344.

Masuda, Y., Haramizu, S., Oki, K., Ohnuki, K., Watanabe, T., Yazawa, S., et al. (2003). Upregulation of uncoupling proteins by oral administration of capsiate, a nonpungent capsaicin analog. *Journal of Applied Physiology, 95*(6), 2408–2415.

McCaskill, M. L., Rogan, E., & Thomas, R. D. (2014). Diallyl sulfide inhibits diethylstilbestrol induced DNA damage in human breast epithelial cells (MCF-10A). *Steroids, 92*, 96–100.

Misawa, K., Hashizume, K., Yamamoto, M., Minegishi, Y., Hase, T., & Shimotoyodome, A. (2015). Ginger extract prevents high-fat diet-induced obesity in mice via activation of the peroxisome proliferator-activated receptor δ pathway. *The Journal of Nutritional Biochemistry, 26*(10), 1058–1067.

Müller, L., Kasper, P., Müller-Tegethoff, K., & Petr, T. (1994). The genotoxic potential in vitro and in vivo of the allyl benzene etheric oils estragole, basil oil and trans-anethole. *Mutation Research Letters, 325*(4), 129–136.

Murase, T., Misawa, K., Haramizu, S., & Hase, T. (2009). Catechin-induced activation of the LKB1/AMP-activated protein kinase pathway. *Biochemical Pharmacology, 78*(1), 78–84.

Nader, M. A., El-Agamy, D. S., & Suddek, G. M. (2010). Protective effects of propolis and thymoquinone on development of atherosclerosis in cholesterol-fed rabbits. *Archives of Pharmacal Research, 33*(4), 637–643.

Naderi-Kalali, B., Allameh, A., Rasaee, M. J., Bach, H. J., Behechti, A., Doods, K., et al. (2005). Suppressive effects of caraway (*Carum carvi*) extracts on 2, 3, 7, 8-tetrachloro-dibenzo-p-dioxin-dependent gene expression of cytochrome P450 1A1 in the rat H4IIE cells. *Toxicology In Vitro, 19*(3), 373–377.

Nagulapalli Venkata, K. C., Swaroop, A., Bagchi, D., & Bishayee, A. (2017). A small plant with big benefits: Fenugreek (*Trigonella foenum-graecum* Linn.) for disease prevention and health promotion. *Molecular Nutrition & Food Research, 61*(6), 1600950.

National Cancer Institute. (2019). Cancer Trends Progress Report. Bethesda, MD: National Cancer Institute, National Institutes of Health, Department of Health and Human Services. https://www.progressreport.cancer.gov. Accessed 7 May 2019.

Niture, S. K., Rao, U. S., & Srivenugopal, K. S. (2006). Chemopreventative strategies targeting the MGMT repair protein: Augmented expression in human lymphocytes and tumor cells by ethanolic and aqueous extracts of several Indian medicinal plants. *International Journal of Oncology, 29*(5), 1269–1278.

Niu, C. S., Wu, H. T., Cheng, K. C., Lin, K. C., Chen, C. T., & Cheng, J. T. (2011). A novel mechanism for decreasing plasma lipid level from imidazoline I-1 receptor activation in high fat diet-fed mice. *Hormone and Metabolic Research, 43*(07), 458–463.

Park, S. Y., Kim, J. Y., Lee, S. M., Jun, C. H., Cho, S. B., Park, C. H., et al. (2014). Capsaicin induces apoptosis and modulates MAPK signaling in human gastric cancer cells. *Molecular Medicine Reports, 9*(2), 499–502.

Pendurthi, U. R., & Rao, L. V. M. (2000). Suppression of transcription factor Egr-1 by curcumin. *Thrombosis Research, 97*(4), 179–189.

Petiwala, S. M., Berhe, S., Li, G., Puthenveetil, A. G., Rahman, O., Nonn, L., & Johnson, J. J. (2014). Rosemary (*Rosmarinus officinalis*) extract modulates CHOP/GADD153 to promote androgen receptor degradation and decreases xenograft tumor growth. *PLoS One, 9*(3), e89772.

Plaisier, C., Cok, A., Scott, J., Opejin, A., Bushhouse, K. T., Salie, M. J., & Louters, L. L. (2011). Effects of cinnamaldehyde on the glucose transport activity of GLUT1. *Biochimie, 93*(2), 339–344.

Power, M. L., & Schulkin, J. (2008). Sex differences in fat storage, fat metabolism, and the health risks from obesity: Possible evolutionary origins. *British Journal of Nutrition, 99*(5), 931–940.

Prasad, S., Gupta, S. C., Tyagi, A. K., & Aggarwal, B. B. (2014). Curcumin, a component of golden spice: From bedside to bench and back. *Biotechnology Advances, 32*(6), 1053–1064.

Priya Rani, M., Padmakumari, K. P., Sankarikutty, B., Lijo Cherian, O., Nisha, V. M., & Raghu, K. G. (2011). Inhibitory potential of ginger extracts against enzymes linked to type 2 diabetes, inflammation and induced oxidative stress. *International Journal of Food Sciences and Nutrition, 62*(2), 106–110.

Pulinilkunnil, T., He, H., Kong, D., Asakura, K., Peroni, O. D., Lee, A., & Kahn, B. B. (2011). Adrenergic regulation of AMP-activated protein kinase in brown adipose tissue in vivo. *Journal of Biological Chemistry, 286*(11), 8798–8809.

Radhakrishnan, E. K., Bava, S. V., Narayanan, S. S., Nath, L. R., Thulasidasan, A. K. T., Soniya, E. V., & Anto, R. J. (2014). [6]-Gingerol induces caspase-dependent apoptosis and prevents PMA-induced proliferation in colon cancer cells by inhibiting MAPK/AP-1 signaling. *PLoS One, 9*(8), e104401.

Renuka, C., Ramesh, N., & Saravanan, K. (2009). Evaluation of the antidiabetic effect of *Trigonella foenum-graecum* seed powder on alloxaninduced diabetic albino rats. *International Journal of Pharm Tech Research, 1*(4), 1580–1584.

Saha, A., Blando, J., Silver, E., Beltran, L., Sessler, J., & DiGiovanni, J. (2014). 6-Shogaol from dried ginger inhibits growth of prostate cancer cells both in vitro and in vivo through inhibition of STAT3 and NF-κB signaling. *Cancer Prevention Research, 7*(6), 627–638.

Samarghandian, S., Borji, A., Farahmand, S. K., Afshari, R., & Davoodi, S. (2013). *Crocus sativus* L.(saffron) stigma aqueous extract induces apoptosis in alveolar human lung cancer cells through caspase-dependent pathways activation. *BioMed Research International, 2013*, 417928.

Samykutty, A., Shetty, A. V., Dakshinamoorthy, G., Bartik, M. M., Johnson, G. L., Webb, B., et al. (2013). Piperine, a bioactive component of pepper spice exerts therapeutic effects on androgen dependent and androgen independent prostate cancer cells. *PLoS One, 8*(6), e65889.

Şanlier, N., & Gencer, F. (2020). Role of spices in the treatment of diabetes mellitus: A minireview. *Trends in Food Science and Technology*. https://doi.org/10.1016/j.tifs.2020.03.018.

Scalbert, A., & Williamson, G. (2000). Dietary intake and bioavailability of polyphenols. *Journal of Nutrition, 130*, 2073S-85S.

Scheckel, K. A., Degner, S. C., & Romagnolo, D. F. (2008). Rosmarinic acid antagonizes activator protein-1–dependent activation of cyclooxygenase-2 expression in human cancer and nonmalignant cell lines. *The Journal of Nutrition, 138*(11), 2098–2105.

Shabtay, A., Sharabani, H., Barvish, Z., Kafka, M., Amichay, D., Levy, J., et al. (2008). Synergistic antileukemic activity of carnosic acid-rich rosemary extract and the 19-nor Gemini vitamin D analogue in a mouse model of systemic acute myeloid leukemia. *Oncology, 75*(3–4), 203–214.

Shanmugam, M. K., Rane, G., Kanchi, M. M., Arfuso, F., Chinnathambi, A., Zayed, M. E., et al. (2015). The multifaceted role of curcumin in cancer prevention and treatment. *Molecules, 20*(2), 2728–2769.

Shen, Y., Fukushima, M., Ito, Y., Muraki, E., Hosono, T., Seki, T., & Ariga, T. (2010). Verification of the antidiabetic effects of cinnamon (*Cinnamomum zeylanicum*) using insulin-uncontrolled type 1 diabetic rats and cultured adipocytes. *Bioscience, Biotechnology, and Biochemistry, 74*(12), 2418–2425.

Shen, Y., Honma, N., Kobayashi, K., Jia, L. N., Hosono, T., Shindo, K., et al. (2014). Cinnamon extract enhances glucose uptake in 3T3-L1 adipocytes and C2C12 myocytes by inducing LKB1-AMP-activated protein kinase signaling. *PLoS One, 9*(2), e87894.

Shidfar, F., Rajab, A., Rahideh, T., Khandouzi, N., Hosseini, S., & Shidfar, S. (2015). The effect of ginger (*Zingiber officinale*) on glycemic markers in patients with type 2 diabetes. *Journal of Complementary and Integrative Medicine, 12*(2), 165–170.

Shwaireb, M. H. (1993). Caraway oil inhibits skin tumors in female BALB/c mice. *Nutrition and Cancer, 19*(3), 321–325.

Subash Babu, P., Prabuseenivasan, S., & Ignacimuthu, S. (2007). Cinnamaldehyde- a potential antidiabetic agent. *Phytomedicine, 14*, 15–22.

Sundaram, G., Theagarajan, R., Gopalakrishnan, K., Babu, G. R., & Murthy, G. D. (2020). Effect of fenugreek consumption with metformin treatment in improving plaque index in diabetic patients. *Journal of Natural Science, Biology and Medicine, 11*(1), 55.

Taher, M., Majid, F. A. A., & Sarmidi, M. R. (2006). A proanthocyanidin from *Cinnamomum zeylanicum* stimulates phosphorylation of insulin receptor in 3t3–l1 adipocytes. *Jurnal Teknologi, 44*(1), 53–68.

Tan, S., Gao, B., Tao, Y., Guo, J., & Su, Z. Q. (2014). Antiobese effects of capsaicin–chitosan microsphere (CCMS) in obese rats induced by high fat diet. *Journal of Agricultural and Food Chemistry, 62*(8), 1866–1874.

Tang, Y., Zheng, S., & Chen, A. (2009). Curcumin eliminates leptin's effects on hepatic stellate cell activation via interrupting leptin signaling. *Endocrinology, 150*(7), 3011–3020.

Tapsell, L. C., Hemphill, I., Cobiac, L., Sullivan, D. R., Fenech, M., Patch, C. S., et al. (2006). Health benefits of herbs and spices: The past, the present, the future. *Review, 185*(S4), S1–S24.

Tung, Y. C., Tsai, M. L., Kuo, F. L., Lai, C. S., Badmaev, V., Ho, C. T., & Pan, M. H. (2015). Se-methyl-L-selenocysteine induces apoptosis via endoplasmic reticulum stress and the death receptor pathway in human colon adenocarcinoma COLO 205 cells. *Journal of Agricultural and Food Chemistry, 63*(20), 5008–5016.

Upasani, S. V., Ingle, P. V., Patil, P. H., Nandedkar, R. Y., Shah, V. S., & Surana, S. J. (2013). Traditional Indian spices useful in Diabetes Mellitus—An updated review. *Journal of Pharmaceutical and BioSciences, 4*, 157–161.

Vanitha, M., Karpagam, T., Varalakshmi, B., & Pandian, R. S. (2012). A comparative study on the anti-diabetic potential of *Aloe vera* gel and fenugreek seeds on experimentally induced diabetic rats. *Pharmacognosy Communications, 2*, 57–61.

Vasconcelos, E. F., Mota, N. F., Gomes-Rochette, D. C. S., Nunes Pinheiro, S. M., & Nabavi, D. F. (2019). Ginger (Zingiber officinale Roscoe). *Nonvitamin and Nonmineral Nutritional Supplements*, 235–9. Elsevier..

Viuda-Martos, M., Ruiz-Navajas, Y., Fernández-López, J., & Pérez-Álvarez, J. A. (2010). Spices as functional foods. *Critical Reviews in Food Science and Nutrition, 51*(1), 13–28.

Wang, J., Zhang, X., Lan, H., & Wang, W. (2017). Effect of garlic supplement in the management of type 2 diabetes mellitus (T2DM): A meta-analysis of randomized controlled trials. *Food & Nutrition Research, 61*(1), 1377571.

Wei, C. K., Tsai, Y. H., Korinek, M., Hung, P. H., El-Shazly, M., Cheng, Y. B., et al. (2017). 6-paradol and 6-shogaol, the pungent compounds of ginger, promote glucose utilization in adipocytes and myotubes, and 6-paradol reduces blood glucose in high-fat diet-fed mice. *International Journal of Molecular Sciences, 18*(1), 168.

World Health Organization. (2000). *Obesity: Preventing and managing the global epidemic (No. 894)*. World Health Organization.

World Health Organization. (2016). World Health Organization obesity and overweight fact sheet.

Yaffe, P. B., Power Coombs, M. R., Doucette, C. D., Walsh, M., & Hoskin, D. W. (2015). Piperine, an alkaloid from black pepper, inhibits growth of human colon cancer cells via G1 arrest and apoptosis triggered by endoplasmic reticulum stress. *Molecular Carcinogenesis, 54*(10), 1070–1085.

Yaheya, Md., & Ismail, Md. (2010). Role of spices in diabetes mellitus. *Research Journal Of Pharmaceutical, Biological And Chemical Sciences, 1*(3), July – September, 30..

Yang, J. Y., Della-Fera, M. A., Nelson-Dooley, C., & Baile, C. A. (2006). Molecular mechanisms of apoptosis induced by ajoene in 3T3-L1 adipocytes. *Obesity, 14*(3), 388–397.

Yi, T., Cho, S. G., Yi, Z., Pang, X., Rodriguez, M., Wang, Y., et al. (2008). Thymoquinone inhibits tumor angiogenesis and tumor growth through suppressing AKT and extracellular signal-regulated kinase signaling pathways. *Molecular Cancer Therapeutics, 7*(7), 1789–1796.

Zandi, P., Basu, S. K., Khatibani, L. B., Balogun, M. O., Aremu, M. O., Sharma, M., et al. (2015). Fenugreek (Trigonella foenum-graecum L.) seed: A review of physiological and biochemical properties and their genetic improvement. *Acta Physiologiae Plantarum, 37*(1), 1714.

Zhai, B., Zhang, C., Sheng, Y., Zhao, C., He, X., Xu, W., et al. (2018). Hypoglycemic and hypolipidemic effect of S-allyl-cysteine sulfoxide (alliin) in DIO mice. *Scientific Reports, 8*(1), 1–7.

Zhang, Y., Xu, L., Ding, M., Su, G., & Zhao, Y. (2019). Anti-obesity effect of garlic oil on obese rats via Shenque point administration. *Journal of Ethnopharmacology, 231*, 486–493.

Zhang, Z., Wang, C. Z., Wen, X. D., Shoyama, Y., & Yuan, C. S. (2013). Role of saffron and its constituents on cancer chemoprevention. *Pharmaceutical Biology, 51*(7), 920–924.

Zheng, G. Q., Kenney, P. M., & Lam, L. K. (1992). Anethofuran, carvone, and limonene: Potential cancer chemoprotective agents from dill weed oil and caraway oil. *Planta Medica, 58*(04), 338–341.

Zhu, J., Chen, H., Song, Z., Wang, X., & Sun, Z. (2018). Effects of ginger (*Zingiber officinale* Roscoe) on type 2 diabetes mellitus and components of the metabolic syndrome: A systematic review and meta-analysis of randomized controlled trials. *Evidence-Based Complementary and Alternative Medicine, 2018*, 5692962.

MicroRNAs as Targets of Dietary Phytochemicals in Obesity and Cancer

Chukwuebuka Egbuna, Muhammad Akram, Kingsley Chukwuemeka Patrick-Iwuanyanwu, Mehwish Iqbal, Eugene N. Onyeike, Chukwuemelie Zedech Uche, and Sadia Hassan

1 Introduction

Plant-based chemicals are naturally occurring substances in botanicals; though, the word is commonly employed to mention chemicals from plants that are organically active. These plant chemicals are found in various plant-based foods for instance

C. Egbuna
Africa Centre of Excellence for Public Health and Toxicological Research (ACE-PUTOR), University of Port Harcourt, Port Harcourt, Nigeria

Department of Biochemistry, Faculty of Natural Science, University of Port Harcourt, Port Harcourt, Choba, Nigeria

Department of Biochemistry, Faculty of Natural Sciences, Chukwuemeka Odumegwu Ojukwu University, Igbaria, Anambra, Nigeria

M. Akram (✉)
Department of Eastern Medicine, Government College University Faisalabad, Faisalabad, Pakistan

K. C. Patrick-Iwuanyanwu · E. N. Onyeike
Africa Centre of Excellence for Public Health and Toxicological Research (ACE-PUTOR), University of Port Harcourt, Port Harcourt, Nigeria

Department of Biochemistry, Faculty of Natural Science, University of Port Harcourt, Port Harcourt, Choba, Nigeria

M. Iqbal
Institute of Management Sciences, Dow University of Health Sciences, Karachi, Pakistan

C. Z. Uche
Department of Medical Biochemistry and Molecular Biology, Faculty of Basic Medical Sciences, University of Nigeria, Enugu Campus, Nsukka, Nigeria

S. Hassan
Department of Nutritional Sciences, Faculty of Science and Technology, Government College Women University, Faisalabad, Pakistan

C. Egbuna, S. Hassan (eds.), *Dietary Phytochemicals*,
https://doi.org/10.1007/978-3-030-72999-8_10

vegetables, fruits, drinks, and grains with various applications. A number of laboratorial and epidemiological researches have proposed that plant chemicals may reduce the effects of some severe or chronic diseases (Reddy et al. 2003; Steinmetz and Potter, 1996; Block et al. 1992). Consequently, more than tens of thousands of chemicals of plants have been recognized (Russo et al. 2010) and an exciting fact is that almost 47% of food and drug administration-endorsed anticarcinogenic medicines are obtained from plants (Newman and Cragg 2007). Alkaloids, polyphenols, organosulfur compounds and terpenoids, are renowned groups of plant chemicals having anticancer properties (Kaur et al. 2018). Few polyphenols for instance, curcumin and resveratrol alter signaling of inflammation, oxidative stress, and demonstrate anti-inflammatory and antioxidant activities (Bishayee 2009; Kasi et al. 2016). *MicroRNAs (miRNAs) are minute regulatory noncoding molecules of ribonucleic acids that influence expression of gene all the way through posttranscription* (Krol et al. 2010). There are over 500 miRNAs that have been recognized in humans (Bhardwaj et al. 2010). MicroRNAs control different biological processes for example differentiation of cell, survival, development, migration, their dysregulation and death of cells which may cause a range of developmental defects (Srivastava et al. 2013; Cai et al. 2009, Sassen et al. 2008). Recent information suggests that some plant chemicals control the expression of a range of miRNAs. Expression of miRNAs is tissue- or developmental phase-specific, and their abnormal expression is related to the development of pathogenic states (Gregory and Shiekhattar 2005). For instance, irregular expression of a number of miRNAs is related to the commencement and advancement of metastasis and carcinogenesis (Zhang et al. 2007). A number of researches have demonstrated that miRNAs had diverse patterns of expression in a range of human ailments, including carcinoma, and are practically involved in processes of disease (Calin and Croce 2006).

2 Phytochemical Modulated miRNAs and Its Role in Obesity

Overweight and obesity are two extremely common ailments in our contemporary societies. According to the WHO, global rates of obesity, determined by the body mass index, BMI (Table 1), have almost increased by two folds since 1980. In 2008, 35% of adults were overweight while 11% was obese. In 2011, over 40 million (40,000,000) children below the age of 5 were overweight. Whereas this fact is

Table 1 The body mass index acceptable values

BMI range	Categorization
BMI < 18.5	Underweight
BMI between 18.5–24.9	Normal weight
BMI between 25–29.9	Overweight
BMI of 30 or more	Obese
BMI of more than 40	Severe obesity

endemic in US, with an adult obesity rate of 35%, obesity and overweight are increasing considerably in other developed and under developed countries (Ogden et al. 2012). Surplus weight gain of the body enhances the possibility of a variety of metabolic diseases for instance type 2 diabetes mellitus, osteoarthritis, cardiac disease, dyslipidaemia, and certain forms of cancer (Haslam and James 2005). Fundamental management of obesity consists of low-fat and low-calorie diets, enhanced physical activity, and various plans contributing to the variation of lifestyle. Though, weight losses attained with intervention in lifestyle are reserved and bounded by increased rates of recidivism and recompensing decelerating metabolism. In this circumstance, it is important to recognize the role of the inherited and epigenetic characteristics in the pathogenesis of obesity. In agreement with this information, various researches have proposed that numerous loci on the human genome are linked with metabolic syndrome and obesity (Berndt et al. 2013; Kunej et al. 2013; Loos and Yeo 2014). Excitingly, omics approaches obscured an association between the expression of a number of miRNAs in different body tissues (e.g., liver, adipose tissue, and pancreas) and metabolic diseases or obesity (Kunej et al. 2013; Jiang et al. 2009; Dumortier et al. 2013). Though the list of miRNAs related with diet is essential, numerous research studies proposed that the miR-203, miR-17/20/93 family,miR-200b/c family, miR-21/590-5p family, miR-221/222 family and let-7/miR-98 family are the most dysregulated in this perspective (Palmer et al. 2014). The metabolic set of symptoms and its persistent diseases are identified to be caused by harmful nutrients and dietary constituents, imbalanced energy from food intake, and spending on inherited factors. Healthy nutrition and way of living life are strongly related with multimodal illness prevention plan. Current researches have revealed the impact of nutrients, plant chemicals, and other bioactive useful foods on methods of epigenetic that have a significant role in the control of many target genes through modulation of miRNA expression. McroRNAs are identified to control homeostasis of lipid, and the irregular expression of miRNAs has been studied in various metabolic ailments, including diabetes and obesity (Rottiers and Naar 2012). Nutritional constituents that modify endogenous micro RNA expression and purpose by means of epigenetic factors, which leads to the control of numerous biological methods, counting apoptosis, invasion, cell proliferation, migration, angiogenesis, resistance of insulin and adipokine synthesis, are the foundation of nutrigenomics science and are authorized to decrease mortality rate of disease (Ross and Davis 2014). Nutritional polyphenols have been made known to improve resistance of insulin and dyslipidemia by the modulation of specific miRNAs. Proanthocyanidins, a polyphenol found in diets, enhanced the expression of adenosine triphosphate-binding cassette transporter1 by downregulating miR-33, which leads to increasing the cholesterol efflux of liver and the manufacturing of new high density lipoprotein cholesterol in liver cells of obese rats. Coffee and Quercetin stimulated the expression of miR-122, which averts diet-provoked hepatic steatosis in rats (Boesch-Saadatmandi et al. 2012; Murase et al. 2011). Lycopene in food downregulates Fabp7 through the upregulation of miR 21 expression, which ultimately leads to the decrease in hepatic intracellular lipid accumulation (Ahn et al. 2012). Extracts of polyphenols from *Hibiscus sabdariffa* plant was reported to control the metabolism of lipid and glucose through the modulation of miR-103, miR-122, and miR-107 in diet-stimulated hyperlipidemic mice (Joven et al. 2012). Jeon et al. (2013) revealed that fisetin, a natural flavonol found in a variety of vegetables and fruits, controls metabolism of lipid by altering expression of miRNA.

3 MicroRNAs as Phytochemicals Targets in Carcinogenesis

MicroRNAs are known to participate in the control of the progression and development of cancer. They either work as growth enhancers (oncomiRs; oncogenic miRNAs) or suppressors (anti-oncomiRs; tumor suppressor miRNAs). Phytochemicals are nonnutritive chemicals in plant that have a variety of applications including anticancer and anti-inflammatory roles. Several researches have revealed that the consumption of vegetable and fruit-rich foods could reduce the incidence of cancer (Block et al. 1992; Reddy et al. 2003; Steinmetz and Potter 1996; Rao et al. 1997). Interestingly, about 47% of food and drug administration approved anticancerous medicines are obtained from plants (Newman and Cragg 2007; Russo 2007). These plant chemicals put forth their anticancer effects through modulation of numerous target proteins influencing various signaling pathways (Russo et al. 2010; Tyagi et al. 2011; Tyagi and Ghosh 2011). Several researches have revealed that the expression of the tumor suppressor Lethal-7 is considerably downregulated in a variety of cancers (Johnson et al. 2005; Yu et al. 2007; Zhang et al. 2007). It has been demonstrated that return of Lethal-7 reduced growth of tumor in a K-RAS (Ki-ras2 Kirsten rat sarcoma viral oncogene homolog) model of lung carcinoma (Kumar et al. 2008). Other miRNAs for instance miR-15a and miR-16 are either removed or downregulated in a large number of patients of the unending lymphocytic leukemia, and their overexpression induced the death of cells (Nicolini et al. 2006). The function of miR-451 as a cancer suppressor miRNA by reducing the invasion, migration and growth of cancer cell, is known (Liu, Liu et al. 2014; Liu, Zhang et al. 2014; Lv et al. 2014). MicroRNAs are perceived as smart targets for therapy and prevention of cancer because of their tumor suppressing or oncogenic activities. Different researchers have proposed that the modulation of miRNAs is one of the key mechanisms in the anticancer phytochemicals.

4 Dietary Phytochemicals and miRNA

4.1 Resveratrol

Resveratrol is a stilbenoid that has been made known to have anticancer activities in opposition to a range of cancers including carcinoma of breast, pulmonary cancer, glioma, carcinoma of prostate, gastric carcinoma, thyroid cancer, colon carcinoma, and neuroblastoma since 1997 (Siddiqui et al. 2015; Bishayee et al. 2010; Jang et al. 1997; Athar et al. 2007). Resveratrol lessened the expression of many oncogenic miRNAs, specifically, miR-21, miR-17, miR-92a-2, miR-103-1, miR-25, andmiR-103-2, in cancerous cells of human colon (Tili et al. 2010). It has been revealed that the anticancer effect of resveratrol on cancer cells of pancreas was because of the inhibition of oncogenic miR-21 (Liu et al. 2013). Furthermore, the interdependent antitumor activity of resveratrol and miR-200c has been revealed in pulmonary carcinoma cells of human (bai et al. 2014).

4.2 Genistein

Genistein is another important polyphenol that demonstrated considerable anticancer effects in the course of the regulation of miRNAs. Management with genistein was proven to increase the death of cells interdependently with miR-16 in chronic lymphocytic leukemia cells of human (Salerno et al. 2009). Xu et al. have illustrated that management of ovarian carcinoma cells with genistein caused a reduction of migration and growth of cells through inhibition of miR-27a (Xu et al. 2013). Additionally, genistein has been revealed to upregulate the tumor suppressor miR-574-3p in cancer cells of prostate (Chiyomaru et al. 2013). Genistein moreover plays a role of tumor suppressor in the course of inhibition of miR-27a in cancer cells of pancreas.

4.3 Conjugated Linoleic Acids

Conjugated linoleic acids in some food supplements has been revealed to decrease fat of the body accumulated in white adipose tissues of abdomen and enhance lean mass of body by increasing the process of fat breakdown, oxidation of fat, apoptosis of fat cells and decreasing the volume of fat cells, the storage and uptake of FAs, and the reduction of enzymes engaged in fat metabolism (Fischer-Posovszky et al. 2007). Conjugated linoleic acids management in mouse considerably enhanced sensitivity of insulin and decreased resistance of insulin by downregulating miR-107 and miR-103 in mouse fed a typical-fat diet, proposing that miRNAs are controlled by dietary fatty acids in obesity (Parra et al. 2010).

4.4 Cinnamic Acid and Cinnamaldehyde

The chief constituents of cinnamon are cinnamic acid and cinnamaldehyde (Kang et al. 2019). Current research studies illustrated biological effects of cinnamaldehyde as having antioxidant, anti-inflammatory, and anti-obesity properties (Huang et al. 2011). Moreover, numerous derivatives of hydroxycinnamic acid for example cinnamic acid, improved the obesity stimulated by a high fat diet in rats (Mnafgui et al. 2015). Cinnamaldehyde (forty micrometer) averts segregation of adipocyte during the downregulation of Ppar-γ and Ccat-enhancer-connecting protein alpha (C/EBPα) (Lu et al. 2018). Likewise, in a bioequivalence study, supplementation of cinnamaldehyde (10 mg per kg) and induced lipolysis, downregulated adipogenic genes and brown adipose tissue activity. These outcomes led to reduced gain in body weight and accumulation of fat after 14 weeks of management (Lu et al. 2018). Cinnamaldehyde controls inflammation in macrophages in the course of the suppression of miR-155 and miR-21 (Qu et al. 2019).

4.5 Ajoene

In this perspective, some complexes have been revealed to exert effects on accumulation of fat, for instance ajoene, in concentrations from 25 μm to 200 μm in 3T3-L1 cells (Ambati et al. 2009). In this view, Kim and Kim (2011) revealed a prospective anti-obesity effect of *Allium sativum* (either 2 or 4%) in obese mouse, downregulating adipogenic genes. Similarly, in one more research with obese mouse managed with powder of garlic (2 or 5%) for 7 weeks, reduces body weight, white adipose tissue, and adipogenic gene expression was establish in white adipose tissue and enhanced Ucps micro ribonucleic acid levels and 5' AMP-activated protein kinase activity in brown adipose tissue.

4.6 Curcumin

Curcumin is an important component of *Curcuma longa* and has been consumed as an essential constituent of spice in Indian cuisine and as a long-established medicine in Asian countries for many decades (Asher and Spelman 2013). It holds chemotherapeutic and chemopreventive activities against many unwanted growths or tumors, it also helps in reversing high blood cholesterol, high blood glucose and other symptoms associated with obesity by targeting nuclear factor-kB, Signal transducer and activator of transcription 3, COX-2, mammalian object of rapamycin indicating cascades, protein kinase B and resulting in reduction of obesity-associated growth of cancer (Asher and Spelman 2013; Khor et al. 2006; Sun et al. 2013; Vyas et al. 2009). Curcumin put forth's its healing effects by regulating miRNAs identified to play a significant role in carcinoma (Reuter et al. 2011). Animal and in vitro studies sustain the proposition that curcumin may encourage the browning process and/or the process of heat production in humans or animals. For example, Wang et al. (2015) stated that mouse treated with curcumin (50 or 100 mg per kg per day) for fifty days exhibited a minimum body weight gain and lower fat accumulation without any effect on food intake, and also demonstrated improved acceptance to cold exposure (Nishikawa et al. 2018). In vitro studies have revealed that curcumin also stimulates the process of browning and mitochondrial biogenesis in 3T3-L1 line of cell and principal inguinal white adipocytes (Kim et al. 2016; Lone et al. 2016).

4.7 *Epigallocatechin-3-Gallate*

Epigallocatechin-3-gallate (EGCG) is a flavonoid of polyphenol that holds considerable anticancer and antioxidant properties. It has been made known that EGCG stimulates apoptosis in liver cell cancer through increased expression of miR-16 (Tsang and Kwok 2010). In pulmonary carcinoma, EGCG upregulated the expression of miR-210, which lead to the reduction in cell growth (Wang et al. 2011). Epigallocatechin-3-gallate increased the efficiency of cisplatin through downregulation of miR-98-5p in A549 non-small cells pulmonary carcinoma.

4.8 *Quercetin*

The consumption of a quercetin-loaded food has been revealed to alter the expression of forty-eight distinctive miRNAs. These miRNAs have been accounted to reduce invasion and metastasis of tumor (miR-146a/b, 194, and 503), reduce proliferation of cell (miR-125a, 302c, lethal-7 family, 195, 26a, 155, 503, and 215), stimulate cell death (miR-125a, 34a, 605, 26b, lethal-7 g, 491, and 16), and upregulate tumor suppressor miRNAs (lethal-7 family, miR-125a, 146a, 183, 106a, 98, 19b, and 381) (Lam et al. 2012). Moreover, flavonols for instance quercetin have been studied extensively for the anti-obesity effects (Peñarrieta et al. 2014). In a current study, the effect of quercetin on miR-155 expression in RAW264.7 macrophages induced with lipopolysaccharide was assessed. Authors found that the management with quercetin (10 micrometer) downregulated the expression of miR-155 in lipopolysaccharide-induced macrophages (Boesch-Saadatmandi et al. 2011).

5 Conclusion

For effective control, prevention and treatment of obesity and cancer, the fundamental knowledge of the role of the various miRNAs in obesity and cancer needs to be understood. Phytochemicals such as resveratrol, genistein, conjugated linoleic acids, cinnamic acid cinnamaldehyde, ajoene, curcumin, epigallocatechin-3-gallate, quercetin have been found to alter the expression of certain miRNAs. It is recommended that food sources of these phytochemicals should be consumed for maximum benefits.

References

Ahn, J., Lee, H., Jung, C. H., & Ha, T. (2012). Lycopene inhibits hepatic steatosis via microRNA-21-induced downregulation of fatty acid-binding protein 7 in mice fed a high-fat diet. *Molecular Nutrition & Food Research, 56*, 1665–1674.

Ambati, S., Yang, J. Y., Rayalam, S., Park, H. J., Della-Fera, M. A., & Baile, C. A. (2009). Ajoene exerts potent effects in 3T3-L1 adipocytes by inhibiting adipogenesis and inducing apoptosis. *Phytotherapy Research, 23*, 513–518.

Asher, G. N., & Spelman, K. (2013). Clinical utility of curcumin extract. *Alternative Therapies in Health and Medicine, 19*(2), 20–22.

Athar, M., Back, J. H., Tang, X., et al. (2007). Resveratrol: A review of preclinical studies for human cancer prevention. *Toxicology and Applied Pharmacology, 224*(3), 274–283.

Bai, T., Dong, D. S., & Pei, L. (2014). Synergistic antitumor activity of resveratrol and miR-200c in human lung cancer. *Oncology Reports, 31*(5), 2293–2297.

Berndt, S. I., Gustafsson, S., Mägi, R., Ganna, A., Wheeler, E., Feitosa, M. F., Justice, A. E., Monda, K. L., Croteau-Chonka, D. C., Day, F. R., et al. (2013). Genome-wide meta-analysis identifies 11 new loci for anthropometrictraits and provides insights into genetic architecture. *Nature Genetics, 45*, 501–512.

Bhardwaj, A., Singh, S., & Singh, A. P. (2010). MicroRNA-based cancer therapeutics: Big hope from small RNAs. *Molecular and Cellular Pharmacology, 2*, 213–219.

Bishayee, A. (2009). Cancer prevention and treatment with resveratrol: From rodent studies to clinical trials. *Cancer Prevention Research, 2*, 409–418.

Bishayee, A., Politis, T., & Darvesh, A. S. (2010). Resveratrol in the chemoprevention and treatment of hepatocellular carcinoma. *Cancer Treatment Reviews, 36*(1), 43–53.

Block, G., Patterson, B., & Subar, A. (1992). Fruit, vegetables, and cancer prevention: A review of the epidemiological evidence. *Nutrition and Cancer, 18*(1), 1–29.

Boesch-Saadatmandi, C., Loboda, A., Wagner, A. E., Stachurska, A., Jozkowicz, A., Dulak, J., Döring, F., Wolram, S., & Rimbach, G. (2011). Effect of quercetin and its metabolites isorhamnetin and quercetin-3-glucuronide on inflammatory gene expression: Role of miR-155. *The Journal of Nutritional Biochemistry, 22*, 293–299.

Boesch-Saadatmandi, C., Wagner, A. E., Wolffram, S., & Rimbach, G. (2012). Effect of quercetin on inflammatory gene expression in mice liver in vivo – Role of redox factor 1, miRNA-122 and miRNA-125b. *Pharmacological Research, 65*, 523–530.

Cai, Y., Yu, X., Hu, S., & Yu, J. (2009). A brief review on the mechanisms of mirna regulation. *Genomics, Proteomics & Bioinformatics, 7*, 147–154.

Calin, G. A., & Croce, C. M. (2006). MicroRNA signatures in human cancers. *Nature Reviews. Cancer, 6*, 857–866.

Chiyomaru, T., Yamamura, S., Fukuhara, S., et al. (2013). Genistein up-regulates tumor suppressor microRNA-574-3p in prostate cancer. *PLoS One, 8*(3), e58929.

Dumortier, O., Hinault, C., & Van Obberghen, E. (2013). MicroRNAs and metabolism crosstalk in energy homeostasis. *Cell Metabolism, 18*, 312–324.

Fischer-Posovszky, P., Kukulus, V., Zulet, M. A., Debatin, K. M., & Wabitsch, M. (2007). Conjugated linoleic acids promote human fat cell apoptosis. *Hormone and Metabolic Research, 39*, 186–191.

Gregory, R. I., & Shiekhattar, R. (2005). Microrna biogenesis and cancer. *Cancer Research, 65*, 3509–3512.

Haslam, D. W., & James, W. P. T. (2005). Obesity. *Lancet, 366*, 1197–1209.

Huang, B., Yuan, H. D., Kim, D. Y., Quan, H. Y., & Chung, S. H. (2011). Cinnamaldehyde prevents adipocyte differentiation and adipogenesis via regulation of Peroxisome Proliferator-Activated Receptor- (PPAR) and AMP-Activated Protein Kinase (AMPK) Pathways. *Journal of Agricultural and Food Chemistry, 59*, 3666–3673.

Jang, M., Cai, L., Udeani, G. O., Slowing, K. V., Thomas, C. F., Beecher, C. W., Fong, H. H., Farnsworth, N. R., Kinghorn, A. D., Mehta, R. G., et al. (1997). Cancer chemopreventive activity of resveratrol, a natural product derived from grapes. *Science, 275*, 218–220.

Jeon, T. I., Park, J. W., Ahn, J., Jung, C. H., & Ha, T. Y. (2013). Fisetin protects against hepatosteatosis in mice by inhibiting mir-378. *Molecular Nutrition & Food Research, 57*, 1931–1937.

Jiang, Q., Wang, Y., Hao, Y., Juan, L., Teng, M., Zhang, X., Li, M., Wang, G., & Liu, Y. (2009). miR2Disease: A manuallycurated database for microRNA deregulation in human disease. *Nucleic Acids Research, 37*, D98–D104.

Johnson, S. M., Grosshans, H., Shingara, J., et al. (2005). RAS is regulated by the let-7 microRNA family. *Cell, 120*(5), 635–647.

Joven, J., Espinel, E., Rull, A., Aragones, G., Rodriguez-Gallego, E., Camps, J., Micol, V., Herranz-Lopez, M., Menendez, J. A., Borras, I., et al. (2012). Plantderived polyphenols regulate expression of miRNA paralogs miR- 103/107 and miR-122 and prevent diet-induced fatty liver disease in hyperlipidemic mice. *Biochimica et Biophysica Acta, 1820*, 894–899.

Kang, N. H., Mukherjee, S., & Yun, J. W. (2019). Trans-cinnamic acid stimulates white fat browning and activates brown adipocytes. *Nutrients, 11*, 577.

Kasi, P. D., Tamilselvam, R., Skalicka-Wozniak, K., Nabavi, S. F., Daglia, M., Bishayee, A., Pazoki-Toroudi, H., & Nabavi, S. M. (2016). Molecular targets of curcumin for cancer therapy: An updated review. *Tumour Biology, 37*, 13017–13028.

Kaur, V., Kumar, M., Kumar, A., Kaur, K., Dhillon, V. S., & Kaur, S. (2018). Pharmacotherapeutic potential of phytochemicals: Implications in cancer chemoprevention and future perspectives. *Biomedicine & Pharmacotherapy, 97*, 564–586.

Khor, T. O., Keum, Y.-S., Lin, W., et al. (2006). Combined inhibitory effects of curcumin and phenethyl isothiocyanate on the growth of human PC-3 prostate xenografts in immunodeficient mice. *Cancer Research, 66*(2), 613–621.

Kim, M., & Kim, H. (2011). Effect of garlic on high fat induced obesity. *Acta Biologica Hungarica, 62*, 244–254.

Kim, S. W., Choi, J. H., Mukherjee, R., Hwang, K. C., & Yun, J. W. (2016). Proteomic identification of fat-browning markers in cultured white adipocytes treated with curcumin. *Molecular and Cellular Biochemistry, 415*, 51–66.

Krol, J., Loedige, I., & Filipowicz, W. (2010). The widespread regulation of microRNA biogenesis, function and decay. *Nature Reviews. Genetics, 11*, 597–610.

Kumar, M. S., Erkeland, S. J., Pester, R. E., et al. (2008). Suppression of non-small cell lung tumor development by the let-7 microRNA family. *Proceedings of the National Academy of Sciences of the United States of America, 105*(10), 3903–3908.

Kunej, T., Jevsinek Skok, D., Zorc, M., Ogrinc, A., Michal, J. J., Kovac, M., & Jiang, Z. (2013). Obesity gene atlas inmammals. *Journal of Genomics, 1*, 45–55.

Lam, T. K., Shao, S., Zhao, Y., et al. (2012). Influence of quercetin-rich food intake on microRNA expression in lung cancer tissues. *Cancer Epidemiology, Biomarkers & Prevention, 21*(12), 2176–2184.

Liu, D., Liu, C., Wang, X., Ingvarsson, S., & Chen, H. (2014). MicroRNA-451 suppresses tumor cell growth by downregulating IL6R gene expression. *Cancer Epidemiology, 38*(1), 85–92.

Liu, P., Liang, H., Xia, Q., et al. (2013). Resveratrol induces apoptosis of pancreatic cancers cells by inhibiting miR-21 regulation of BCL-2 expression. *Clinical and Translational Oncology, 15*(9), 741–746.

Liu, X., Zhang, X., Xiang, J., Lv, Y., & Shi, J. (2014). miR-451: Potential role as tumor suppressor of human hepatoma cell growth and invasion. *International Journal of Oncology, 45*, 739–745.

Lone, J., Choi, J. H., Kim, S. W., & Yun, J. W. (2016). Curcumin induces brown fat-like phenotype in 3T3-L1 and primary white adipocytes. *Journal of Nutritional Biochemistry, 27*, 193–202.

Loos, R. J. F., & Yeo, G. S. H. (2014). The bigger picture of FTO: The first GWAS-identified obesity gene. *Nature Reviews. Endocrinology, 10*, 51–61.

Lu, M., Cao, Y., Xiao, J., Song, M., & Ho, C. T. (2018). Molecular mechanisms of the anti-obesity effect of bioactive ingredients in common spices: A review. *Food & Function, 9*, 4569–4581.

Lv, G., Hu, Z., Tie, Y., et al. (2014). MicroRNA-451 regulates activating transcription factor 2 expression and inhibits liver cancer cell migration. *Oncology Reports, 32*(3), 1021–1028.

Mnafgui, K., Derbali, A., Sayadi, S., Gharsallah, N., Elfeki, A., & Allouche, N. (2015). Anti-obesity and cardioprotective effects of cinnamic acid in high fat diet-induced obese rats. *Journal of Food Science and Technology, 52*, 4369–4377.

Murase, T., Misawa, K., Minegishi, Y., Aoki, M., Ominami, H., Suzuki, Y., Shibuya, Y., & Hase, T. (2011). Coffee polyphenols suppress diet-induced body fat accumulation by downregulating SREBP-1c and related molecules in C57BL/6J mice. *American Journal of Physiology. Endocrinology and Metabolism, 300*, E122–E133.

Newman, D. J., & Cragg, G. M. (2007). Natural products as sources of new drugs over the last 25 years. *Journal of Natural Products, 70*, 461–477.

Nicolini, A., Carpi, A., & Rossi, G. (2006). Cytokines in breast cancer. *Cytokine & Growth Factor Reviews, 17*, 325–337.

Nishikawa, S., Kamiya, M., Aoyama, H., Nomura, M., Hyodo, T., Ozeki, A., Lee, H., Takahashi, T., Imaizumi, A., & Tsuda, T. (2018). Highly dispersible and bioavailable curcumin but not native curcumin induces brown-like adipocyte formation in mice. *Molecular Nutrition & Food Research, 62*, 1700731.

Ogden, C. L., Carroll, M. D., Kit, B. K., & Flegal, K. M. (2012). Prevalence of obesity in the United States, 2009–2010. *NCHS Data Brief, 82*, 1–8.

Palmer, J. D., Soule, B. P., Simone, B. A., Zaorsky, N. G., Jin, L., & Simone, N. L. (2014). MicroRNA expression altered by diet: Can food be medicinal? *Ageing Research Reviews, 17*, 16–24.

Parra, P., Serra, F., & Palou, A. (2010). Expression of adipose microRNAs is sensitive to dietary conjugated linoleic acid treatment in mice. *PLoS One, 5*, e13005.

Peñarrieta, M., Tejeda, L., Mollinedo, P., Vila, J., & Bravo, J. (2014). Phenolic compounds in food. *Revista Boliviana de Química, 31*, 68–81.

Qu, S., Shen, Y., Wang, M., Wang, X., & Yang, Y. (2019). Suppression of miR-21 and miR-155 of macrophage by cinnamaldehyde ameliorates ulcerative colitis. *International Immunopharmacology, 67*, 22–34.

Rao, C. V., Wang, C.-X., Simi, B., et al. (1997). Enhancement of experimental colon cancer by genistein. *Cancer Research, 57*(17), 3717–3722.

Reddy, L., Odhav, B., & Bhoola, K. D. (2003). Natural products for cancer prevention: A global perspective. *Pharmacology & Therapeutics, 99*, 1–13.

Reuter, S., Gupta, S. C., Park, B., Goel, A., & Aggarwal, B. B. (2011). Epigenetic changes induced by curcumin and other natural compounds. *Genes & Nutrition, 6*(2), 93–108.

Ross, S. A., & Davis, C. D. (2014). The emerging role of microRNAs and nutrition in modulating health and disease. *Annual Review of Nutrition, 34*, 305–336.

Rottiers, V., & Naar, A. M. (2012). Micrornas in metabolism and metabolic disorders. *Nature Reviews Molecular Cell Biology, 13*, 239–250.

Russo, G. L. (2007). Ins and outs of dietary phytochemicals in cancer chemoprevention. *Biochemical Pharmacology, 74*(4), 533–544.

Russo, M., Spagnuolo, C., Tedesco, I., & Russo, G. L. (2010). Phytochemicals in cancer prevention and therapy: Truth or dare? *Toxins (Basel), 2*, 517–551.

Salerno, E., Scaglione, B. J., Coffman, F. D., et al. (2009). Correcting miR- 15a/16 genetic defect in New Zealand Black mouse model of CLL enhances drug sensitivity. *Molecular Cancer Therapeutics, 8*(9), 2684–2692.

Sassen, S., Miska, E. A., & Caldas, C. (2008). MicroRNA: Implications for cancer. *Virchows Archiv, 452*, 1–10.

Siddiqui, I. A., Sanna, V., Ahmad, N., Sechi, M., & Mukhtar, H. (2015). Resveratrol nanoformulation for cancer prevention and therapy. *Annals of the New York Academy of Sciences, 1348*, 20–31.

Srivastava, S., Bhardwaj, A., Leavesley, S., Grizzle, W., Singh, S., & Singh, A. (2013). MicroRNAs as potential clinical biomarkers: Emerging approaches for their detection. *Biotechnic & Histochemistry*.

Steinmetz, K. A., & Potter, J. D. (1996). Vegetables, fruit, and cancer prevention: A review. *Journal of the American Dietetic Association, 96*(10), 1027–1039.

Sun, X.-D., Liu, X.-E., & Huang, D.-S. (2013). Curcumin reverses the epithelial-mesenchymal transition of pancreatic cancer cells by inhibiting the Hedgehog signaling pathway. *Oncology Reports, 29*(6), 2401–2407.

Tili, E., Michaille, J. J., Alder, H., et al. (2010). Resveratrol modulates the levels of microRNAs targeting genes encoding tumor suppressors and effectors of TGFβsignaling pathway in SW480 cells. *Biochemical Pharmacology, 80*(12), 2057–2065.

Tsang, W. P., & Kwok, T. T. (2010). Epigallocatechin gallate upregulation of miR-16 and induction of apoptosis in human cancer cells. *The Journal of Nutritional Biochemistry, 21*(2), 140–146.

Tyagi, N., & Ghosh, P. C. (2011). Folate receptor mediated targeted delivery of ricin entrapped into sterically stabilized liposomes to human epidermoid carcinoma (KB) cells: Effect of monensin intercalated into folate-tagged liposomes. *European Journal of Pharmaceutical Sciences, 43*(4), 343–353.

Tyagi, N., Rathore, S. S., & Ghosh, P. C. (2011). Enhanced killing of human epidermoid carcinoma (KB) cells by treatment with ricin encapsulated into sterically stabilized liposomes in combination with monensin. *Drug Delivery, 18*(6), 394–404.

Vyas, H. K., Pal, R., Vishwakarma, R., Lohiya, N. K., & Talwar, G. P. (2009). Selective killing of leukemia and lymphoma cells ectopically expressing hCG$\beta$by a conjugate of curcumin with an antibody against hCGβ subunit. *Oncology, 76*(2), 101–111.

Wang, H., Bian, S., & Yang, C. S. (2011). Green tea polyphenol EGCG suppresses lung cancer cell growth through upregulating miR- 210 expression caused by stabilizing HIF-1α. *Carcinogenesis, 32*(12), 1881–1889.

Wang, S., Wang, X., Ye, Z., Xu, C., Zhang, M., Ruan, B., Wei, M., Jiang, Y., Zhang, Y., Wang, L., et al. (2015). Curcumin promotes browning of white adipose tissue in a norepinephrine-dependent way. *Biochemical and Biophysical Research Communications, 466*, 247–253.

Xu, L., Xiang, J., Shen, J., et al. (2013). Oncogenic microRNA-27a is a target for genistein in ovarian cancer cells. *Anti-Cancer Agents in Medicinal Chemistry, 13*(7), 1126–1132.

Yu, F., Yao, H., Zhu, P., et al. (2007). Let-7 regulates self-renewal and tumorigenicity of breast cancer cells. *Cell, 131*(6), 1109–1123.

Zhang, B., Pan, X., Cobb, G. P., & Anderson, T. A. (2007). microRNAs as oncogenes and tumor suppressors. *Developmental Biology, 302*(1), 1–12.

Natural Phenolic Compounds as Anti-obesity and Anti-cardiovascular Disease Agent

Hefei Zhao and Changmou Xu

1 Introduction

"Plant phenolics", "polyphenols", and "natural phenolic compounds" (NPC) are secondary natural metabolites derived from the shikimate/phenylpropanoid pathway synthesizing monomeric and polymeric phenols and polyphenols, which have high variability among their structure and occurrence and undertake wide physiological roles in plants (Lattanzio 2013; Quideau et al. 2011; Tomás-Barberán and Espín 2001). Phenolic compounds are the defense on the front-line for plants (Zhang and Tsao 2016). Polyphenols consist of two and more phenolic hydroxyl groups linked to more than one aromatic ring (Vermerris and Nicholson 2008). They usually contain two aromatic rings are usually named as ring A and B, also oxygenated heterocycle (ring C) (Vermerris and Nicholson 2008). They may also connect to glucosides to form polyphenol glucosides (Tapiero et al. 2002). Most of the phenolic compounds are derived from plant tissue of fruits and vegetables and accumulated in places of grapes and berries skins and seeds (Poudel et al. 2008; Rodríguez Montealegre et al. 2006), tea leaves (Zderic and Zondervan 2017), nut shell (Moccia et al. 2020), roots, and barks (Panja 2018).

The distribution of NPC profile is also different among plants and tissues, some compounds are commonly shared in different plants while others may be relatively unique in specific parts of plants. Gallic acid and its esterified-derivatives naturally exist in almost all tissue parts of the plant (Daglia et al. 2014). Catechin can be commonly found both in grapes (Yilmaz and Toledo 2003) and green tea (Elbling et al. 2005). Based on the hydroxyl groups and esterification with gallate moiety on catechin backbone, catechin can be characterized to epicatechin (EC), epicatechin

H. Zhao (✉) · C. Xu
Food Processing Center, Department of Food Science and Technology, University of Nebraska-Lincoln, Lincoln, NE, USA
e-mail: hzhao@huskers.unl.edu; cxu13@unl.edu

C. Egbuna, S. Hassan (eds.), *Dietary Phytochemicals*,
https://doi.org/10.1007/978-3-030-72999-8_11

gallate (ECG), epigallocatechin (EGC) and epigallocatechin gallate (EGCG) (Babu et al. 2008). Anthocyanins are colored flavonoids widely distributed in plants (Cao et al. 2001; Pool-Zobel et al. 1999). Ellagic acid is also universally present in fruits (Talcott et al. 2002), such as raspberry (Bobinaitė et al. 2016), strawberry (Schulenburg et al. 2016), blackberry, and blueberry (Sellappan et al. 2002). Theaflavin, theabrownin can be found in dark fermented teas such as Puerh tea (Huang et al. 2019).

These phenolic compounds undertake multiple health-promoting benefits (Cueva et al. 2017; Xu et al. 2010; Zhang and Tsao 2016) which are primarily related to their antioxidant as well as anti-inflammatory properties (de Souza et al. 2018). And those compounds have antioxidant effects such as DPPH scavenging (Roy et al. 2010), ORAC (Oxygen radical absorbance capacity) (Kasote et al. 2015), and ABTS (Prior et al. 2005) in vitro. The anti-obesity effect through polyphenols intake has been found based on mechanisms including but not limited to lowering lipids intake by binding bile salt and inhibiting the activity of pancreatic lipase, reducing lipogenesis, augmenting lipolysis, enhancing fatty acids (FA) b-oxidation, inhibiting differentiation and growth of adipocyte, and anti-inflammatory and suppress oxidative stress (Hamauzu and Suwannachot 2019; Wang et al. 2014). This chapter summarized the mechanisms of Anti-obesity effect of NPC, and for each of the mechanisms, in vitro studies, in silico studies, plasma models, cell models, animal model studies and clinical research, etc., were proposed to be discussed to furnish evidence and supports. However, not all type of studies was provided for some of mechanism. For instance, lowering lipids intake by binding bile salt is not a mechanism in the cell but in the intestinal tract, therefore no cell model is necessary for this mechanism. Also, the body fat decrement and weight loss in clinical research are possibly attributed to the synergistic effect of the combination of various mechanism actions.

2 Effects of Natural Phenolic Compounds on Oil and Fat Metabolism

The aqueous continuous phase essentially undertakes the transportation of various nutrients in the human body, however, due to the water insolubility of lipids, triglycerides (TG) which is the common dietary fats have to be decomposed via enzymolysis which enable water-soluble molecules to be transported in the human body. As can be seen in Fig. 1, the bulk oil is firstly encompassed by surfactant molecules and emulsifiers. Secondly, bile salts and phospholipids are secreted in the duodenum, then attach on the surface of oil droplets, therefore those surfactants stabilized the emulsions. The finally, the oil and water surface enables pancreatic lipase and colipase, secreted by pancreatic acinar cells, adsorb onto the interface, then hydrolyze triglycerides, complete fat digestion in the proximal small intestine (Lowe 1997).

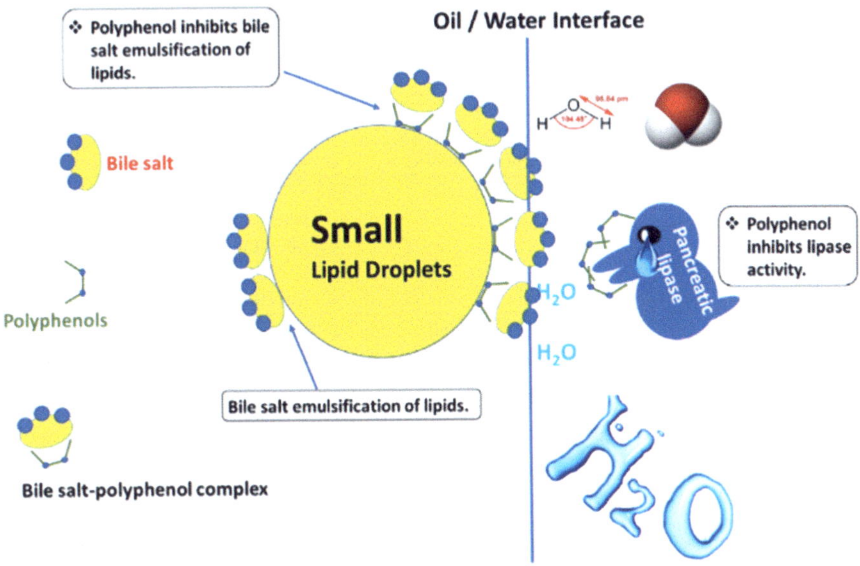

Fig. 1 Mechanism of regiospecific interaction between bile salt and polyphenols to reduce cholesterol micellar solubility and inhibit pancreatic lipase activities

As the interaction between bile salt and oil droplets as well as pancreatic lipase play essential roles in the stage of digestion, interference of these interactions would be potentially helpful for anti-obesity. For instance, a lab-synthesized micelle sequestrant polymers (MSPs), could isolate lipid micelles and significantly eliminate fecal triglycerides and bile acids, therefore reduce lipid absorption via a pH-triggered flocculation process (Qian et al. 2015). The mechanism of lowering oil and fat intake in the digestion and absorption stage were all listed in Table 1. Since the first step of lipid absorption happens around the stomach and duodenum, mechanism and clinical studies targeting this stage by natural phenolic compounds were discussed as follows.

3 Binding Bile Salt to Inhibit Emulsification of Lipids

3.1 In Vitro Studies

The formation of a micellar emulsion of lipid in duodenum provides a large surface for the lipolysis by pancreatic lipase, therefore, reduction of micellar solubility potentially decreases the chances of interaction between pancreatic lipase and lipids, so that inhibit lipid absorption and subsequent obesity.

Table 1 Mechanisms of lowering oil and fat intake in digestion and absorption stage

Mechanism	Model	Potential acting site in the human body	Compounds	Key finding	References
Binding bile salt	In vitro	Duodenum	Theaflavins	• Theaflavins precipitate bile salt micellar solution and reduce cholesterol micellar solubility	Ikeda et al. (2010)
			EGCG and oolongtheanins-3′-O-gallate	• NMR confirmed the regiospecific interaction of EGCG and oolongtheanins-3′-O-gallate with sodium taurocholate	Ogawa et al. (2016)
			Catechin	• Catechin–taurocholic acid complex was confirmed at a 1:1 binding ratio by using electrospray ionization-mass spectrometry	Sakakibara et al. (2019)
Inhibit the activity of pancreatic lipase	In vitro	Proximal small intestine	Seaweed (*Ascophyllum nodosum*) polyphenol-rich extract (PRE)	• Inhibited lipase activity with IC_{50} of 200μg gallic acid equivalents (GAE) per assay	Austin et al. (2018)
			Citrus unshiu peel mixture extract (PCM) and *Diospyros kaki* fruit	• Inhibited lipase activity with an IC_{50} value of 507.01μg/mL	Kim et al. (2016)
	In silico		Black tea theaflavins	• Theaflavin-3,3′-digallate (TFdiG) showed the strongest inhibition due to the hydroxyl groups on galloyl ester moiety increases inhibitory potency	Glisan et al. (2017)
Increase fecal lipid excretion	Animal models	Gut	Black tea polyphenols and polysaccharides	• Increased body composition, fatty acids content in fecal, and regulate fat metabolism	Wu et al. (2016)

(continued)

Table 1 (continued)

Mechanism	Model	Potential acting site in the human body	Compounds	Key finding	References
			(−)-Epigallocatechin-3-gallate (EGCG)	• Increase fecal excretion of bile acids, cholesterol, and total oils and fats, but decrease intestinal bile acid content	Huang et al. (2018)
	Clinical research		Polyphenol-enriched oolong tea	• Polyphenol group had more fecal lipid excretion of than that of the placebo group, significantly	Hsu et al. (2006)
			Black tea polyphenol (BTP)	• Restrained postprandial increase of triacylglycerol	Ashigai et al. (2016)

Ikeda et al. (2010) found that cholesterol micellar solubility was decreased by the key component of black tea polyphenols (BTPs), theaflavins in vitro. Theaflavin (TF)-monogallates (TFMGs), TF-3-gallate (TF3G), and TF-3′-gallate (TF3′G) effectively removed cholesterol out of bile salt micelles in vitro, however, the micellar solubility were limitedly changed by TF and TF-3,3′-digallate (TFDG) which indicate a structure-function relationship. It was observed the solution turned turbid immediately under the condition of adding BTPs or TFMGs. However, the mechanism of actions among BTPs, bile salt, and cholesterol was not demonstrated at that time, and the authors provide speculation that there is a hydrophobic interaction between theaflavins and cholesterol, which was an uncertain presumption at that time. Green-tea catechins which had one galloyl moiety were also reported to decrease the cholesterol micellar solubility in vitro (Ikeda et al. 2003).

To reveal the mechanism of action between bile acid and polyphenols from teas for inhibiting micellar cholesterol solubility, Ogawa et al. (2016) implemented nuclear magnetic resonance (NMR) to indicate a new mechanism that is regiospecific interaction of EGCG and oolongtheanins-3’-O-gallate with sodium taurocholate (a compound of bile salts). Furthermore, Sakakibara et al. (2019) confirmed the complex was formed by one catechin linked by one taurocholic acid by using electrospray ionization-mass spectrometry. Also, binding functional groups and positions on (−)-epigallocatechin gallate (EGCG) were indicated by the strong chemical-shift difference at the carbonyl carbon and the 1′- and 4′-positions in the functional group of the galloyl moiety on ^{13}C NMR spectra. And, the carbonyl carbon atom and the aromatic ring of the galloyl moiety of (−)-EGCG derivatives contributed strong effects by ^{1}H NMR, but the amount of hydroxyl groups on the B ring had no relations on these interactions. They also proposed the change of micellar cholesterol solubility is attributed to EGCG inserting in the micelle by interacting with taurocholic acid hydrophobically.

4 Pancreatic Lipase Inhibition

4.1 In Vitro Studies

The activity of pancreatic lipase has been reported to be inhibited by an edible seaweed (*Ascophyllum nodosum*) polyphenol-rich extract (PRE), by using an oil-based turbidimetric IC_{50} assay at ~230μg in dry weight. This also indicated that the PRE extracts could furnish phenolic components into the gut, therefore possibly reduce lipid digestion in situ. *Diospyros kaki* fruit and *Citrus unshiu* peel mixture extract (PCM) was also found to inhibit lipase activity, and the IC_{50} value was 507.01μg/mL (Kim et al. 2016).

4.2 In Silico Modeling Studies

Based on In silico modeling study, Glisan et al. (2017) reported theaflavins to show pocket-adjacent binding activity to the active position on Asn263 and Asp206, and theaflavins with galloyl are estimated to intervene His264 protonation. Theaflavin-3,3′-digallate (TFdiG) showed the strongest inhibition due to hydroxyl groups on galloyl ester moiety increases inhibitory potency.

4.3 Increase Fecal Lipid Excretion

As previous mentions, polyphenols have anti-obesity effect based on synergistic mechanisms of binding bile salt to inhibit emulsification of lipids as well as inhibiting activity of pancreatic lipase, which cause to incomplete digested lipids in GI tract and increased fecal lipid content. Animal models and clinical trials have been conducted to study fecal lipid content under the interference of polyphenol intakes. The Increase fecal lipid excretion may be attributed to both binding bile salt to inhibit the emulsification of lipids and inhibit the pancreatic lipase activity, therefore it can be considered as Animal models and Clinical research of the two mechanisms.

4.4 *Animal Models*

Based on models of high-fat diet-induced obese rats, Wu et al. (2016) found that body composition was increased by black tea polyphenols and polysaccharides. The mixture also showed an increment of fatty acids content in fecal, regulation of fat metabolism. Huang et al. (2018) found bile acid in mice intestinal system was significantly reduced by (−)-epigallocatechin-3-gallate (EGCG), but bile acids, cholesterol, and total lipids were higher in fecal excretion.

4.5 *Clinical Research*

Hsu et al. (2006) measured the excretion of lipids of twelve healthy adults who had placebo control or polyphenol-enriched oolong tea in 750 mL beverages base and found that the group with the polyphenol-enriched oolong tea period by19.3 ± 12.9 g/3day had significantly higher fecal lipid excretion than the group with the placebo period which was 9.4 ± 7.3 g/3day ($P < 0.01$). Ashigai et al. (2016) reported black tea polyphenol (BTP) intake inhibited postprandial triacylglycerol by a randomized, placebo-controlled, double-blind, crossover study. And for the BTP intake group, the total lipid excretion of was from 5.51 ± 1.73 to 6.87 ± 1.91 g/3 higher than the control beverage group.

5 Gut Microbiota as Potential Targets

The small intestine does not or poorly absorb a major part of dietary phenolic compounds due to not hydrolyzed by the human intestinal enzymes, therefore they can go through gastrointestinal (GI) tract then reach large intestine in which they interact with colonic microbiota through biochemical reaction such as

(1) hydrolysis of O-deglycosylations and ester, for instance, the hydrolysis of ellagitannins, esters of hexahydroxydiphenic acid (HHDP) by *A. muciniphila* produces ellagic acid to stimulate the growth of this beneficial gut microbe (Henning et al. 2017).
(2) cleavage of C-ring, delactonization, demethylation, for instance, One of the lactones of ellagic acid derived from hydrolysis of ellagitannins is cleaved by *G. urolithinfaciens* and *G. pamelaeae* through deleptonization, to form luetic

acid, then the carboxyl group removed by decarboxylation to form urolithins M-5 (Espín et al. 2013; Selma et al. 2014).

(3) reductions of dehydroxylation and a double bond (Espín et al. 2017; Scalbert and Williamson 2000), for example, urolithins M-5 is converted to urolithin M-6 by subtracting the hydroxyl via dihydroxylation by *G. urolithinfaciens* and *G. pamelaeae* to generate urolithins M-5 followed by urolithins C (Selma et al. 2014). Other urolithins A, B, D, M6, M7, and their forms may also be produced by other gut microbes (Cerdá et al. 2005; Espín et al. 2013). Phenolic compounds have been found to have a 'prebiotic-like' effect which enhances beneficial bacteria growth, e.g. lactobacilli and bifidobacteria (Dueñas et al. 2015).

It is emerging that the modulation of gut microbiota by natural bioactive molecules such as polyphenols derived by plants, indicating that polyphenols could alter brain functions that characterize neurodegenerative diseases (Filosa et al. 2018). Chen et al. (2018) reported that the fecal levels of obese rats by treatment of epigallocatechin gallate (EGCG), epicatechin gallate (ECG), and gallocatechin gallate (GCG) were significantly increased. These results suggested an incomplete catechins metabolism by the gut microflora, due to high-fat and high-sugar (HFHS) diet might impact on the tea catechins excretion. IgA-coated bacteria, related to low-grade inflammation in the obesity group, has been reported to be increased by olive oil phenolic compounds which indicates the facilitation of the mucosal immunity at the intestinal level (Martín-Peláez et al. 2016). The ratio Firmicutes/Bacteroidetes, which is an indicator at a high ratio for obesity and overweight status, was reported to be reduced by quercetin supplementation (Etxeberria et al. 2015). These discoveries indicated anti-inflammation and anti-obesity effect due to the gut microbiota modulation induced by phenolic compounds and their metabolic products. However, not all evidence confirms health benefits. An important inflammation marker (C-reactive protein (CRP) was increased in the experimental group consuming phenolic compounds at 500 mg/kg dosages in extra virgin olive oil (Martín-Peláez et al. 2016). The Firmicutes/Bacteroidetes ratio also showed no statistically significant differences instead of reduction as expected.

The gut microbiota can be positively modulated by prebiotics-act molecules such as oligosaccharides and phenolic compounds, which results in increasing saccharolytic fermentation and short-chain fatty acid (SCFA), therefore improves gut barrier function (Bianchi et al. 2019; Li et al. 2018). A higher SCFA is a good indication of the generation of anorectic gut hormones, the peptide YY (PYY) and glucagon-like peptide-1 (GLP-1), which essentially influence on satiety, hunger, insulin sensitivity, and lipolysis. SCFA also inhibits the synthesis of triglycerides (TG) and cholesterol and enhances the function of the gut barrier. The improvement of gut barrier function possibly decrease the translocation of bacteria and the blood LPS, therefore finally reduces tumor necrosis factor (TNF), pro-inflammatory markers which include interleukin-6 (IL-6), and C-reactive protein (CPR), meanwhile increases anti-inflammatory markers including interleukin-4 (IL4), interleukin-10 (IL-10).

6 Reduction of Lipogenesis and Inflammation in Adipose Tissue and Liver

After fatty acid absorption by the digestive system, they are transported in blood vessels, form adipose tissue and liver fat tissue. Lipogenesis, synthesis of fatty acid, attributes to the complete synthesis of palmitate from acetyl CoA in the cytosol, which has been well represented in adipose tissue and liver (Lin and Lin-Shiau 2006). However, the intake of polyphenols for plant sources has an interference effect on lipogenesis. The mechanisms of action of anti-obesity by intake of natural compounds are related to alkaloid and polyphenol targeting lipid metabolism (Cheng et al. 2020). The metabolites compounds of resveratrol were found to be the largest amounts in liver, intermediate concentration in adipose tissue, and the lowest level in the muscle (Andres-Lacueva et al. 2012). Oxidized tea polyphenols have been found to inhibit lipid content in the liver and visceral white adipose tissue in rats (Sumin Wang et al. 2017).

6.1 *In Vitro Studies by Cell Models*

3T3-L1 cell has been utilized as one of the standards for investigating preadipocytes into adipocytes conversion. It has been reported that all fractions of the polyphenol-enriched organic fraction (CPEF) from the crude extract of *Cyclopia intermedia* all inhibited accumulative lipid in 3T3-L1 pre-adipocytes as well as decrease lipid content in the mature cell, however, the effects depended on concentrations (Jack et al. 2018). Lipolysis in mature adipocytes was stimulated by *β*-D-glucoside which is a kind of benzophenone marked as Fraction 1 (F1), hesperidin which is kind of flavanone marked as F2, mangiferin which is kind of xanthone marked as F3. F1 and F2 result in more expression of the messenger RNA (mRNA) for producing hormone-sensitive lipase in mature adipocytes, but uncoupling protein 3 expressions were increased by F1 and neoponcirin which is kind of flavone marked as F4. Phenolic compounds derive from different fractions by high-performance counter-current chromatography (HPCCC) performed anti-obesity effects differently. Crude extracts have higher activities than purified fractions, therefore they illustrate synergistic effects in different compounds in plant extracts. However, the synergistic effect of the extracts, which usually are mixtures of individual phenolic compounds, are still not fully investigated.

6.2 *Animal Models*

Resveratrol has been found to decrease body fat, total cholesterol, triacylglycerol, transaminases, and level of insulin plasma in mice, which also proved by a significant decreasing expression of TNF-α, IL-6, and NF-κB mRNA in the liver. ACC, PPAR-γ, and SREBP-1 mRNA expression were significantly suppressed by resveratrol in high-fat diet (HFD) + RSV mice by analyzing genes relating to liver adipogenesis. Resveratrol augments lipid metabolism and inhibits nonalcoholic fatty liver disease (NAFLD) and pro-inflammation in the liver of HFD mice. *Solanum nigrum polyphenol extract* (SNPE) has been found to reduce hepatic lipid (oleic acid) accumulation (Chang et al. 2017). Therefore, NAFL-induced hepatitis is potentially alleviated by SNE, also the major components of SNE are polyphenolic compounds.

7 Suppression of Lipogenic Enzyme Fatty Acid Synthase (FAS) in Cell

Tea polyphenols promote the expenditure of cellular energy in mitochondria so that possibly decreases the body weight gain. Also, the expression of FAS in the nucleus may be inhibited through the downregulation of the signal transduction pathway of EGF-receptor/ PI3K/Akt/Sp-1, therefore the growth of tissue and lipogenesis in the cell may be also suppressed. Tea polyphenols, e.g. EGCG and theaflavin-3,39-digallate (TF-3) showed inhibition of the growth factor EGF binding to EGFR, hamper the PI3K/Akt signal pathway activity, and decrease the DNA-binding of nuclear transcription factor Sp-1, therefore results in inhibiting the expression of FAS DNA (Lin and Lin-Shiau 2006).

7.1 Cell Model

Polyphenols from Extra-virgin olive oil (EVOO) primarily include single phenols including tyrosol, hydroxytyrosol, vanillin, phenolic acids including caffeic acid, *p*-coumaric acid, vanillic acid, ferulic acid, elenolic acid, lignans including pinoresinol, acetoxy-pinoresinol, flavonoids including apigenin and luteolin, or secoiridoids including deacetoxyoleuropein aglycone, ligstroside aglycone, oleuropein glycoside, oleuropein aglycone. EVOO polyphenols with lignans, flavonoids, and secoiridoids showed suppression of the expression of fatty acid synthase (FASN) protein in HER2 gene-amplified SKBR3 breast cancer cells. However, single phenols and phenolic acids had little effect on FASN expression in SKBR3 and MCF-7/HER2 cells (Menendez et al. 2008).

7.2 Animal Models

Based on Swiss male mice, gallic acid (GA) was found to reduce liver steatosis, body weight, and plasma insulin levels. The liver ACC and FAS mRNA of HFD + GA mice were significantly suppressed by analyzing hepatic steatosis related gene expression (Sousa et al. 2020).

8 Increase of Lipolysis in Cell

By using differentiated 3T3-L1 adipocytes model, almond skin polyphenol extract (ASP) has shown significant promotion of phosphorylation of AMP-activated protein kinase (AMPK), and an increment of the activity of adipose triglyceride lipase and hormone-sensitive lipase, and inhibition of adipogenesis-related transcription factors (Huang et al. 2017). The study indicates that ASP modulates lipolysis through AMPK activation, results in the reduction of adipogenesis, as well as suppression of adipocytes proinflammatory cytokines.

9 Inhibition of Adipocyte Differentiation and Growth

Aranaz et al. (2019) compared the inhibiting effect of different standard phenolic compounds including phenolic acids on adipogenesis by using mouse 3T3-L1 preadipocytes model from different family and within different stages of differentiation. Results showed that accumulating lipid was inhibited by quercetin and resveratrol within the entire differentiating process, but apigenin and myricetin only had their activity during the early and latest differentiating periods other than intermediate, however, inhibition by hesperidin was just opposite. The phenolic acids were confined to have the activity on the early stages, but not for ellagic acids and *p*-coumaric. This anti-adipogenic effect showed up together with the down-regulating Scd1 and Lpl. At the early stages, the inhibition to cell growth of the phenolic compounds correlated significantly with their binding affinity to the PPAR ligand-binding set by using molecular docking analysis, with the significance of $r = 0.7034$ and $p = 0.005$. Results show that, during the early differentiation stages, polyphenols and phenolic acids possibly interact with specific residues of the receptor, and they could be potentially used for the determination of anti-adipogenicity. Phe264, His266, Ile281, Cys285, and Met348 are the most involved amino acid residues which indicate may play a key role in modulating the receptor activity.

This study is very important to show different polyphenols have a different effect in vivo, although the polyphenols almost all perform antioxidant effects such as free radical DPPH scavenging (Roy et al. 2010), Oxygen radical absorbance capacity (ORAC) (Kasote et al. 2015) and ABTS (Prior et al. 2005) in vitro. But, the mechanism on the molecular level for understanding the structure-function relationship of the antioxidant effect in vitro is still necessary for conducting a further study.

10 Anti-atherosclerosis by Natural Phenolic Compounds

Atherosclerosis is one of the cardiovascular disorders involved in a multifactorial disease with inflammatory pathogenesis (Bahramsoltani et al. 2019). The increment of HDL/LDL ratio and plasma total triacylglycerol (TG) level are important index indications for atherogenic disease (Matralis et al. 2016). It also accompanies ischemic heart disease and peripheral arterial disorders (Barquera et al. 2015). Many characteristics of the atherosclerosis inflammation are recapitulated by pathological mechanisms of obesity, which considered as a subacute inflammation of blood vessel and arterial wall cells, with characteristics of the infiltration of macrophages and T cells (Rocha and Libby 2009). The reducing risk of heart disease by the consumption of polyphenols have supported by epidemiology studies (Cassidy et al. 2013). Red blood cells (RBCs), albumin, and lipoproteins are bound with polyphenols. Polyphenols-LDL complex is an antioxidant that inhibits the oxidation of LDL in the blood and the endothelium at the early stages.

11 Conclusion

Based on experimental data, the anti-obesity effect of NPC is a synergetic outcome of a series of mechanisms at the different locus in the human body, and the mechanism of the anti-obesity effect of natural phenolic compounds rests on four major stages and locations. (1) Inhibition of lipid absorption by binding bile salts and pancreatic lipase in digestion stage primarily in the duodenum, which results in fecal lipid excretion; (2) 'prebiotic-like' effect positively altering Gut microbes to enhance gut barrier function; (3) a series of anti-inflammation and modulation of fat metabolism in the cell; (4) Anti-atherosclerosis and suppression of plasma total triacylglycerol (TG) level through binding red blood cell (RBCs) and lipoproteins, and modulation of and HDL/LDL ratio link to atherosclerotic plaque in the cardiovascular system.

References

Andres-Lacueva, C., Macarulla, M. T., Rotches-Ribalta, M., Boto-Ordóñez, M., Urpi-Sarda, M., Rodríguez, V. M., & Portillo, M. P. (2012). Distribution of resveratrol metabolites in liver, adipose tissue, and skeletal muscle in rats fed different doses of this *polyphenol. Journal of Agricultural and Food Chemistry, 60*(19), 4833–4840. https://doi.org/10.1021/jf3001108.

Aranaz, P., Navarro-Herrera, D., Zabala, M., Miguéliz, I., Romo-Hualde, A., López-Yoldi, M., et al. (2019). Phenolic compounds inhibit 3T3-L1 adipogenesis depending on the stage of differentiation and their binding affinity to PPARγ. *Molecules (Basel, Switzerland), 24*(6), 1045. https://doi.org/10.3390/molecules24061045.

Ashigai, H., Taniguchi, Y., Suzuki, M., Ikeshima, E., Kanaya, T., Zembutsu, K., et al. (2016). Fecal lipid excretion after consumption of a black tea polyphenol containing beverage-randomized, placebo-controlled, double-blind, crossover study. *Biological & Pharmaceutical Bulletin, 39*(5), 699–704. https://doi.org/10.1248/bpb.b15-00662.

Austin, C., Stewart, D., Allwood, J. W., & McDougall, G. J. (2018). Extracts from the edible seaweed: *Ascophyllum nodosum*, inhibit lipase activity in vitro: Contributions of phenolic and polysaccharide components. *Food & Function, 9*, 502–510. Royal Society of Chemistry. https://doi.org/10.1039/c7fo01690e.

Babu, P. V. A., Liu, D., & Liu, D. (2008). Green tea catechins and cardiovascular health: An update. *Current Medicinal Chemistry, 15*(18), 1840–1850. Retrieved 9 September 2018 from http://www.ncbi.nlm.nih.gov/pubmed/18691042.

Bahramsoltani, R., Ebrahimi, F., Farzaei, M. H., Baratpourmoghaddam, A., Ahmadi, P., Rostamiasrabadi, P., et al. (2019). Dietary polyphenols for atherosclerosis: A comprehensive review and future perspectives. *Critical Reviews in Food Science and Nutrition, 59*(1), 114–132. https://doi.org/10.1080/10408398.2017.1360244.

Barquera, S., Pedroza-Tobías, A., Medina, C., Hernández-Barrera, L., Bibbins-Domingo, K., Lozano, R., & Moran, A. E. (2015). Global overview of the epidemiology of atherosclerotic cardiovascular disease. *Archives of Medical Research*. https://doi.org/10.1016/j.arcmed.2015.06.006.

Bianchi, F., Duque, A. L. R. F., Saad, S. M. I., & Sivieri, K. (2019). Gut microbiome approaches to treat obesity in humans. *Applied Microbiology and Biotechnology, 103*(3), 1081–1094. https://doi.org/10.1007/s00253-018-9570-8.

Bobinaitė, R., Viskelis, P., Bobinas, Č., Mieželienė, A., Alenčikienė, G., & Venskutonis, P. R. (2016). Raspberry marc extracts increase antioxidative potential, ellagic acid, ellagitannin and anthocyanin concentrations in fruit purees. *LWT- Food Science and Technology, 66*(460), 467. Retrieved 19 June 2018 from. https://doi.org/10.1016/J.LWT.2015.10.069.

Cao, G., Muccitelli, H. U., Sánchez-Moreno, C., & Prior, R. L. (2001). Anthocyanins are absorbed in glycated forms in elderly women: A pharmacokinetic study. *The American Journal of Clinical Nutrition, 73*(5), 920–926. Retrieved 11 September 2018 from. https://doi.org/10.1093/ajcn/73.5.920.

Cassidy, A., Mukamal, K. J., Liu, L., Franz, M., Eliassen, A. H., & Rimm, E. B. (2013). High anthocyanin intake is associated with a reduced risk of myocardial infarction in young and middle-aged women. *Circulation, 127*(2), 188–196.

Cerdá, B., Periago, P., Espín, J. C., & Tomás-Barberán, F. A. (2005). Identification of urolithin A as a metabolite produced by human colon microflora from ellagic acid and related compounds. *Journal of Agricultural and Food Chemistry, 53*(14), 5571–5576. Retrieved 29 June 2019 from. https://doi.org/10.1021/JF050384I.

Chang, J.-J., Chung, D.-J., Lee, Y.-J., Wen, B.-H., Jao, H.-Y., & Wang, C.-J. (2017). Solanum nigrum polyphenol extracts inhibit hepatic inflammation, oxidative stress, and lipogenesis in high-fat-diet-treated mice. *Journal of Agricultural and Food Chemistry, 65*(42), 9255–9265. https://doi.org/10.1021/acs.jafc.7b03578.

Chen, B., Zhou, J., Meng, Q., Zhang, Y., Zhang, S., & Zhang, L. (2018). Comparative analysis of fecal phenolic content between normal and obese rats after oral administration of tea polyphenols. *Food & Function, 9*(9), 4858–4864. Retrieved 22 June 2020 from. https://doi.org/10.1039/c8fo00609a.

Cheng, C., Li, Z., Zhao, X., Liao, C., Quan, J., Bode, A. M., et al. (2020). Natural alkaloid and polyphenol compounds targeting lipid metabolism: Treatment implications in metabolic diseases. *European Journal of Pharmacology*. Elsevier B.V. Retrieved from. https://doi.org/10.1016/j.ejphar.2020.172922.

Cueva, C., Gil-Sánchez, I., Ayuda-Durán, B., González-Manzano, S., González-Paramás, A., Santos-Buelga, C., et al. (2017). An integrated view of the effects of wine polyphenols and their relevant metabolites on gut and host health. *Molecules, 22*(1), 99. Retrieved 19 June 2018 from. https://doi.org/10.3390/molecules22010099.

Daglia, M., Di Lorenzo, A., Nabavi, S. F., Talas, Z. S., & Nabavi, S. M. (2014). Polyphenols: Well beyond the antioxidant capacity: Gallic acid and related compounds as neuroprotective agents: You are what you eat! *Pharmaceutical Biotechnology, 15*(4), 362–372. Retrieved 9 September 2018 from https://www.ingentaconnect.com/content/ben/cpb/2014/00000015/00000004/art00008.

de Souza, E. L., de Albuquerque, T. M. R., dos Santos, A. S., Massa, N. M. L., & de Brito Alves, J. L. (2018). Potential interactions among phenolic compounds and probiotics for mutual boosting of their health-promoting properties and food functionalities – A review. *Critical Reviews in Food Science and Nutrition, 59*(10), 1–15. Retrieved 19 June 2018 from. https://doi.org/10.1080/10408398.2018.1425285.

Dueñas, M., Muñoz-González, I., Cueva, C., Jiménez-Girón, A., Sánchez-Patán, F., Santos-Buelga, C., et al. (2015). A survey of modulation of gut microbiota by dietary polyphenols. *BioMed Research International, 2015*, 1–15. Retrieved 28 June 2019 from. https://doi.org/10.1155/2015/850902.

Elbling, L., Weiss, R.-M., Teufelhofer, O., Uhl, M., Knasmueller, S., Schulte-Hermann, R., et al. (2005). Green tea extract and (–)-epigallocatechin-3-gallate, the major tea catechin, exert oxidant but lack antioxidant activities. *The FASEB Journal, 19*(7), 807–809. Retrieved 9 September 2018 from. https://doi.org/10.1096/fj.04-2915fje.

Espín, J. C., González-Sarrías, A., & Tomás-Barberán, F. A. (2017). The gut microbiota: A key factor in the therapeutic effects of (poly)phenols. *Biochemical Pharmacology, 139*, 82–93. Retrieved 23 June 2019 from. https://doi.org/10.1016/J.BCP.2017.04.033.

Espín, J. C., Larrosa, M., García-Conesa, M. T., & Tomás-Barberán, F. (2013). Biological significance of urolithins, the gut microbial ellagic acid-derived metabolites: The evidence so far. *Evidence-based Complementary and Alternative Medicine*, (2013), 270418. Retrieved 29 June 2019 from. https://doi.org/10.1155/2013/270418.

Etxeberria, U., Arias, N., Boqué, N., Macarulla, M. T., Portillo, M. P., Martínez, J. A., & Milagro, F. I. (2015). Reshaping faecal gut microbiota composition by the intake of trans-resveratrol and quercetin in high-fat sucrose diet-fed rats. *The Journal of Nutritional Biochemistry, 26*(6), 651–660. Retrieved 29 June 2019 from. https://doi.org/10.1016/J.JNUTBIO.2015.01.002.

Filosa, S., Di Meo, F., & Crispi, S. (2018). Polyphenols-gut microbiota interplay and brain neuromodulation. *Neural Regeneration Research, 13*, 12. Wolters Kluwer Medknow Publications. Retrieved 22 June 2020 from. https://doi.org/10.4103/1673-5374.241429.

Glisan, S. L., Grove, K. A., Yennawar, N. H., & Lambert, J. D. (2017). Inhibition of pancreatic lipase by black tea theaflavins: Comparative enzymology and in silico modeling studies. *Food Chemistry, 216*, 296–300. https://doi.org/10.1016/j.foodchem.2016.08.052.

Hamauzu, Y., & Suwannachot, J. (2019). Non-extractable polyphenols and in vitro bile acid-binding capacity of dried persimmon (*Diospyros kaki*) fruit. *Food Chemistry, 293*, 127–133. https://doi.org/10.1016/j.foodchem.2019.04.092.

Henning, S. M., Summanen, P. H., Lee, R.-P., Yang, J., Finegold, S. M., Heber, D., & Li, Z. (2017). Pomegranate ellagitannins stimulate the growth of Akkermansia muciniphila in vivo. *Anaerobe, 43*, 56–60. Retrieved 28 June 2019 from. https://doi.org/10.1016/J.ANAEROBE.2016.12.003.

Hsu, T.-F., Kusumoto, A., Abe, K., Hosoda, K., Kiso, Y., Wang, M.-F., & Yamamoto, S. (2006). Polyphenol-enriched oolong tea increases fecal lipid excretion. *European Journal of Clinical Nutrition, 60*(11), 1330–1336. https://doi.org/10.1038/sj.ejcn.1602464.

Huang, F., Zheng, X., Ma, X., Jiang, R., Zhou, W., Zhou, S., et al. (2019). Theabrownin from Pu-erh tea attenuates hypercholesterolemia via modulation of gut microbiota and bile acid metabolism. *Nature Communications, 10*(1), 4971. https://doi.org/10.1038/s41467-019-12896-x.

Huang, J., Feng, S., Liu, A., Dai, Z., Wang, H., Reuhl, K., et al. (2018). Green tea polyphenol EGCG alleviates metabolic abnormality and fatty liver by decreasing bile acid and lipid absorption in mice. *Molecular Nutrition & Food Research, 62*(4), 1700696. https://doi.org/10.1002/mnfr.201700696.

Huang, W.-C., Chen, C.-Y., & Wu, S.-J. (2017). Almond skin polyphenol extract inhibits inflammation and promotes lipolysis in differentiated 3T3-L1 adipocytes. *Journal of Medicinal Food, 20*(2), 103–109. Retrieved 29 June 2020 from. https://doi.org/10.1089/jmf.2016.3806.

Ikeda, I., Kobayashi, M., Hamada, T., Tsuda, K., Goto, H., Imaizumi, K., et al. (2003). Heat-epimerized tea catechins rich in gallocatechin gallate and catechin gallate are more effective to inhibit cholesterol absorption than tea catechins rich in epigallocatechin gallate and epicatechin gallate. *Journal of Agricultural and Food Chemistry, 51*(25), 7303–7307. https://doi.org/10.1021/jf034728l.

Ikeda, I., Yamahira, T., Kato, M., & Ishikawa, A. (2010). Black-tea polyphenols decrease micellar solubility of cholesterol in vitro and intestinal absorption of cholesterol in rats. *Journal of*

Agricultural and Food Chemistry, 58(15), 8591–8595. Retrieved 8 May 2017 from. https://doi.org/10.1021/jf1015285.

Jack, B. U., Malherbe, C. J., Willenburg, E. L., de Beer, D., Huisamen, B., Joubert, E., et al. (2018). Polyphenol-enriched fractions of cyclopia intermedia selectively affect lipogenesis and lipolysis in 3T3-L1 adipocytes. *Planta Medica, 84*(02), 100–110. https://doi.org/10.1055/s-0043-119463.

Kasote, D. M., Katyare, S. S., Hegde, M. V., & Bae, H. (2015). Significance of antioxidant potential of plants and its relevance to therapeutic applications. *International Journal of Biological Sciences, 11*(8), 982. Retrieved 1 July 2019 from. https://doi.org/10.7150/IJBS.12096.

Lattanzio, V. (2013). Phenolic compounds: Introduction. In *Natural products* (pp. 1543–1580). Berlin, Heidelberg: Springer. Retrieved 28 June 2019 from. https://doi.org/10.1007/978-3-642-22144-6_57.

Li, W., Zhang, K., & Yang, H. (2018). Pectin alleviates high fat (Lard) diet-induced nonalcoholic fatty liver disease in mice: Possible role of short-chain fatty acids and gut microbiota regulated by pectin. Journal of Agricultural and Food Chemistry, 66(30), 8015–8025. Retrieved from doi: https://doi.org/10.1021/acs.jafc.8b02979

Lin, J.-K., & Lin-Shiau, S.-Y. (2006). Mechanisms of hypolipidemic and anti-obesity effects of tea and tea polyphenols. *Molecular Nutrition & Food Research, 50*(2), 211–217. Retrieved 28 April 2017 from. https://doi.org/10.1002/mnfr.200500138.

Lowe, M. E. (1997). Structure and function of pancreatic lipase and colipase. *Annual Review of Nutrition, 17*(1), 141–158. Retrieved from. https://doi.org/10.1146/annurev.nutr.17.1.141.

Martín-Peláez, S., Castañer, O., Solà, R., Motilva, M., Castell, M., Pérez-Cano, F., et al. (2016). Influence of phenol-enriched olive oils on human intestinal immune function. *Nutrients, 8*(4), 213. Retrieved 29 June 2019 from. https://doi.org/10.3390/nu8040213.

Matralis, A. N., Bavavea, E.-I., Incerpi, S., Pedersen, J. Z., & Kourounakis, A. P. (2016). Balancing antioxidant, hypolipidemic and anti-inflammatory activity in a single agent: The example of 2-hydroxy-2-substituted morpholine, 1,4-benzoxazine and 1,4-benzothiazine derivatives as a rational therapeutic approach against atherosclerosis. *Current Medicinal Chemistry, 24*(12). https://doi.org/10.2174/0929867323666160814001803.

Menendez, J. A., Vazquez-Martin, A., Oliveras-Ferraros, C., Garcia-Villalba, R., Carrasco-Pancorbo, A., Fernandez-Gutierrez, A., & Segura-Carretero, A. (2008). Analyzing effects of extra-virgin olive polyphenols on breast cancer-associated fatty acid synthase protein expression using reverse-phase protein microarrays. *International Journal of Molecular Medicine, 22*(4), 433–439. Retrieved 27 June 2020 from. https://doi.org/10.3892/ijmm_00000039.

Moccia, F., Agustin-Salazar, S., Berg, A.-L., Setaro, B., Micillo, R., Pizzo, E., et al. (2020). Pecan (Carya illinoinensis (Wagenh.) K. Koch) Nut shell as an accessible polyphenol source for active packaging and food colorant stabilization. *ACS Sustainable Chemistry & Engineering, 8*(17), 6700–6712. https://doi.org/10.1021/acssuschemeng.0c00356.

Ogawa, K., Hirose, S., Nagaoka, S., & Yanase, E. (2016). Interaction between tea polyphenols and bile acid inhibits micellar cholesterol solubility. *Journal of Agricultural and Food Chemistry, 64*(1), 204–209. Retrieved 8 May 2017 from. https://doi.org/10.1021/acs.jafc.5b05088.

Panja, P. (2018). Green extraction methods of food polyphenols from vegetable materials. *Current Opinion in Food Science*. https://doi.org/10.1016/j.cofs.2017.11.012.

Pool-Zobel, B. L., Bub, A., Schröder, N., & Rechkemmer, G. (1999). Anthocyanins are potent antioxidants in model systems but do not reduce endogenous oxidative DNA damage in human colon cells. *European Journal of Nutrition, 38*(5), 227–234. Retrieved 11 September 2018 from. https://doi.org/10.1007/s003940050065.

Poudel, P. R., Tamura, H., Kataoka, I., & Mochioka, R. (2008). Phenolic compounds and antioxidant activities of skins and seeds of five wild grapes and two hybrids native to Japan. *Journal of Food Composition and Analysis, 21*(8), 622–625. Retrieved 9 September 2018 from. https://doi.org/10.1016/J.JFCA.2008.07.003.

Prior, R. L., Wu, X., & Schaich, K. (2005). Standardized methods for the determination of antioxidant capacity and phenolics in foods and dietary supplements. *Journal of Agricultural and*

Food Chemistry, 53(10), 4290–4302. Retrieved 2 July 2019 from. https://doi.org/10.1021/JF0502698.

Qian, J., Sullivan, P. B., & Berkland, C. (2015). pH-Responsive Micelle sequestrant polymers inhibit fat absorption. *Biomacromolecules, 16*(8), 2340–2346. https://doi.org/10.1021/acs.biomac.5b00560.

Quideau, S., Deffieux, D., Douat-Casassus, C., & Pouységu, L. (2011). Plant polyphenols: Chemical properties, biological activities, and synthesis. *Angewandte Chemie International Edition, 50*(3), 586–621. Retrieved 28 April 2017 from. https://doi.org/10.1002/anie.201000044.

Rocha, V. Z., & Libby, P. (2009). Obesity, inflammation, and atherosclerosis. *Nature Reviews Cardiology, 6*(6), 399–409. https://doi.org/10.1038/nrcardio.2009.55.

Rodríguez Montealegre, R., Romero Peces, R., Chacón Vozmediano, J. L., Martínez Gascueña, J., & García Romero, E. (2006). Phenolic compounds in skins and seeds of ten grape Vitis vinifera varieties grown in a warm climate. *Journal of Food Composition and Analysis, 19*(6–7), 687–693. Retrieved 9 September 2018 from. https://doi.org/10.1016/J.JFCA.2005.05.003.

Roy, M. K., Koide, M., Rao, T. P., Okubo, T., Ogasawara, Y., & Juneja, L. R. (2010). ORAC and DPPH assay comparison to assess antioxidant capacity of tea infusions: Relationship between total polyphenol and individual catechin content. *International Journal of Food Sciences and Nutrition, 61*(2), 109–124. Retrieved 2 July 2019 from. https://doi.org/10.3109/09637480903292601.

Sakakibara, T., Sawada, Y., Wang, J., Nagaoka, S., & Yanase, E. (2019). Molecular mechanism by which tea catechins decrease the micellar solubility of cholesterol. *Journal of Agricultural and Food Chemistry, 67*(25), 7128–7135. https://doi.org/10.1021/acs.jafc.9b02265.

Scalbert, A., & Williamson, G. (2000). Dietary intake and bioavailability of polyphenols. *The Journal of Nutrition, 130*(8), 2073S–2085S. Retrieved 28 June 2019 from. https://doi.org/10.1093/jn/130.8.2073S.

Schulenburg, K., Feller, A., Hoffmann, T., Schecker, J. H., Martens, S., & Schwab, W. (2016). Formation of β-glucogallin, the precursor of ellagic acid in strawberry and raspberry. *Journal of Experimental Botany, 67*(8), 2299–2308. Retrieved 19 June 2018 from. https://doi.org/10.1093/jxb/erw036.

Sellappan, S., Akoh, C. C., & Krewer, G. (2002). Phenolic compounds and antioxidant capacity of Georgia-grown blueberries and blackberries. *Journal of Agricultural and Food Chemistry, 50*(8), 2432–2438. Retrieved 19 June 2018 from. https://doi.org/10.1021/JF011097R.

Selma, M. V., Beltrán, D., García-Villalba, R., Espín, J. C., & Tomás-Barberán, F. A. (2014). Description of urolithin production capacity from ellagic acid of two human intestinal Gordonibacter species. *Food & Function, 5*(8), 1779–1784. Retrieved 29 June 2019 from. https://doi.org/10.1039/C4FO00092G.

Sousa, J. N., Paraíso, A. F., Andrade, J. M. O., Lelis, D. F., Santos, E. M., Lima, J. P., et al. (2020). Oral gallic acid improve liver steatosis and metabolism modulating hepatic lipogenic markers in obese mice. *Experimental Gerontology, 134*, 110881. https://doi.org/10.1016/j.exger.2020.110881.

Talcott, S. T., Lee, J.-H., Talcott, S. T., & Lee, J.-H. (2002). Ellagic acid and flavonoid antioxidant content of muscadine wine and juice. *Journal of Agricultural and Food Chemistry, 50*(11), 3186–3192. Retrieved 19 June 2018 from. https://doi.org/10.1021/jf011500u.

Tapiero, H., Tew, K., Nguyen Ba, G., & Mathé, G. (2002). Polyphenols: Do they play a role in the prevention of human pathologies? *Biomedicine & Pharmacotherapy, 56*(4), 200–207. Retrieved 29 June 2019 from. https://doi.org/10.1016/S0753-3322(02)00178-6.

Tomás-Barberán, F. A., & Espín, J. C. (2001). Phenolic compounds and related enzymes as determinants of quality in fruits and vegetables. *Journal of the Science of Food and Agriculture, 81*(9), 853–876. Retrieved 28 June 2019 from. https://doi.org/10.1002/jsfa.885.

Vermerris, W., & Nicholson, R. (2008). Families of phenolic compounds and means of classification. In *Phenolic compound biochemistry* (pp. 1–34). Springer.

Wang, S., Moustaid-Moussa, N., Chen, L., Mo, H., Shastri, A., Su, R., et al. (2014). Novel insights of dietary polyphenols and obesity. *Journal of Nutritional Biochemistry*. https://doi.org/10.1016/j.jnutbio.2013.09.001.

Wang, S., Huang, Y., Xu, H., Zhu, Q., Lu, H., Zhang, M., et al. (2017). Oxidized tea polyphenols prevent lipid accumulation in liver and visceral white adipose tissue in rats. *European Journal of Nutrition, 56*(6), 2037–2048. Retrieved 23 June 2020 from. https://doi.org/10.1007/s00394-016-1241-x.

Wu, T., Guo, Y., Liu, R., Wang, K., & Zhang, M. (2016). Black tea polyphenols and polysaccharides improve body composition, increase fecal fatty acid, and regulate fat metabolism in high-fat diet-induced obese rats. *Food & Function, 7*(5), 2469–2478. Retrieved 22 June 2020 from. https://doi.org/10.1039/c6fo00401f.

Xu, C., Zhang, Y., Cao, L., & Lu, J. (2010). Phenolic compounds and antioxidant properties of different grape cultivars grown in China. *Food Chemistry, 119*(4), 1557–1565. https://doi.org/10.1016/j.foodchem.2009.09.042.

Yilmaz, Y., & Toledo, R. T. (2003). Major flavonoids in grape seeds and skins: Antioxidant capacity of catechin, epicatechin, and gallic acid. *Journal of Agricultural and Food Chemistry, 52*(2), 255–260. Retrieved 9 September 2018 from. https://doi.org/10.1021/JF030117H.

Zderic, A., & Zondervan, E. (2017). Product-driven process synthesis: Extraction of polyphenols from tea. *Journal of Food Engineering, 196*, 113–122. https://doi.org/10.1016/j.jfoodeng.2016.10.019.

Zhang, H., & Tsao, R. (2016). Dietary polyphenols, oxidative stress and antioxidant and anti-inflammatory effects. *Current Opinion in Food Science*. Retrieved 19 June 2018 from https://www.sciencedirect.com/science/article/pii/S2214799316300133.

Harnessing the Potential of Phytochemicals for Breast Cancer Treatment

Manvi Singh, Sradhanjali Mohapatra, Sanskriti, Navneet Kaur, Abeeda Mushtaq, Sheikh Zahid, Arshad A. Pandith, Sheikh Mansoor, and Zeenat Iqbal

1 Introduction

Breast cancer (BC) is a heterogenous disease which remains the most commonly detected cancer; a major public health issue worldwide. Globally, it is the main cause of death among females, and some effect can be seen in males as well as transgender (Bray et al. 2018). BC has been a complex phenomenon with regards to biochemistry, pathophysiology, pathogenesis and it is categorized as uncontrollable cell proliferation triggered by malfunctioning of genes involved in the production of proteins like tumor suppressers, growth factor receptors, anti-apoptotic proteins, growth factors and, transcription factors (Khan et al. 2016). BC is difficult to treat due to the presence of several types of receptors which respond differently to the BC treatment. Some examples are estrogen receptor positive (T47D and MCF-7), estrogen receptor negative (MDA-MB-468, SKBR3 etc.), human epidermal growth factor receptor-2 and progesterone receptor (Hudis and Gianni 2011).

M. Singh · S. Mohapatra · Z. Iqbal (✉)
Department of Pharmaceutics, School of Pharmaceutical Education and Research, Jamia Hamdard, New Delhi, Delhi, India
e-mail: zeenatiqbal@jamiahamdard.ac.in

Sanskriti
Department of Biotechnology, Jamia Hamdard, New Delhi, Delhi, India

N. Kaur · A. Mushtaq · S. Zahid
Division of Biochemistry, Faculty of Basic Sciences, Sher-e-Kashmir University of Agricultural Sciences and Technology of Jammu, Jammu, Jammu and Kashmir, India

A. A. Pandith · S. Mansoor (✉)
Department of Advanced Centre for Human Genetics, Sher-I-Kashmir Institute of Medical Sciences, Srinagar, Jammu and Kashmir, India

C. Egbuna, S. Hassan (eds.), *Dietary Phytochemicals*,
https://doi.org/10.1007/978-3-030-72999-8_12

BC is caused by many related factors such as gender, age, reproductive factors, weight, family history of the patient, mammographic density, hormonal exposure, exposure to carcinogenic agents like alcohol, tobacco, UV rays, benzene, asbestos, etc. Other factors can be lifestyle conditions like early menarche, lactation, use of oral contraceptives, parity, nulliparity (Bhatt et al. 2017). Only 5–10% of BC cases are caused by mutation in genes like BRCA1 and BRCA2 (Li et al. 2017).

Early detection of BC can significantly reduce death rates in near future. For this achievement, it is important to fasten the process of discovery, diagnosis and treatment. Regular checkup of the clinical condition like hormone balance, age, receptor status, presence of metastasis can provide patient with the best possible treatment. Several BC diagnostic methods are developed over the years for early detection and quality treatment. Some of them are mammogram, magnetic resonance imaging, computerized tomography, ultrasound, breast self-examination etc. which are effective and efficient but have disadvantages like time consuming, inappropriate for young and lactating females, too expensive. Hence, the need of the hour is to develop a rapid and highly sensitive diagnostic method (Ng et al. 2017).

For the treatment of BC, methods such as radiotherapy, hormonal therapy, chemotherapy, surgery, monoclonal antibody therapy, mastectomy are available. They too have certain drawbacks such as their side effect, drug resistance (hormonal treatment for triple negative breast cancer has shown resistance), high incidence of tumor recurrence and disease progression. This has led scientists and researchers to think on how to find appropriate treatment for BC which will have less side effect with reduced chances of drug resistance. In this scenario, phytochemicals have proved themselves to be the influential preventive method for treatment of BC. Today, people are more interested in phytochemicals when compared with the synthetic ones because phytochemicals contain naturally active compounds supporting human health (Dmitri et al. 2015). Figure 1 shows the mechanism of breast cancer development.

Meta-analysis study of the phytochemicals shows that they have protective and preventive effect on the normal cells and they decrease the risk of BC in hormonal receptor negative and post-menopausal women (Zheng et al. 2016). Research studies have shown that many medicinal plants and their biochemically active compounds have inhibitory effects on BC cells by inhibiting angiogenesis, down-regulating ER-α expression, arresting of the cell cycle metastasis, proliferation and inducing caspase mediated cell death (apoptosis). A thorough literature survey has been done and the studies are summarized to define the role of phytochemicals in effective treatment and prevention of breast cancer.

2 Statistical Evidence Including the Indian Scenario

Breast cancer is considered the most common cancer in female with nearly 1.67 million new cases representing 25% of all cancers in 2012. Epidemiological study suggests that women in less developed region had a higher number of cases

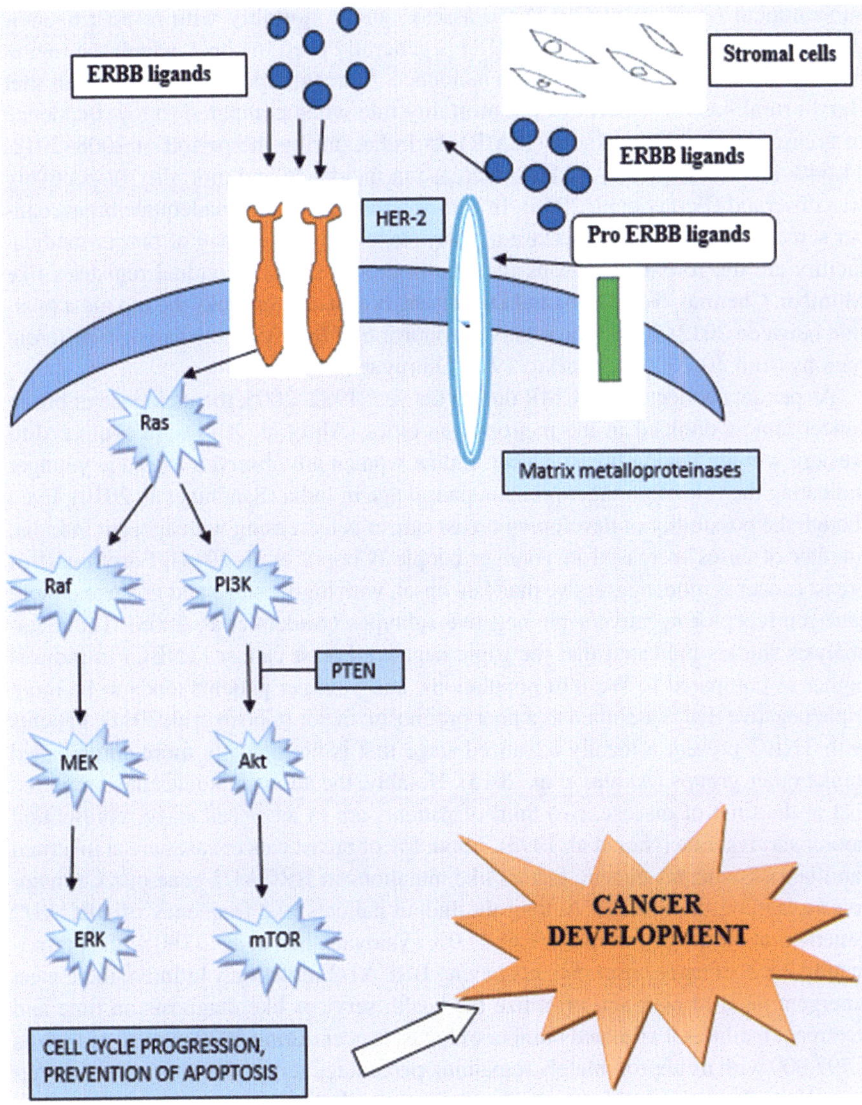

Fig. 1 ERBB ligands are released from stromal cells and/or produced by activation of estrogen receptor, which causes the matrix metalloproteinase-mediated cleavage and release of pro-EGF. Ras, Raf, PI3k, PTEN (gene alterations) activate further Raf/MEK/ERK and PI3K/Akt cascade leads to cancer

(883,000) than those in more developed regions (794,000) (Ferlay et al. 2015). The incidence rate of breast cancer varies among different parts of world based on race and ethnicity (DeSantis et al. 2014) and by the year 2050, it is estimated to reach 3.2 million (Hortobagyi et al. 2005). As per global and Indian data, there is a significant increase in both incidence and cancer associated morbidity and mortality in Indian subcontinent (Babu et al. 2013). To assess cancer mortality with respect to incidence, mortality/incidence ratio (MIR) is generally used to check whether a region has higher/lower mortality based on incidence. A recent report of ICMR reveals that Barshi rural has 66.3 MIR as high mortality rate when compared to low incidence of breast cancer in rural India (ICMR). In India, during the period of 2008–2012, 11.54% and subsequently 13.82% increase in incidence and mortality respectively was observed (Ferlay et al. 2015). In addition to this, lack of inadequate breast cancer screening, detection of disease in early stage and inadequacy of proper medical facility are the foremost reasons for hike in mortality. In individual registries like Mumbai, Chennai, New Delhi and Dibrugarh, breast cancer ranks the top most position between 2012–2014. The relative proportion of breast cancer varies in different regions from 30.7% (Chennai) to 19% (Dibrugarh).

As per data collected by ICMR during the year 1982–2005, the incidence of breast cancer almost doubled in the metropolitan cities (Ali et al. 2011). As compared to western women having breast cancer, Indian women are observed a decade younger indicating the role of younger premenopausal age in India (Sandhu et al. 2010). Even though the possibility of developing breast cancer is increasing with age, but in India, number of cases increased in younger people (Chopra et al. 2014). Early onset of breast cancer is more aggressive than late onset, with higher stage and grade and more estrogen receptor negative/triple-negative subtypes (Anders et al. 2008). The meta-analysis studies indicated that the triple negative breast cancer (TNBC) in India is higher as compared to Western populations, and younger patients tends to be more triple negative that is assumed as a poor prognostic factor (Ghosh et al. 2011). Patients with TNBC present a locally advanced stage that is biologically more complicated unlike other groups (Akhtar et al. 2015). Notably, the survival studies have reported that at the time of disease, two-third of patients are in advanced stage representing poorer survival rate (Nair et al. 1993). About 5% of breast cancer cases are at inherited familial risk whereas genetic factors like mutations in BRCA1/2 gene plays a major role in leading the disease. Among the Indian patients, the frequency of BRCA1/2 genetic mutations vary from 2.9% to 24.0% (Vaidyanathan et al. 2009) with approximately 2.8% of early onset patients having BRCA1/2 mutations. In India, there is an emergent need to plan and prioritize the health services like diagnosis on time and treatment facilities. Estimated number of breast cancer during 2020 may go as high as 1,797,900 with its approximately remaining percentage same 10% among all the cancers. Nowadays, the incidence of breast cancer in Indian women has surpassed the cervical cancer, previously considered as most common cancer, with still its existence as most common in rural India.

3 Current Treatments Strategies for Breast Cancer

This section explains the types of treatments that are the standard of care for early-stage and locally advanced breast cancer. It comprises 10.4% (Breast Cancer Treatment 2020, https://www.webmd.com/breast-cancer/breast-cancer-treatment)

of all cancer cases among women which makes it the second most common type of non-skin cancer (after lung cancer) and the fifth most common cause of death due to cancer. Breast cancer is more prevalent in women than in men, as males tend to have poor outcome due to slow rate of diagnosis. The treatment plan is affected by the behavior and biology of breast cancer. Some tumors are larger but grow slowly, while others are small but grow quickly. Treatment options and recommendations are very personalized and depend on several factors, including:

- The subtype of tumor together with hormone receptor status (ER, PR), HER2 and nodal status.
- The stage of the tumor.
- Genomic markers, such as Oncotype DX™ or MammaPrint™, if appropriate
- The age of patient, general health, status of menopause and preferences.
- The presence of known mutations in inherited breast cancer genes, such as BRCA1 or BRCA2 (Vaidyanathan et al. 2009).

3.1 Chemotherapy

Hormone therapy drugs are used to prevent hormones, especially estrogen, for enhancing the growth of breast cancer cells. Medicines comprise tamoxifen (Nolvadex) for women before and after menopause and aromatase inhibitors including anastrozole (Arimidex), exemestane (Aromasin), and letrozole (Femara) for postmenopausal women. Side effects can include hot flashes and vaginal dryness. Other types of this therapy work by inhibiting hormones from the ovaries, either through medication or by surgery. Fulvestrant (Faslodex) is an injection that prevents estrogen from sticking to cancer cells. ("Breast Cancer Treatment" 2020). Targeted therapy such as lapatinib (Tykerb), fam-trastuzumab-deruxtecan-nxki (Enhertu), trastuzumab (Herceptin) and pertuzumab (Perjeta) boost the immune system of body to help destroy cancer cells. These medicines target breast cancer cells that have high levels of a protein called HER2. T-DM1 or ado-trastuzumab emtansine (Kadcyla) is a medicine that combines Herceptin and the chemotherapy medicine emtansine to target HER2 positive cancer cells. Abemaciclib (Verzenio), palbociclib (Ibrance) and ribociclib (Kisqali) are frequently used with an aromatase inhibitor or fulvestrant (Faslodex), in women with various types of advanced cancer. Abemaciclib (Verzenio) can be used alone in women who have been already treated with chemotherapy and hormone therapy. HER2-positive breast cancer can also be inhibited by Neratinib (Nerlynx) which blocks the cancer cells from receiving growth signals. A latest class of drugs called PARP (poly ADP ribose polymerase) inhibitors attacks an enzyme that fuels cancer cells. PARP inhibitors include talazoparib (Talzenna) and olaparib (Lynparza). Alpelisib (Piqray) a PI3K inhibitor treats breast cancer in men and women having mutations due to treatment with hormonal therapy ("Breast Cancer Treatment" 2020).

3.2 Surgery

The most common form of non-invasive breast cancer is Ductal carcinoma in situ (DCIS) and Lobular carcinoma in situ (LCIS) not so common form is considered as a marker for increased breast cancer risk. For both DCIS and early-stage invasive breast cancer, generally doctors recommend surgery for the removal of tumor. A small area of healthy tissue around the tumor, called as margin is also removed by the surgeon to make sure that the entire tumor is removed. Lumpectomy (removal of the lump only), or surgical removal of the entire breast (mastectomy) is performed according to the type and stage of the tumor. Standard practice requires the surgeon to establish that the tissue removed in the operation has margins clear of cancer, indicating that the cancer has been completely excised. If in case the excised tissue does not show clear margins, further operations may be essential to remove more tissue. This sometimes requires the removal of part of the pectoralis major muscle (main muscle of the anterior chest wall). More recent technique of sentinel lymph node (SLN) dissection has become common as it requires the removal of fewer lymph nodes, resulting in low side effects. Over the past few decade advances in sentinel lymph node mapping has increased the accuracy of detecting sentinel lymph node from 80% using blue dye alone to between 92% and 98% using combined modalities (Sharma et al. 2010).

3.3 Gene Therapy

Generally, cancer arises because of an accumulation of multiple molecular genetic defects that end up in a cellular phenotype characterized by unregulated cellular growth. A variety of gene therapies have been developed as potential strategy for treating cancer. Current knowledge of tumor suppresser genes and proto-oncogene in the malignancy genesis has encouraged the development of gene therapy tactics aimed at restoring or ablating such genes, respectively. In other strategies, cancer cells are capable of converting a systemically delivered prodrug to a toxic metabolite, or a target for destruction by replicating viral vectors conversely transfer of drug resistance genes into normal cells may provide chemoprotection during high dose antineoplastic treatment. Finally, immune system modulation can activate anticancer drug defense mechanisms (Osborne et al. 2004).

3.4 Oncogenes Inactivation

Numerous oncogenic proteins have been identified which are associated with various malignancies. The most commonly applied approach to date in clinical trials has been use of antisense strategies. Transcription of oncogenes also can be suppressed by the use of adenoviral gene E1A, which interferes with the transcription of erbB-2, a useful approach in treating cancer that overexpresses this oncogenic protein, such as breast and ovarian cancer (Osborne et al. 2004).

3.5 Augmentation of Tumor Suppresser Genes

3.5.1 BRCA1 and BRCA2

BRCA1 and BRCA2 are hypothesized to play a critical role in maintaining integrity of DNA and are tumor suppressor genes. Inherited mutation in the BRCA1 and BRCA2 genes are caused by hereditary BC. Maximum mutations of the BRCA1/BRCA2 genes in breast and/or ovarian cancer families are point mutations or small deletions and insertions extended over the coding sequence and splice site junctions. Large number of rearrangements in BRCA1and BRCA2 account for less than 1% of examined patients suffering from hereditary breast and ovarian cancer (Jackson et al. 2014). All of these mutations are adequate for shortening BRCA1 protein, leading to weakening of its physiologic function (Varley 2003). Furthermore, aggressive triple-negative breast cancer (TNBC) has also been found associated with sporadic mutations in BRCA1 (Gonzalez-Angulo et al. 2011).

3.5.2 Androgen Receptor

The human androgen receptor (AR) is a member of nuclear receptor superfamily and is involved in hormone action. Non-transcriptional/non-genomic mechanism can also be used for activation of this receptor. AR receptor is present in 70–90% of all breast cancers and is an effective therapeutic target in ER-negative breast cancers as it is widely expressed in most of the ER-positive breast cancers (67–88% of cases) with apocrine differentiation, leading to a reduction in mortality (Gucalp and Traina 2016) . The anti-androgen agents (e.g., bicalutamide, enzalutamide) provided new gateway for development of therapeutic strategies in advanced breast cancer expressing AR. Evidence also suggest that AR is able to interact with a set of signaling pathways including MAPK pathways, PI3K/AKT/mTOR and cell cycle, or Ras-dependent pathways and many crucial molecules, such as pioneering factors (FOXA1), PTEN, BRCA1, and BRCA2 are involved in carcinogens and progression of TNBC by interaction with AR (Shi et al. 2018). Targetable agents such as PI3-K, Src, and/or MEK inhibitors or their rational combinations with therapies depending on hormone and chemotherapeutic agents are presently provided for further assessment in breast cancer (Giovannelli et al. 2019).

3.5.3 Immunomodulation

A range of cytokines have the ability to boost immunity against cancer cells, and this observation has led to the development of gene-based technique to regulate the immune reaction in malignancy.

- **Ectopic Cytokine Expression**: Many cytokines have led to the decline in tumor growth when expressed ectopically in tumor cells or in their microenvironment. Initially, various immunostimulatory factors have no effect on the growth rate of the tumor, but later on provide immunity against tumor growth if the animal is treated with wild type tumor cells.
- **Enhancement of Immune**: Medicines are used to boost a person's own immune system to identify and destroy cancer cells more efficiently. It is a type of biological therapy in which elements made from living organisms are used to treat cancer. One such method is to express a highly immunogenic molecule allotype MHC antigen on the surface of cancer cells (Disis and Park 2009).

3.5.4 Endocrine (Antihormonal) Treatment

At present, endocrine treatment is the first line of therapy for patients with metastatic breast cancer and positive or unknown hormone receptor status whose disease is not at a life-threatening stage. A delay of 10–12 weeks is necessary before the endocrine treatment starts to take effect. Extensive visceral metastases, CNS (central nervous system) metastases and/or an urgent need for remission (pain) are main reasons not to undergo endocrine treatment. Concurrent chemoendocrine therapy must be avoided, as this is no more effective than the separate therapies but has more side effects (Sledge Jr et al. 2000). Endocrine maintenance therapy following a response to chemotherapy improves not only disease-free survival but also overall survival (Gerber et al. 2010).

3.5.5 Inclination Towards Phytochemicals

With a surge in cancer incidence and high rates of mortality, keen understandings of dietary combinations as potential chemo-preventive agents is certainly an attractive approach against cancer. The main disadvantage of conventional therapies is the recurrence, as most of the cancer stem cells are not eliminated from the body. Moreover, multi-drug resistance (MDR) is a significant clinical challenge. Besides, in many countries, economic and social burdens due to cost of treatments remain among many problems to be conquered for reduction in cancer incidences and mortality rate globally. As a result, there is a vital need to discover novel approaches and therapies that may offer comparatively cost-effective regimens with fewer side effects (Rizeq et al. 2020). Several studies have focused on bioactive molecules which have been isolated, characterized and identified as phytochemicals which are non-essential non-nutritive plant chemicals meaning they are not required by the human body for sustaining life having disease preventive or protective properties (Mai et al. 2007: Carriere et al. 2016). These phytochemicals are arranged on the basis of their chemical composition and structure from organosulfur compounds, carotenoids to phenolics. Results have shown that many of these phytochemicals have the ability to target a wide variety of signaling molecules associated with cellular growth, proliferation, differentiation, and death in a sequence of cell types (Lee et al. 2011: Howes and Simmonds 2014). Chemicals within these plants target numerous molecular subtypes of breast cancer and breast CSCs (Dandawate et al. 2013). Phytochemicals are not toxic and perform a wide variety of biological activities including antiproliferative, anti-inflammatory, antioxidant and anticancer properties. Many studies have focused on the use of phytochemicals as adjuvant therapies. A variety of phytochemicals behave as SERMs (selective estrogen receptor modulator), including isoflavones, coumestans, lignans and prenylflavonoids. Isoflavones inhibit breast cancer growth, metastasis and induce apoptosis by the regulation of different signaling pathways. Genistein and daidzein isoflavones inhibit cell proliferation, angiogenesis and induce apoptosis by cell cycle arrest in ER(+) breast cancer cells (Israel et al. 2018). Epigallocatechin gallate (EGCG) a polyphenolic catechin, is a phytochemical present in green tea (Mai et al. 2007). It is utilized to suppress the growth and angiogenesis and to increase breast cancer cells apoptosis (Gu et al. 2013). Due to the drug-resistant nature of BCSC (breast cancer surveillance consortium) utilization of natural dietary supplements as alternative and/or adjuvant therapies for existing breast tumors has gained interest (Oh et al. 2016). Phytochemicals have shown effective results at targeting various signaling pathways and BCSCs in breast cancer. Utilizing well-formulated phytochemicals in nanoparticles, in combination with currently accepted endocrine and/or standard chemotherapies, can prove a feasible technique of overcoming resistance and targeting BCSCs in metastatic breast cancer (Israel et al. 2018). More than 5000 phytochemicals have been found in vegetables, grains and fruits with a huge quantity still undiscovered (Meybodi et al. 2017). The inclination of phytochemicals towards cancer treatment, maybe due to their ability in regulating different signaling pathways that inhibit cell proliferation, advance cell death, and invasion, sensitize cancerous cells, and enhance the immune system. Cytotoxic anti-tumour agents and inhibitors from phytochemicals may work together producing inhibitory action against tumor growth (Knobloch et al. 2016).

4 Phytochemicals for Breast Cancer Therapy

4.1 Plumbagin

Plumbagin is a secondary metabolite found in Plumbaginaceae, Ebenceae, Dioncophyllaceae Ancestrocladaceae and Droseraceae family plants. It is mainly extracted from root, leaf and bark of the plant. Chemically it is 5-hydroxy-2-methyl-1,4-naphthoquinone, a metabolite of 1,4-naphthoquinone class of quinone family and an analogue of vitamin K3 (Liu et al. 2017). It is known for its anti-cancerous, anti-inflammatory, anti-oxidant, antimicrobial, neuroprotective and hypolipidemic properties. It activates proapoptotic proteins like Bax and Bak and downregulate anti apoptotic proteins like Bcl-xL and Bcl-2. Plumbagin can reduce breast cancer cell proliferation by inducing cell cycle arrest (G2/M and G1) and autophagy by regulating different signaling pathways and proteins (Kuo et al. 2006; Zhang et al. 2016). Signaling Pathways such as PI3K/AKT, mTOR, RANKL signaling, STAT3 signaling are found to be downregulated in breast cancer cell line in mouse models (Sung et al. 2012; Yan et al. 2013). Plumbagin inhibits progression, invasion, metastasis and even angiogenesis by regulating different proteins like p53, p21, NF-κB, CXCL12,IL-1α, TGF-β, MMP-2,MMP-9 (Manu et al. 2011; Lee et al. 2019). ER and Her2 overexpressed breast cancer cells and endothelial progenitor cells (angiogenesis) are also suppressed by plumbagin (Kawiak et al. 2012).

4.2 Apigenin

Apigenin (5,7,4′-trihydroxyflavone) is a flavonoid commonly found in the dietary food like parsley, peppermint, thyme, celery, citrus fruits, apple wine, beer, honey etc. It possess antioxidant, anti-inflammatory and anticarcinogenic properties (Liu et al. 2013). The reason being more permeability of the compound through lipid bilayer in the cell membrane. It lowers the cancer cell survival by inhibiting nuclear factor kappa beta (NF-kB), JAK2 and STAT3 phosphorylation and CoCl-2-induced VEGF secretion in HER2 overexpressing cancerous cells hence, inhibits angiogenesis. It elevates apoptosis by upregulating caspase 8, caspase 3, poly (ADP-ribose) polymerase cleavage (PARP) cleavage, p53, phospho-p53 and p21 (Seo et al. 2012; Choi and Kim 2009). Antimetastatic property of Apigenin can be identified by downregulation of matrix metalloproteinase-9 (MMP-9) in MCF-7 cell line as well as in estrogen-insensitive breast tumor cells (Lindenmeyer et al. 2001). Modification of certain genes such as nonsteroidal anti-inflammatory drug (NSAID) activated gene-1 (NAG-1), aromatase or cytochrome P450 (CYP19) also decrease the cancer progression (Yang et al. 2014).

4.3 Isothiocyanate

Role of isothiocyanate for cancer treatment is well known from last two decade (Warin et al. 2009). Cruciferous vegetables like broccoli, cabbage, turnip, etc., have abundant source of isothiocyanate (Wu et al. 2009). Isothiocyanate acts as a histone deacetylase (HDAC) inhibitor and decreases the growth factor receptors expression like estrogen receptor. Its treatment inhibits cell growth, causes cell cycle arrest (G2-M phase) and induce DNA fragmentation. By targeting Fas, it induces cell apoptosis by activating caspases, releasing cytochrome c into cytosol, increasing the expression of proapoptotic proteins like bax (Tseng et al. 2004) and decreasing antiapoptotic proteins (Bcl-2). Benzyl isothiocyanate inhibits breast cancer stem-like cells renewal both in vitro and in vivo (Kim et al. 2013a, b).

4.4 *Quercetin*

The idea of using quercetin as anticancer flavonoid is derived from its ability to prevent free radicals. It is mostly found in its glycoside form in wine, vegetables, fruit and tea (Zhang and Du 2012). It acts as an antioxidant, inhibits proliferation, metastasis, angiogenesis and cause cell death. Different breast cancer subtypes respond differently upon quercetin treatment. In basal subtype of breast cancer, it causes cell cycle arrest, apoptosis, alter cell morphology and inhibits cell migration by targeting P53, JNK signaling, PARP and β-catenin (Nguyen et al. 2017). Using xenograft models, it has been shown that it's antiangiogenetic effect is comparable to that of tamoxifen (Aggarwal et al. 2016). It also compliments with other drugs like doxorubicin and cisplatin by increasing the bioavailability, reversing the resistance of breast cancer (DOX) and enhancing their effect and reducing the side effects like renal toxicity (CP) (Choudhuri et al. 2005).

4.5 *Curcumin*

Curcumin is a polyphenol product isolated from rhizome of *Curcuma longa.* It has been used for medicinal purpose like infection, inflammation, arthritis, viral infection, and amyloid aggregation. Recent studies have reported the anti-cancerous property of curcumin by activating apoptosis and inhibiting pre-cancer processes. It inhibits cell proliferation by causing cell cycle arrest at G1 stage by proteasomal degradation of cyclin E and upregulation of CDK inhibitors, p53, p21 and p27. It also inhibits the initiation by blocking the alliance of CDK 4 and CDK 6 and inhibiting the phosphorylation of Retinoblastoma (Rb). Curcumin induces apoptosis via activating p53 and proapoptotic (Bax, caspase-3) proteins and downregulating Bcl-xL (Choudhuri et al. 2002). It also targets PI3K/Akt/mTOR signalling and Wnt/β catenin signaling pathways, NF-κB pathway and VEGF to obstruct the proliferation, EMT, metastasis and angiogenesis in Breast cancer (Park 2017; Prasad et al. 2009).

4.6 *Catechins*

There are four major Catechins (flavan-3-oils) present in green tea which are epigallocatechin-3-gallate (EGCG, 59%), epigallocatechin (EGC, 19%), epicatechin3-gallate (ECG 13.6%), and epicatechin (EC 6.4%) (Cabrera et al. 2006). It enhances the effect of drugs in chemotherapy like tamoxifen and also reduces the required dose amount. It enhances the cytotoxic effect of 4-hydroxytamoxifen by earlier induction of apoptosis in the cell by downregulation of S-phase kinase protein 2 (Skp2), component of the Skp1-cullin 1-Fbox protein (SCF) ubiquitin ligase complex, which ultimately target G1-to-S phase progression regulator (Stuart et al. 2007). Similar results were reported with the combination of raloxifene associated with reduced phosphorylation of EGFR- and AKT-signaling proteins (Stuart and Rosengren 2008). mTOR, Akt and EGFR phosphorylation was reduced and SAPKs, JNK1/2 and P38 phosphorylation was induced. It also reverses the tamoxifen drug resistant breast cancer (Farabegoli et al. 2011).

4.7 Lycopene

Lycopene is a carotenoid lacking provitamin A activity and is found in plant food like tomato, watermelon, pink grapefruit, apricot, papaya etc. It has a strong antioxidant activity and causes interference in insulin like growth factor-1 receptor signaling which leads to disruption of cell cycle progression (Karas et al. 2000). Lycopene inhibits both ER dependent and independent breast cancer cells. Simultaneously, it regulates apoptosis, cell communication, MAPK, cell cycle and DNA repair mechanism pathways (King-Batoon et al. 2008). It also attenuates tumorigenesis by promoter methylation modulation of gene expression (Gloria et al. 2014).

4.8 Hesperidin

Hesperidin is a flavonoid commonly found in citrus plants such as orange, lemon etc. It eliminates cancerous cells by apoptosis without harming normal cells and increases the intrinsic antioxidant activity (Singh et al. 2020). Studies have shown that hesperidin has antiproliferative effect in breast cancer cell lines (Lee et al. 2010). It suppresses activation of NF-κB and Akt pathway and decrease the Programmed death ligand 1(PD-L1) protein which is elevated in aggressive breast cancer. In this way, it reduces the growth, migration and ultimately the aggressiveness of the cancer. It also targets ER signaling in ER positive breast cancer. Complementary effect of hesperidin with chemotherapy drugs like Tamoxifen and Doxorubicin has also been reported (Febriansah et al. 2014).

4.9 Anthocyanin

Anthocyanin is a type of flavonoid which provides dense pigment to many fruits, vegetables and even pigmented cereals for example grapes, blueberries, purple sweet potatoes, red cabbage, etc. Earlier it was known for its effects in lipid profile improvement, inflammation and antioxidant activity. Recent studies have shown its role in controlling cancer progression (Viskelis et al. 2009). In an experiment with anthocyanin extracted from black rice, the breast cancer cells [MDA-MB-453 (ER−, HER2/neu+)] exhibited increased apoptosis upon activation of caspases, cleavage of PARP, depolarisation of mitochondrial membrane and release of cytochrome C. The same results were also depicted in Xenograft model mice which showed suppressed tumor growth and angiogenesis upon oral administration of the drug (Hui et al. 2010). Expression of certain proteins like COX-2, iNOS, c-Jun, MMP-9, MMP-2 and uPA was also found to be decreased. Studies have also reported that moderate consumption of wine is beneficial for breast cancer patient undergoing radiotherapy in order to reduce skin toxicity. Black rice anthocyanin also suppresses the metastasis by targeting RAS/RAF/MAPK pathway (Chen et al. 2015). Grape seed extract contained with plenty of proanthocyanin has been used in clinical trials for breast cancer in post menstrual staged women (Lin et al. 2017).

4.10 Colchicine Alkaloid

Colchicine is an alkaloid and a natural tubulin inhibitor which hampers cell division and has been in use for curing diseases such as Mediterranean fever, Bechet's syndrome, Sweet's syndrome, scleroderma, amyloidosis, liver cirrhosis and gout (Lidar and Livneh 2007). On treatment to breast cancer cell line MCF-7, more cells were found in G2/M phase which implicated the inhibitory effect of colchicine against cell growth & is dependent on dosage and time variance (Sun et al. 2016). Colchicine is a toxic substance, hence numerous colchicine analogues which are less toxic are used. CH-35, an analogue of colchicine, was found to be more effective than taxol in controlling breast cancer cells overexpressing βIII isotype of tubulin which is more aggressive, metastatic and drug resistant. Certain derivatives of colchicine are also used for cancer treatment like carbamate derivatives, Naphthalene chalcone derivatives etc., (Urbaniak et al. 2020).

4.11 Polyphyllin D

Polyphyllin D is a steroidal saponin which is found majorly in Paris polyphylla, a Liliaceae family plant used traditionally in Chinese medicine to cure mastitis, convulsion, tuberculous, sore throat, meningitis and tumors of liver, pancreas, urinary bladder, respiratory system, digestive tract, and brain as well as leukemia. It also possesses antibacterial, hemostatic, anti-inflammatory, spermicidal, hemostatic, sedative, and analgesic properties (Lee and Lin 1988). The anti-cancerous property of polyphyllin D is linked with the mitochondrial dysfunction (independent of ER) and the in vivo study has projected that the intravenous dose reduces the tumor weight and size by 50% without harming heart and liver functioning of the host (Lee et al. 2005).

4.12 Genistein

Genistein is a soybean isoflavone often consumed in Asian food. Number of findings have reported that early exposure (before puberty) to genistein reduces the susceptibility towards breast cancer, probably because of differentiation of mammary gland during this period. The mechanism supporting this is related to activation of ER-α and enhanced expression of BRCA1 and PTEN protein (Warri et al. 2008). Effect of Genistein depends upon the dose, type of tissue and level of ER isoform. When ERβ is present, the high dose of drug is beneficial and leads to reduced proliferation whereas in case of ERα, high concentration of genistein is related to high proliferation and worse prognosis. It specifically targets EGFR, PDGFR, IR, Abl, Fgr, Fyn and Src and inhibits NF-kB signaling. In triple negative breast cancer, genistein affect the cell growth in dose and time dependent manner. Induction of apoptosis and G2/M cell cycle arrest was the effect of NF-kB inhibition via NOTCH-1 signaling, ultimately targeting antiapoptotic gene expression (Bcl-2 and Bcl-xL) (Pan et al. 2012).

4.13 Resveratrol

Resveratrol is a phytoalexin present in many plants such as grapes, peanuts and berries and provides resistance to microbes, fungi etc. It epigenetically reduces the expression of certain genes involved in breast cancer like RASSF-1α (Zhu et al. 2012). In combination with resveratrol, herceptin encourages a reduction in HER-2 receptor cells and resistance to doxorubicin which is reversed in the presence of resveratrol (Abdel-Latif et al. 2015). It enhances p21 expression and reduces the expression of antiapoptotic proteins (PI3K, AKT, NFκB) and ultimately reduces the solid tumor size. Mixed micelle loaded Resveratrol demonstrated a better treatment approach for doxorubicin resistant breast cancer cells (Wang et al. 2015).

4.14 Berberine

Berberine is a nitrogenous cyclic compound. Its structure is like intercalating agent (for example: ethidium) and possess fluorescent properties. Berberine decreases the ROS level by inhibiting xanthine oxidase, cyclooxygenase-2 and superoxide dismutase enzyme activity (Liu et al. 2008). However, this effect of berberine seems to depend upon the cell condition and type, in case of cancerous cell it induces ROS formation during the activation of different apoptotic pathways like MAPK, AKT, ERK1/2 and JNK. Berberine induces apoptosis by caspase activation, DNA fragmentation and mitochondrial collapse (Patil et al. 2010). It inhibits TNF-α induced MMP-9 activity and interrupts the invasion mechanism by suppression of AP-1 DNA binding activity (Kim et al. 2008). 17β-estradiol is metabolically transformed into two forms i.e. 2-hydroxyestradiol and 4-hydroxyestradiol which are further converted into non mutagenic and genotoxic quinone, respectively. The ratio between 2-hydroxyestradiol and 4-hydroxyestradiol is essential for the carcinogenic activity of estradiol. The ratio of the enzymes involved CYP1A1 (2-OH E2) and CYP1B1 (4-OH E2) are regulated by berberine which increases CYP1A1 without altering CYP1B1 (Wen et al. 2014).

5 Pharmaceutical Compounding of Phytochemicals

Phytochemicals are small non-nutritive compounds extracted from plants. They are source of many micronutrients, vitamins, minerals, and fibers and have potential of protecting humans from diseases. A broad classification of phytochemical is based on the chemical structure as phenolic, alkaloids, carotenoids, nitrogen containing compounds and organosulfur (Liu 2003; Akhtar Siddiqui et al. 2015). Due to the drawbacks and side effects of conventional chemotherapy treatment, finding a new therapeutic agent is indispensable. The area of phytochemicals is barely explored, and researchers are still trying to examine the core of this domain. In the following section the emerging phytochemicals with potential of anti-cancerous activity against breast cancer were discussed.

Stages in cancer development include inflammation, initiation, progression, invasion, migration and angiogenesis. Inflammation is the elementary steps of cancer, as it provides a suitable environment for tumor initiation, progression, and metastasis. Hence, inflammation control has proved to be a pivotal step. Major involvement in this process is related to proinflammatory transcription factors, cytokines, and NF-kB. The most challenging part for the treatment is controlling the cancer invasion and metastasis. Once a cancer is metastasized, it is unattainable to control it. Currently, we are lacking such drugs which can be used to cure metastasize cancer. Both these steps are very crucial for the prospective of controlling cancer progression with dietary elements.

Polyphenol extracted from pomegranate juice shows anti-inflammatory effect by attenuating macrophages response to M1 pro inflammatory activation in macrophage like cell line (Aharoni et al. 2015). Recently, the antineoplastic role of *Chlorella pyrenoids* has been discovered by Kubatka et al. (2015) in both in vitro and in vivo study. It has also been reported that chlorella powder activates caspase-7 and decreases vascular endothelial growth factor receptor-2 expression in animal model. In vitro study showed increase in apoptosis and reactive oxygen species whereas decrease in mitochondrial membrane potential (Kubatka et al. 2016a, b). The same study was done on young barley mixture, oregano and fruit peel contained

Flavin-7. Lyophilized oregano (rich in phenolic compounds and terpenoids) decrease tumor frequency, tumor incidence and volume by almost 50% in vivo and results in decreased proliferative marker (Ki67) and cancer stem cell markers (CD24 and EpCAM) (Kubatka et al. 2017). Lyophilized substance of Flavin-7 oral administration showed significant decrease in Bcl-2, Ki67, and VEGFR-2 expression, which implies antiproliferative, apoptotic and antimetastatic role of it (Kubatka et al. 2016a, b). Clove buds (*Syzygium aromaticum*) also possess antineoplastic property increasing Bax, caspase-3 and ALDH1 expression and RASSF1A promoter methylation. All these phytochemicals inhibit anti apoptotic pathways and activate mitochondrial apoptotic pathways. Modulation of tumor growth by *Aloe vera* and honey by elevating apoptosis susceptibility and decreasing cell growth has also been reported in some studies (Tomasin and Cintra Gomes-Marcondes 2011). Propolis or bee glue is a resin like material collected by honey bees from various plants and contains many bioactive compounds. Chrysin (a flavonoid) is an active compound present in propolis, is a histone deacetylase inhibitor. It inhibits histone deacetylase 8 (HDAC8) enzymatic activity which deacetylate the histone and result into tumor growth. The data was also validated in xenograft animal model (Kim et al. 2013a, b). Chrysin also inhibits the hypoxial survival of tumor cells stopping activation of STAT3 and downregulating VEGF expression that ultimately inhibits angiogenesis and metastasis. It is also a potential adjuvant with other antimetastatic drugs (Lirdprapamongkol et al. 2013). Cannabinoid based medicine were found to be useful in the treatment of all three subtypes of breast cancer i.e. hormone receptor positive, HER2 positive and triple negative breast cancer. In hormone receptor positive cancer, activated cannabinoid receptor acts via inhibition of adenyl cyclase that further leads to deactivation of Raf-1/ERK/MAP pathway and thus inhibits cell proliferation. Tetrahydrocannabinol (THC) major active compound in cannabinoids activate transcription factor JunD which eventually activate the apoptosis process. Cannabinoid also arrest cell cycle at different stages and also inhibit migration and invasion by inhibiting ERK. The fatty acid amide hydrolase (FAAH) and monoacylglycerol lipase (MGL) level are elevated in cancerous cells which interfere with the activation of nuclear factor (erythroid-derived 2)-like 2 (Nrf2)/antioxidant responsive element (ARE) pathways which finally leads to apoptosis. Cannabinoid decreases the level of FAAH and MGL in cancer cell and induce apoptosis (Chakravarti et al. 2014). In case of HER-2 positive and triple negative breast cancer, cannabinoid inhibits protein like ERK, AKT, COX-2 and FAK which are responsible for elevating cell proliferation, metastasis and decreasing apoptosis (Nasser et al. 2011). It activates Wnt/catenin signaling pathway, cell cycle arrest, autophagy and inhibits angiogenesis through MMP-2 and MMP-9 (Shrivastava et al. 2011). Table 1 depicts the various pharmaceutical formulation available in clinical trials data having phytochemicals as key component.

Table 1 List of clinical trials relating to the management of breast cancer using a variety of botanicals

Study design	No. of patient (women)	Study length	Extract/formulation/dosage form	Study outcome	Reference
Pilot human clinical trials	6 breast cancer patients	12 weeks	Gold nanoparticles-based Nano Swarna Bhasma Two capsules thrice a day after food	The study succeeded in achieving tremendous clinical benefits of Nano Swarna Bhasma drug in adjuvant therapy. The study helped to treat human metastatic breast cancer patients by using a proprietary combination of gold nanoparticles and phytochemicals to develop the Nano-Ayurvedic drug: Nano Swarna Bhasma, through innovative green nanotechnology	Khoobchandani et al. (2020)
Randomized controlled trial	19 breast cancer patients	From biopsy-confirmed diagnosis to surgery (6 ± 2 days)	Dietary polyphenols Capsules contained pomegranate, orange, lemon, olive, cocoa, and grapeseed extracts plus resveratrol: 3 capsules per day	The study comprehensively describes the metabolic profiling of dietary phenolics and methylxanthines in malignant tissue as well as normal tissue. It was observed that the metabolites, detected in breast tissues, did not exert antiproliferative or estrogenic/antiestrogenic activities in MCF-7 breast cancer cells. Thus, the study concluded that even if phase-II conjugation might obstruct a direct anticancer activity, long-term tumor-senescent chemoprevention cannot be discarded	Ávila-Gálvez et al. (2019)

(continued)

Table 1 (continued)

Study design	No. of patient (women)	Study length	Extract/formulation/dosage form	Study outcome	Reference
Randomized controlled trial	62 breast cancer patients	Entire period of radiotherapy	Nigella sativa: 5% gel medications twice daily during radiotherapy period	The study showed that there is a significant reduction in the severity of acute radiation dermatitis and interruption in the onset of moist desquamation in breast cancer patients after application of Nigella sativa gel	Rafati et al. (2019)
Randomized controlled trial	93 patients	24 months	Chemotherapy and Xiaoaiping	The study showed that patients in the combined group had a better quality of life than the control patients and concluded that Xiaoaiping can improve alopecia, WBC count, AST levels, nausea, vomiting, and diarrhea symptoms, and the quality of life in breast cancer patients	Yu et al. (2019)
Randomized controlled trial	85 patients	12th weeks	Remifemin	The study showed that Cimicifuga racemosa (Remifemin) is effective, reliable, and oncological safe for treatment of menopausal syndrome caused by luteinizing-hormone-releasing hormone analogue in breast cancer	Wang et al. (2019)

Study design	No. of patient (women)	Study length	Extract/formulation/dosage form	Study outcome	Reference
Prospective open-label randomized phase III study	161 female breast cancer patients	20 months	Mebo ointment and Biafine cream	A comparative study was carried out to assess the efficacy of two topical agents used for management of acute radiation dermatitis during radiation therapy in breast cancer patients. This study concluded that there was no significant difference in grades 2 and 3 dermatitis between Mebo and Biafine but the incidence of severe pruritus and severe local skin pain was significantly reduced by the Mebo ointment, which might positively affect the quality of life of cancer patients	Geara et al. (2018)
Randomized, double-blinded, placebo-controlled phase II clinical trial	1075 healthy postmenopausal women (932 completers)	12 months	Decaffeinated green tea extract—capsules containing 1315 mg total catechins, including 843 mg epigallocatechin-3-gallate (EGCG): 4 capsules per day	The study showed that there is no significant effect on mammographic density (MD) measures in all women after 1-year supplementation with a high dose of EGCG, but reduced percent of MD was observed in younger women. An age-dependent chemopreventive effect of green tea intake on breast cancer risk was observed, which is similar to those of tamoxifen	Samavat et al. (2017)

(continued)

Table 1 (continued)

Study design	No. of patient (women)	Study length	Extract/formulation/dosage form	Study outcome	Reference
Pilot randomized controlled trial	60 patients	6 months	Dry extract of Propolis with sodium bicarbonate	The research showed that Propolis plus bicarbonate was safe, well-tolerated, and promisingly effectual in the prevention of ophthalmic metastases in patients with breast cancer	Piredda et al. (2017)
Randomized controlled trials	150 female patients with breast cancer selected 119 patients completed the study	3 chemotherapy cycles	Ginger: 500 mg ginger powder, twice a day for 3 days	The study showed that in all sessions, ginger decreased the severity of vomiting in patients as compared to those who received a placebo	Ansari et al. (2016)
Single-center, double-blinded, randomized, placebo-controlled, parallel-group trial	120 patients	20 weeks	*P. grandiflorum* granules: twice daily for 12 weeks	The study showed that in patients with early breast cancer receiving anthracycline-based chemotherapy, *P. grandiflorum* shows cardioprotective action. Further this is effective as well as safe to use in breast cancer patients	Hao et al. (2017)
Controlled clinical trial	30 cancer patients	8 months	Capilen® cream (a topical phytotherapic product): two times daily two weeks before and during radiotherapy	The study shows that there is a delay in the onset of radiation dermatitis in breast cancer patients treated with Capilen® cream after radiotherapy and warrant further investigation	Stefanelli et al. (2014)
Open-label prospective nonrandomized comparative trial	100 breast cancer patients in all stages	26 months	*Withania somnifera* root extract: 2 g at every 8 h, throughout the course of chemotherapy. Either a combination of chemotherapy with oral *Withania somnifera* or chemotherapy alone	The study showed that *Withania somnifera* has potential against cancer-related fatigue, additionally improving the quality of life. However, further study with a larger sample size is warranted	Biswal et al. (2012)

Study design	No. of patient (women)	Study length	Extract/formulation/dosage form	Study outcome	Reference
Randomized, double-blind, placebo-controlled study	58 breast cancer patients who received chemotherapy or radiotherapy	Before, and following, 6 weeks of cancer treatment	Mixture of Rose geranium (*Pelargonium graveolens*), Ganoderma tsugae (Ganodermataceae), Codonopsis pilosula (Campanulaceae), and Angelica sinensis (Apiaceae) [RG-CMH]	The study showed that cancer patients receiving chemotherapy/radiotherapy during cancer treatment may have the capacity to delay, or ease, the reduction in levels of leucocytes and neutrophils after administration of RG-CMH	Zhuang et al. (2011)
Randomized controlled trial	75 patients with progressive fatigue after their first cycle of chemotherapy	50 days	Guarana: 50 mg by mouth twice daily for 21 days	The study showed that breast cancer patients receiving systemic chemotherapy can use Guarana as an effective, inexpensive, and nontoxic alternative for the short-term treatment of fatigue but needs further evaluation for its generalizability to chronic cancer-related fatigue and other types of cancer	de Oliveira Campos et al. (2011)
A prospective observational study	50 breast cancer patients with tamoxifen treatment	6 months	Isopropanolic extract of black cohosh: 1–4 tablets, 2.5 mg	The study showed that black cohosh extract is well tolerated in breast cancer patients treated with tamoxifen. Hot flashes, sleep problems, sweating, and anxiety improved, whereas urogenital and musculoskeletal complaints did not change. The study concluded that black cohosh extract seems to be a reasonable treatment option in tamoxifen-treated breast cancer patients with predominantly psychovegetative symptoms	Rostock et al. (2011)

6 Conclusion

Phytochemicals obtained from medicinal plants have shown potentials against BC cells in vitro as well as in vivo. Scientific approaches have shown an inverse relationship between dietary phytochemicals such as soy, fruits, mushrooms, etc., and the risk of breast cancer. Phytochemicals act as natural antioxidants by scavenging ROS, regulating cell signaling pathways, increasing the expression level of antioxidant enzymes, interacting with target proteins via hydrogen bonding have contributed to the treatment of BC. Phytochemicals prevent the production of cancer cells and induce apoptosis in BC cells by causing down-regulation of anti-apoptotic factors cell cycle arrest, induces cell death, inhibits FAS, and inhibits metastasis in BC cells by modulating protein degrading enzymes. Overall, the phytochemicals have a great role in suppressing BC. In the years to come more phytochemicals will be extracted, identified, and analyzed for medicinal research purposes to understand the molecular and cellular mechanisms of action and their adverse side effects to provide BC patients with the best possible treatment.

References

Abdel-Latif, G. A., Al-Abd, A. M., Tadros, M. G., Al-Abbasi, F. A., Khalifa, A. E., & Abdel-Naim, A. B. (2015). The chemomodulatory effects of resveratrol and didox on herceptin cytotoxicity in breast cancer cell lines. *Scientific Reports, 5*, 12054.

Aggarwal, B. B., Banerjee, S., Bharadwaj, U., Sung, B., Shishodia, S., & Sethi, G. (2016). Erratum: Curcumin induces the degradation of cyclin e expression through ubiquitin-dependent pathway and up-regulates cyclin-dependent kinase inhibitors p21 and p27 in multiple human tumor cell lines (Biochem. Pharmacol.(2007) 73 (1024–1032)). *Biochemical Pharmacology, 102*.

Aharoni, S., Lati, Y., Aviram, M., & Fuhrman, B. (2015). Pomegranate juice polyphenols induce a phenotypic switch in macrophage polarization favoring a M2 anti-inflammatory state. *BioFactors, 41*(1), 44–51.

Akhtar, M., Dasgupta, S., & Rangwala, M. (2015). Triple negative breast cancer: An Indian perspective. *Breast Cancer: Targets and Therapy, 7*, 239.

Akhtar Siddiqui, J., Singh, A., Chagtoo, M., Singh, N., Madhav Godbole, M., & Chakravarti, B. (2015). Phytochemicals for breast cancer therapy: Current status and future implications. *Current Cancer Drug Targets, 15*(2), 116–135.

Ali, I., Wani, W. A., & Saleem, K. (2011). Cancer scenario in India with future perspectives. *Cancer Therapy, 8*, 56–70.

Anders, C. K., Hsu, D. S., Broadwater, G., Acharya, C. R., Foekens, J. A., Zhang, Y., Wang, Y., Marcom, P. K., Marks, J. R., Febbo, P. G., & Nevins, J. R. (2008). Young age at diagnosis correlates with worse prognosis and defines a subset of breast cancers with shared patterns of gene expression. *Journal of Clinical Oncology, 26*(20), 3324–3330.

Ansari, M., Porouhan, P., Mohammadianpanah, M., Omidvari, S., Mosalaei, A., Ahmadloo, N., Nasrollahi, H., & Hamedi, S. (2016). Efficacy of ginger in control of chemotherapy induced nausea and vomiting in breast cancer patients receiving doxorubicin-based. *Asian Pacific Journal of Cancer Prevention; Chemotherapy, 17*(8), 3877–3880.

Ávila-Gálvez, M. Á., García-Villalba, R., Martínez-Díaz, F., Ocaña-Castillo, B., Monedero-Saiz, T., Torrecillas-Sánchez, A., Abellán, B., González-Sarrías, A., & Espín, J. C. (2019). Metabolic profiling of dietary polyphenols and methylxanthines in normal and malignant mammary tissues from breast cancer patients. *Molecular Nutrition & Food Research, 63*(9), 1801239.

Babu, G. R., Lakshmi, S. B., & Thiyagarajan, J. A. (2013). Epidemiological correlates of breast cancer in South India. *Asian Pacific Journal of Cancer Prevention, 14*(9), 5077–5083.

Bhatt, A. P., Redinbo, M. R., & Bultman, S. J. (2017). The role of the microbiome in cancer development and therapy. *CA: A Cancer Journal for Clinicians, 67*(4), 326–344.

Biswal, B., Sulaiman, S., Ismail, H., Zakaria, H., & Musa, K. (2012). Effect of *Withania somnifera* (Ashwagandha) on the development of chemotherapy-induced fatigue and quality of life in breast cancer patients. *Integrative Cancer Therapies, 12*(4), 312–322.

Bray, F., Ferlay, J., Soerjomataram, I., Siegel, R. L., Torre, L. A., & Jemal, A. (2018). Global cancer statistics 2018: GLOBOCAN estimates of incidence and mortality worldwide for 36 cancers in 185 countries. *CA: A Cancer Journal for Clinicians, 68*(6), 394–424.

Breast Cancer Treatment. (2020). *WebMD Medical Reference*. https://www.webmd.com/breast-cancer/breast-cancer-treatment

Cabrera, C., Artacho, R., & Giménez, R. (2006). Beneficial effects of green tea—A review. *Journal of the American College of Nutrition, 25*(2), 79–99.

Carriere, P. P., Llopis, S. D., Naiki, A. C., Nguyen, G., Phan, T., Nguyen, M. M., Preyan, L. C., Yearby, L., Pratt, J., Burks, H., & Davenport, I. R. (2016). Glyceollin I reverses epithelial to mesenchymal transition in letrozole resistant breast cancer through ZEB1. *International Journal of Environmental Research and Public Health, 13*(1), 10.

Chakravarti, B., Ravi, J., & Ganju, R. K. (2014). Cannabinoids as therapeutic agents in cancer: Current status and future implications. *Oncotarget, 5*(15), 5852.

Chen, X. Y., Zhou, J., Luo, L. P., Han, B., Li, F., Chen, J. Y., Zhu, Y. F., Chen, W., & Yu, X. P. (2015). Black rice anthocyanins suppress metastasis of breast cancer cells by targeting RAS/RAF/MAPK pathway. *BioMed Research International, 2015*.

Choi, E. J., & Kim, G. H. (2009). Apigenin causes G2/M arrest associated with the modulation of p21Cip1 and Cdc2 and activates p53-dependent apoptosis pathway in human breast cancer SK-BR-3 cells. *The Journal of Nutritional Biochemistry, 20*(4), 285–290.

Chopra, B., Kaur, V., Singh, K., Verma, M., Singh, S., & Singh, A. (2014). Age shift: Breast cancer is occurring in younger age groups-Is it true? *Clinical Cancer Investigation Journal, 3*(6), 526.

Choudhuri, T., Pal, S., Agwarwal, M. L., Das, T., & Sa, G. (2002). Curcumin induces apoptosis in human breast cancer cells through p53-dependent Bax induction. *FEBS Letters, 512*(1-3), 334–340.

Choudhuri, T., Pal, S., Das, T., & Sa, G. (2005). Curcumin selectively induces apoptosis in deregulated cyclin D1-expressed cells at G2 phase of cell cycle in a p53-dependent manner. *Journal of Biological Chemistry, 280*(20), 20059–20068.

Dandawate, P., Padhye, S., Ahmad, A., & Sarkar, F. H. (2013). Novel strategies targeting cancer stem cells through phytochemicals and their analogs. *Drug Delivery and Translational Research, 3*(2), 165–182.

de Oliveira Campos, M., Riechelmann, R., Martins, L., Hassan, B., Casa, F., & Giglio, A. (2011). Guarana (*Paullinia cupana*) improves fatigue in breast cancer patients undergoing systemic chemotherapy. *The Journal of Alternative and Complementary Medicine, 17*(6), 505–512.

DeSantis, C., Ma, J., Bryan, L., & Jemal, A. (2014). Breast cancer statistics, 2013. *CA: A Cancer Journal for Clinicians, 64*(1), 52–62.

Disis, M. L., & Park, K. H. (2009). Immunomodulation of breast cancer via tumor antigen specific Th1. *Cancer Research and Treatment: Official Journal of Korean Cancer Association, 41*(3), 117.

Dmitri, O., Levitsky, V., & Dembitsky, M. (2015). Anti-breast cancer agents derived from plants. *Natural Products Bioprospecting, 5*, 1–16.

Farabegoli, F., Papi, A., & Orlandi, M. (2011). (−)-Epigallocatechin-3-gallate down-regulates EGFR, MMP-2, MMP-9 and EMMPRIN and inhibits the invasion of MCF-7 tamoxifen-resistant cells. *Bioscience Reports, 31*(2), 99–108.

Febriansah, R., Dyaningtyas, D. P., Nurulita, N. A., Meiyanto, E., & Nugroho, A. E. (2014). Hesperidin as a preventive resistance agent in MCF–7 breast cancer cells line resistance to doxorubicin. *Asian Pacific Journal of Tropical Biomedicine, 4*(3), 228–233.

Ferlay, J., Soerjomataram, I., Dikshit, R., Eser, S., Mathers, C., Rebelo, M., Parkin, D. M., Forman, D., & Bray, F. (2015). Cancer incidence and mortality worldwide: Sources, methods and major patterns in GLOBOCAN 2012. *International Journal of Cancer, 136*(5), E359–E386.

Geara, F., Eid, T., Zouain, N., Thebian, R., Andraos, T., Chehab, C., Ramia, P., Youssef, B., & Zeidan, Y. (2018). Randomized, prospective, open-label phase III trial comparing mebo ointment with biafine cream for the management of acute dermatitis during radiotherapy for breast cancer. *American Journal of Clinical Oncology, 41*(12), 1257–1262.

Gerber, B., Freund, M., & Reimer, T. (2010). Recurrent breast cancer: Treatment strategies for maintaining and prolonging good quality of life. *Deutsches Arzteblatt International, 107*(6), 85.

Ghosh, J., Gupta, S., Desai, S., Shet, T., Radhakrishnan, S., Suryavanshi, P., Parmar, V., Jalali, R., Goyal, G., Hawaldar, R., & Patil, A. (2011). Estrogen, progesterone and HER2 receptor expression in breast tumors of patients, and their usage of HER2-targeted therapy, in a tertiary care centre in India. *Indian Journal of Cancer, 48*(4), 391.

Giovannelli, P., Di Donato, M., Galasso, G., Di Zazzo, E., Medici, N., Bilancio, A., Migliaccio, A., & Castoria, G. (2019). Breast cancer stem cells: The role of sex steroid receptors. *World Journal of Stem Cells, 11*(9), 594.

Gloria, N. F., Soares, N., Brand, C., Oliveira, F. L., Borojevic, R., & Teodoro, A. J. (2014). Lycopene and beta-carotene induce cell-cycle arrest and apoptosis in human breast cancer cell lines. *Anticancer Research, 34*(3), 1377–1386.

Gonzalez-Angulo, A. M., Timms, K. M., Liu, S., Chen, H., Litton, J. K., Potter, J., Lanchbury, J. S., Stemke-Hale, K., Hennessy, B. T., Arun, B. K., & Hortobagyi, G. N. (2011). Incidence and outcome of BRCA mutations in unselected patients with triple receptor-negative breast cancer. *Clinical Cancer Research, 17*(5), 1082–1089.

Gu, J. W., Makey, K. L., Tucker, K. B., Chinchar, E., Mao, X., Pei, I., Thomas, E. Y., & Miele, L. (2013). EGCG, a major green tea catechin suppresses breast tumor angiogenesis and growth via inhibiting the activation of HIF-1α and NFκB, and VEGF expression. *Vascular Cell, 5*(1), 1–10.

Gucalp, A., & Traina, T. A. (2016). Targeting the androgen receptor in triple-negative breast cancer. *Current Problems in Cancer, 40*(2–4), 141–150.

Hao, W., Liu, S., Qin, Y., Sun, C., Chen, L., Wu, C., & Bao, Y. (2017). Cardioprotective effect of *Platycodon grandiflorum* in patients with early breast cancer receiving anthracycline-based chemotherapy: Study protocol for a randomized controlled trial. *Trials, 18*(1).

Hortobagyi, G. N., de la Garza Salazar, J., Pritchard, K., Amadori, D., Haidinger, R., Hudis, C. A., Khaled, H., Liu, M. C., Martin, M., Namer, M., & O'Shaughnessy, J. A. (2005). The global breast cancer burden: Variations in epidemiology and survival. *Clinical Breast Cancer, 6*(5), 391–401.

Howes, M. J. R., & Simmonds, M. S. (2014). The role of phytochemicals as micronutrients in health and disease. *Current Opinion in Clinical Nutrition & Metabolic Care, 17*(6), 558–566.

Hudis, C. A., & Gianni, L. (2011). Triple-negative breast cancer: An unmet medical need. *The Oncologist, 16.*

Hui, C., Bin, Y., Xiaoping, Y., Long, Y., Chunye, C., Mantian, M., & Wenhua, L. (2010). Anticancer activities of an anthocyanin-rich extract from black rice against breast cancer cells in vitro and in vivo. *Nutrition and Cancer, 62*(8), 1128–1136.

Israel, B. E. B., Tilghman, S. L., Parker-Lemieux, K., & Payton-Stewart, F. (2018). Phytochemicals: Current strategies for treating breast cancer. *Oncology Letters, 15*(5), 7471–7478.

Jackson, S. A., Davis, A. A., Li, J., Yi, N., McCormick, S. R., Grant, C., Fallen, T., Crawford, B., Loranger, K., Litton, J., & Arun, B. (2014). Characteristics of individuals with breast cancer rearrangements in BRCA1 and BRCA2. *Cancer, 120*(10), 1557–1564.

Karas, M., Amir, H., Fishman, D., Danilenko, M., Segal, S., Nahum, A., Koifmann, A., Giat, Y., Levy, J., & Sharoni, Y. (2000). Lycopene interferes with cell cycle progression and insulin-like growth factor I signaling in mammary cancer cells. *Nutrition and Cancer, 36*(1), 101–111.

Kawiak, A., Zawacka-Pankau, J., & Lojkowska, E. (2012). Plumbagin induces apoptosis in Her2-overexpressing breast cancer cells through the mitochondrial-mediated pathway. *Journal of Natural Products, 75*(4), 747–751.

Khan, M., Maryam, A., Zhang, H., Mehmood, T., & Ma, T. (2016). Killing cancer with platycodin D through multiple mechanisms. *Journal of Cellular and Molecular Medicine, 20*(3), 389–402.

Khoobchandani, M., Katti, K. K., Karikachery, A. R., Thipe, V. C., Srisrimal, D., Mohandoss, D. K. D., Darshakumar, R. D., Joshi, C. M., & Katti, K. V. (2020). New approaches in breast cancer therapy through green nanotechnology and nano-ayurvedic medicine–pre-clinical and pilot human clinical investigations. *International Journal of Nanomedicine, 15*, 181.

Kim, S., Choi, J. H., Kim, J. B., Nam, S. J., Yang, J. H., Kim, J. H., & Lee, J. E. (2008). Berberine suppresses TNF-α-induced MMP-9 and cell invasion through inhibition of AP-1 activity in MDA-MB-231 human breast cancer cells. *Molecules, 13*(12), 2975–2985.

Kim, S. H., Sehrawat, A., & Singh, S. V. (2013a). Dietary chemopreventative benzyl isothiocyanate inhibits breast cancer stem cells in vitro and in vivo. *Cancer Prevention Research, 6*(8), 782–790.

Kim, S. Y., Lee, I. S., & Moon, A. (2013b). 2-Hydroxychalcone and xanthohumol inhibit invasion of triple negative breast cancer cells. *Chemico-Biological Interactions, 203*(3), 565–572.

King-Batoon, A., Leszczynska, J. M., & Klein, C. B. (2008). Modulation of gene methylation by genistein or lycopene in breast cancer cells. *Environmental and Molecular Mutagenesis, 49*(1), 36–45.

Knobloch, T. J., Uhrig, L. K., Pearl, D. K., Casto, B. C., Warner, B. M., Clinton, S. K., Sardo-Molmenti, C. L., Ferguson, J. M., Daly, B. T., Riedl, K., & Schwartz, S. J. (2016). Suppression of proinflammatory and prosurvival biomarkers in oral cancer patients consuming a black raspberry phytochemical-rich troche. *Cancer Prevention Research, 9*(2), 159–171.

Kubatka, P., Kapinová, A., Kello, M., Kruzliak, P., Kajo, K., Výbohová, D., Mahmood, S., Murin, R., Viera, T., Mojžiš, J., & Zulli, A. (2016b). Fruit peel polyphenols demonstrate substantial anti-tumour effects in the model of breast cancer. *European Journal of Nutrition, 55*(3), 955–965.

Kubatka, P., Kapinová, A., Kružliak, P., Kello, M., Výbohová, D., Kajo, K., Novák, M., Chripková, M., Adamkov, M., Péč, M., & Mojžiš, J. (2015). Antineoplastic effects of *Chlorella pyrenoidosa* in the breast cancer model. *Nutrition, 31*(4), 560–569.

Kubatka, P., Kello, M., Kajo, K., Kruzliak, P., Výbohová, D., Mojžiš, J., Adamkov, M., Fialová, S., Veizerová, L., Zulli, A., & Péč, M. (2017). Oregano demonstrates distinct tumour-suppressive effects in the breast carcinoma model. *European Journal of Nutrition, 56*(3), 1303–1316.

Kubatka, P., Kello, M., Kajo, K., Kruzliak, P., Výbohová, D., Šmejkal, K., Maršík, P., Zulli, A., Gönciová, G., Mojžiš, J., & Kapinová, A. (2016a). Young barley indicates antitumor effects in experimental breast cancer in vivo and in vitro. *Nutrition and Cancer, 68*(4), 611–621.

Kuo, P. L., Hsu, Y. L., & Cho, C. Y. (2006). Plumbagin induces G2-M arrest and autophagy by inhibiting the AKT/mammalian target of rapamycin pathway in breast cancer cells. *Molecular Cancer Therapeutics, 5*(12), 3209–3221.

Lee, C. J., Wilson, L., Jordan, M. A., Nguyen, V., Tang, J., & Smiyun, G. (2010). Hesperidin suppressed proliferations of both Human breast cancer and androgen-dependent prostate cancer cells. *Phytotherapy Research, 24*(S1), S15–S19.

Lee, H., & Lin, J. Y. (1988). Antimutagenic activity of extracts from anticancer drugs in Chinese medicine. *Mutation Research/Genetic Toxicology, 204*(2), 229–234.

Lee, H. P., Chen, P. C., Wang, S. W., Fong, Y. C., Tsai, C. H., Tsai, F. J., Chung, J. G., Huang, C. Y., Yang, J. S., Hsu, Y. M., & Li, T. M. (2019). Plumbagin suppresses endothelial progenitor cell-related angiogenesis in vitro and in vivo. *Journal of Functional Foods, 52*, 537–544.

Lee, K. W., Bode, A. M., & Dong, Z. (2011). Molecular targets of phytochemicals for cancer prevention. *Nature Reviews Cancer, 11*(3), 211–218.

Lee, M. S., Chan, J. Y. W., Kong, S. K., Yu, B., Eng-Choon, V. O., Nai-Ching, H. W., Mak Chung-Wai, T., & Fung, K. P. (2005). Effects of polyphyllin D, a steroidal saponin in Paris polyphylla, in growth inhibition of human breast cancer cells and in xenograft. *Cancer Biology & Therapy, 4*(11), 1248–1254.

Li, Y., Li, S., Meng, X., Gan, R. Y., Zhang, J. J., & Li, H. B. (2017). Dietary natural products for prevention and treatment of breast cancer. *Nutrients, 9*(7), 728.

Lidar, M., & Livneh, A. (2007). Familial Mediterranean fever: Clinical, molecular and management advancements. *The Netherlands Journal of Medicine, 65*(9), 318–324.

Lin, B. W., Gong, C. C., Song, H. F., & Cui, Y. Y. (2017). Effects of anthocyanins on the prevention and treatment of cancer. *British Journal of Pharmacology, 174*(11), 1226–1243.

Lindenmeyer, F., Li, H., Menashi, S., Soria, C., & Lu, H. (2001). Apigenin acts on the tumor cell invasion process and regulates protease production. *Nutrition and Cancer, 39*(1), 139–147.

Lirdprapamongkol, K., Sakurai, H., Abdelhamed, S., Yokoyama, S., Maruyama, T., Athikomkulchai, S., Viriyaroj, A., Awale, S., Yagita, H., Ruchirawat, S., & Svasti, J. (2013). A flavonoid chrysin suppresses hypoxic survival and metastatic growth of mouse breast cancer cells. *Oncology Reports, 30*(5), 2357–2364.

Liu, R., Zhang, H., Yuan, M., Zhou, J., Tu, Q., Liu, J. J., & Wang, J. (2013). Synthesis and biological evaluation of apigenin derivatives as antibacterial and antiproliferative agents. *Molecules, 18*(9), 11496–11511.

Liu, R. H. (2003). Health benefits of fruit and vegetables are from additive and synergistic combinations of phytochemicals. *The American Journal of Clinical Nutrition, 78*(3), 517S–520S.

Liu, W., Liu, P., Tao, S., Deng, Y., Li, X., Lan, T., Zhang, X., Guo, F., Huang, W., Chen, F., & Huang, H. (2008). Berberine inhibits aldose reductase and oxidative stress in rat mesangial cells cultured under high glucose. *Archives of Biochemistry and Biophysics, 475*(2), 128–134.

Liu, Y., Cai, Y., He, C., Chen, M., & Li, H. (2017). Anticancer properties and pharmaceutical applications of plumbagin: A review. *The American Journal of Chinese Medicine, 45*(03), 423–441.

Mai, Z., Blackburn, G. L., & Zhou, J. R. (2007). Genistein sensitizes inhibitory effect of tamoxifen on the growth of estrogen receptor-positive and HER2-overexpressing human breast cancer cells. *Molecular Carcinogenesis, 46*(7), 534–542.

Manu, K. A., Shanmugam, M. K., Rajendran, P., Li, F., Ramachandran, L., Hay, H. S., Kannaiyan, R., Swamy, S. N., Vali, S., Kapoor, S., & Ramesh, B. (2011). Plumbagin inhibits invasion and migration of breast and gastric cancer cells by downregulating the expression of chemokine receptor CXCR4. *Molecular Cancer, 10*(1), 107.

Meybodi, N. M., Mortazavian, A. M., Monfared, A. B., Sohrabvandi, S., & Meybodi, F. A. (2017). Phytochemicals in cancer prevention: A review of the evidence. *Iranian Journal of Cancer Prevention, 10*(1), e7219.

Nair, M. K., Sankaranarayanan, R., Nair, K. S., Amma, N. S., Varghese, C., Padmakumari, G., & Cherian, T. (1993). Overall survival from breast cancer in Kerala, India, in relation to menstrual, reproductive, and clinical factors. *Cancer, 71*(5), 1791–1796.

Nasser, M. W., Qamri, Z., Deol, Y. S., Smith, D., Shilo, K., Zou, X., & Ganju, R. K. (2011). Crosstalk between chemokine receptor CXCR4 and cannabinoid receptor CB 2 in modulating breast cancer growth and invasion. *PLoS One, 6*(9), e23901.

Ng, Z. X., Ong, M. S., Jegadeesan, T., Deng, S., & Yap, C. T. (2017). Breast cancer: Exploring the facts and holistic needs during and beyond treatment. *Healthcare, 5*(2), 26. Multidisciplinary Digital Publishing Institute.

Nguyen, L. T., Lee, Y. H., Sharma, A. R., Park, J. B., Jagga, S., Sharma, G., Lee, S. S., & Nam, J. S. (2017). Quercetin induces apoptosis and cell cycle arrest in triple-negative breast cancer cells through modulation of Foxo3a activity. *The Korean Journal of Physiology & Pharmacology, 21*(2), 205–213.

Oh, J., Hlatky, L., Jeong, Y. S., & Kim, D. (2016). Therapeutic effectiveness of anticancer phytochemicals on cancer stem cells. *Toxins, 8*(7), 199.

Osborne, C., Wilson, P., & Tripathy, D. (2004). Oncogenes and tumor suppressor genes in breast cancer: Potential diagnostic and therapeutic applications. *The Oncologist, 9*(4), 361–377.

Pan, H., Zhou, W., He, W., Liu, X., Ding, Q., Ling, L., Zha, X., & Wang, S. (2012). Genistein inhibits MDA-MB-231 triple-negative breast cancer cell growth by inhibiting NF-κB activity via the Notch-1 pathway. *International Journal of Molecular Medicine, 30*(2), 337–343.

Park, Y. H. (2017). The nuclear factor-kappa B pathway and response to treatment in breast cancer. *Pharmacogenomics, 18*(18), 1697–1709.

Patil, J. B., Kim, J., & Jayaprakasha, G. K. (2010). Berberine induces apoptosis in breast cancer cells (MCF-7) through mitochondrial-dependent pathway. *European Journal of Pharmacology, 645*(1-3), 70–78.

Piredda, M., Facchinetti, G., Biagioli, V., Giannarelli, D., Armento, G., Tonini, G., & De Marinis, M. (2017). Propolis in the prevention of oral mucositis in breast cancer patients receiving adjuvant chemotherapy: A pilot randomised controlled trial. *European Journal of Cancer Care, 26*(6), e12757.

Prasad, C. P., Rath, G., Mathur, S., Bhatnagar, D., & Ralhan, R. (2009). Potent growth suppressive activity of curcumin in human breast cancer cells: Modulation of Wnt/β-catenin signaling. *Chemico-Biological Interactions, 181*(2), 263–271.

Rafati, M., Ghasemi, A., Saeedi, M., Habibi, E., Salehifar, E., Mosazadeh, M., & Maham, M. (2019). Nigella sativa L. for prevention of acute radiation dermatitis in breast cancer: A randomized, double-blind, placebo-controlled, clinical trial. *Complementary Therapies in Medicine, 47*, 102205.

Rizeq, B., Gupta, I., Ilesanmi, J., AlSafran, M., Rahman, M. M., & Ouhtit, A. (2020). The power of phytochemicals combination in cancer chemoprevention. *Journal of Cancer, 11*(15), 4521.

Rostock, M., Fischer, J., Mumm, A., Stammwitz, U., Saller, R., & Bartsch, H. (2011). Black cohosh (*Cimicifuga racemosa*) in tamoxifen-treated breast cancer patients with climacteric complaints: A prospective observational study. *Gynecological Endocrinology, 27*(10), 844–848.

Samavat, H., Ursin, G., Emory, T., Lee, E., Wang, R., Torkelson, C., Dostal, A., Swenson, K., Le, C., Yang, C., Yu, M., Yee, D., Wu, A., Yuan, J., & Kurzer, M. (2017). A randomized controlled trial of green tea extract supplementation and mammographic density in postmenopausal women at increased risk of breast cancer. *Cancer Prevention Research, 10*(12), 710–718.

Sandhu, D. S., Sandhu, S., Karwasra, R. K., & Marwah, S. (2010). Profile of breast cancer patients at a tertiary care hospital in North India. *Indian Journal of Cancer, 47*(1), 16.

Seo, H. S., Choi, H. S., Kim, S. R., Choi, Y. K., Woo, S. M., Shin, I., Woo, J. K., Park, S. Y., Shin, Y. C., & Ko, S. K. (2012). Apigenin induces apoptosis via extrinsic pathway, inducing p53 and inhibiting STAT3 and NFκB signaling in HER2-overexpressing breast cancer cells. *Molecular and Cellular Biochemistry, 366*(1–2), 319–334.

Sharma, G. N., Dave, R., Sanadya, J., Sharma, P., & Sharma, K. K. (2010). Various types and management of breast cancer: An overview. *Journal of Advanced Pharmaceutical Technology & Research, 1*(2), 109.

Shi, Y., Yang, F., Huang, D., & Guan, X. (2018). Androgen blockade based clinical trials landscape in triple negative breast cancer. *Biochimica et Biophysica Acta (BBA): Reviews on Cancer, 1870*(2), 283–290.

Shrivastava, A., Kuzontkoski, P. M., Groopman, J. E., & Prasad, A. (2011). Cannabidiol induces programmed cell death in breast cancer cells by coordinating the cross-talk between apoptosis and autophagy. *Molecular Cancer Therapeutics, 10*(7), 1161–1172.

Singh, N., Kushwaha, P., Gupta, A., & Prakash, O. (2020). Recent advances of novel therapeutic agents from botanicals for prevention and therapy of breast cancer: An updated review. *Current Cancer Therapy Reviews, 16*(1), 5–18.

Sledge, G. W., Jr., Hu, P., Falkson, G., Tormey, D., Abeloff, M., & Eastern Cooperative Oncology Group. (2000). Comparison of chemotherapy with chemohormonal therapy as first-line therapy for metastatic, hormone-sensitive breast cancer: An Eastern Cooperative Oncology Group study. *Journal of Clinical Oncology, 18*(2), 262–262.

Stefanelli, A., Forte, L., Medoro, S., Sgualdo, A., Lombardo, D., Zini, G., Maronta, D., Rainieri, E., Pascale, G., Bagnolatti, P., Colella, M., Princivalle, S., & Fiorica, F. (2014). Topical use of phytotherapic cream (Capilen® cream) to prevent radiodermatitis in breast cancer: A prospective historically controlled clinical study. *Giornale Italiano di Dermatologia e Venereologia, 149*(1), 107–113.

Stuart, E. C., Larsen, L., & Rosengren, R. J. (2007). Potential mechanisms for the synergistic cytotoxicity elicited by 4-hydroxytamoxifen and epigallocatechin gallate in MDA-MB-231 cells. *International Journal of Oncology, 30*(6), 1407–1412.

Stuart, E. C., & Rosengren, R. J. (2008). The combination of raloxifene and epigallocatechin gallate suppresses growth and induces apoptosis in MDA-MB-231 cells. *Life Sciences, 82*(17–18), 943–948.

Sun, Y., Lin, X., & Chang, H. (2016). Proliferation inhibition and apoptosis of breast cancer MCF-7 cells under the influence of colchicine. *Journal of BUON, 3*, 570–575.

Sung, B., Oyajobi, B., & Aggarwal, B. B. (2012). Plumbagin inhibits osteoclastogenesis and reduces human breast cancer-induced osteolytic bone metastasis in mice through suppression of RANKL signaling. *Molecular Cancer Therapeutics, 11*(2), 350–359.

Togni, S., Maramaldi, G., Bonetta, A., Giacomelli, L., & Di Pierro, F. (2015). Clinical evaluation of safety and efficacy of Boswellia-based cream for prevention of adjuvant radiotherapy skin damage in mammary carcinoma: A randomized placebo controlled trial. *European Review for Medical and Pharmacological Sciences, 19*(8), 1338–1344.

Tomasin, R., & Cintra Gomes-Marcondes, M. C. (2011). Oral administration of Aloe vera and honey reduces walker tumour growth by decreasing cell proliferation and increasing apoptosis in tumour tissue. *Phytotherapy Research, 25*(4), 619–623.

Tseng, E., Ramsay, E. A. S., & Morris, M. E. (2004). Dietary organic isothiocyanates are cytotoxic in human breast cancer MCF-7 and mammary epithelial MCF-12A cell lines. *Experimental Biology and Medicine, 229*(8), 835–842.

Urbaniak, A., Jousheghany, F., Piña-Oviedo, S., Yuan, Y., Majcher-Uchańska, U., Klejborowska, G., Moorjani, A., Monzavi-Karbassi, B., Huczyński, A., & Chambers, T. C. (2020). Carbamate derivatives of colchicine show potent activity towards primary acute lymphoblastic leukemia and primary breast cancer cells—In vitro and ex vivo study. *Journal of Biochemical and Molecular Toxicology, 34*(6), e22487.

Vaidyanathan, K., Lakhotia, S., Ravishankar, H. M., Tabassum, U., Mukherjee, G., & Somasundaram, K. (2009). BRCA1 and BRCA2 germline mutation analysis among Indian women from south India: Identification of four novel mutations and high-frequency occurrence of 185delAG mutation. *Journal of Biosciences, 34*(3), 415–422.

Varley, J. M. (2003). Germline TP53 mutations and Li-Fraumeni syndrome. *Human Mutation, 21*(3), 313–320.

Viskelis, P., Rubinskienė, M., Jasutienė, I., Šarkinas, A., Daubaras, R., & Česonienė, L. (2009). Anthocyanins, antioxidative, and antimicrobial properties of American cranberry (*Vaccinium macrocarpon* Ait.) and their press cakes. *Journal of Food Science, 74*(2), C157–C161.

Wang, C., Huang, Q., Liang, C., Zhang, Y., Deng, D., Yu, Y., Chen, D., Yang, H., & Yu, X. (2019). Effect of *Cimicifuga racemosa* on menopausal syndrome caused by LHRH-a in breast cancer. *Journal of Ethnopharmacology, 238*, 111840.

Wang, S., Chen, R., Morott, J., Repka, M. A., Wang, Y., & Chen, M. (2015). mPEG-b-PCL/TPGS mixed micelles for delivery of resveratrol in overcoming resistant breast cancer. *Expert Opinion on Drug Delivery, 12*(3), 361–373.

Warin, R., Chambers, W. H., Potter, D. M., & Singh, S. V. (2009). Prevention of mammary carcinogenesis in MMTV-neu mice by cruciferous vegetable constituent benzyl isothiocyanate. *Cancer Research, 69*(24), 9473–9480.

Warri, A., Saarinen, N. M., Makela, S., & Hilakivi-Clarke, L. (2008). The role of early life genistein exposures in modifying breast cancer risk. *British Journal of Cancer, 98*(9), 1485–1493.

Wen, C. J., Wu, L. X., Fu, L. J., Shen, D. Y., Zhang, X., Zhang, Y. W., Yu, J., & Zhou, H. H. (2014). Preferential induction of CYP1A1 over CYP1B1 in human breast cancer MCF-7 cells after exposure to berberine. *Asian Pacific Journal of Cancer Prevention, 15*(1), 495–499.

Wu, X., Zhou, Q. H., & Xu, K. (2009). Are isothiocyanates potential anti-cancer drugs? *Acta Pharmacologica Sinica, 30*(5), 501–512.

Yan, W., Tu, B., Liu, Y. Y., Wang, T. Y., Qiao, H., Zhai, Z. J., Li, H. W., & Tang, T. T. (2013). Suppressive effects of plumbagin on invasion and migration of breast cancer cells via the inhibition of STAT3 signaling and down-regulation of inflammatory cytokine expressions. *Bone Research, 1*(1), 362–370.

Yang, M. H., Kim, J., Khan, I. A., Walker, L. A., & Khan, S. I. (2014). Nonsteroidal anti-inflammatory drug activated gene-1 (NAG-1) modulators from natural products as anti-cancer agents. *Life Sciences, 100*(2), 75–84.

Yu, F., Li, Y., Zou, J., Jiang, L., Wang, C., Tang, Y., Gao, B., Luo, D., & Jiang, X. (2019). The Chinese herb Xiaoaiping protects against breast cancer chemotherapy-induced alopecia and other side effects: A randomized controlled trial. *Journal of International Medical Research, 47*(6), 2607–2614.

Zhang, F., & Du, G. (2012). Dysregulated lipid metabolism in cancer. *World Journal of Biological Chemistry, 3*(8), 167.

Zhang, X. Q., Yang, C. Y., Rao, X. F., & Xiong, J. P. (2016). Plumbagin shows anti-cancer activity in human breast cancer cells by the upregulation of p53 and p21 and suppression of G1 cell cycle regulators. *European Journal of Gynaecological Oncology, 37*(1), 30.

Zheng, J., Zhou, Y., Li, Y., Xu, D. P., Li, S., & Li, H. B. (2016). Spices for prevention and treatment of cancers. *Nutrients, 8*(8), 495.

Zhu, W., Qin, W., Zhang, K., Rottinghaus, G. E., Chen, Y. C., Kliethermes, B., & Sauter, E. R. (2012). Trans-resveratrol alters mammary promoter hypermethylation in women at increased risk for breast cancer. *Nutrition and Cancer, 64*(3), 393–400.

Zhuang, S., Chiu, H., Chen, S., Tsai, J., Lee, M., Lee, H., Shen, Y., Yan, Y., Shane, G., & Wang, C. (2011). Effects of a Chinese medical herbs complex on cellular immunity and toxicity-related conditions of breast cancer patients. *British Journal of Nutrition, 107*(5), 712–718.

Index

C. Egbuna, S. Hassan (eds.), *Dietary Phytochemicals*,
https://doi.org/10.1007/978-3-030-72999-8

O

U

V

W

X

Y

Z

The manufacturer's authorised representative in the EU is Springer Nature Customer Service Centre GmbH, Europaplatz 3, 69115 Heidelberg, Germany. If you have any concerns regarding our products, please contact ProductSafety@springernature.com

Printed and bound by CPI Group (UK) Ltd, Croydon, CR0 4YY
22/09/2026
02221062-0012